AF900878

The Forest Certification Handbook

SECOND EDITION

THE EARTHSCAN FORESTRY LIBRARY

Plantations, Privatization, Poverty and Power: Changing Ownership and Management of State Forests
Mike Garforth and James Mayers (eds)

The Sustainable Forestry Handbook 2nd edition
Sophie Higman, James Mayers, Stephen Bass, Neil Judd and Ruth Nussbaum

The Forest Certification Handbook 2nd edition
Ruth Nussbaum and Markku Simula

The Forest Certification Handbook

Second Edition

Ruth Nussbaum and Markku Simula

ProForest

earthscan
from Routledge

iied
International
Institute for
Environment and
Development

This publication is an output from a research project funded by the United Kingdom Department for International Development (DFID) for the benefit of developing countries. The views expressed are not necessarily those of DFID.

R6370E – Forestry Research Programme

First published by Earthscan in the UK and USA in 2005

For a full list of publications please contact:
Earthscan
2 Park Square, Milton Park, Abingdon, Oxfordshire OX14 4RN
52 Vanderbilt Avenue, New York, NY 10017

Routledge is an imprint of the Taylor & Francis Group, an informa business

First issued in paperback 2018

Copyright © ProForest, 2005

All rights reserved. No part of this book may be reprinted or reproduced or utilised in any form or by any electronic, mechanical, or other means, now known or hereafter invented, including photocopying and recording, or in any information storage or retrieval system, without permission in writing from the publishers.

Notices
Practitioners and researchers must always rely on their own experience and knowledge in evaluating and using any information, methods, compounds, or experiments described herein. In using such information or methods they should be mindful of their own safety and the safety of others, including parties for whom they have a professional responsibility.

Product or corporate names may be trademarks or registered trademarks, and are used only for identification and explanation without intent to infringe.

Typesetting by Mapset Ltd, Gateshead, UK

Cover design by Ruth Bateman

Earthscan publishes in association with the International Institute for Environment and Development

A catalogue record for this book is available from the British Library

Library of Congress Cataloging-in-Publication Data

Nussbaum, Ruth, 1964–
The forest certification handbook / Ruth Nussbaum and Markku Simula – 2nd ed.
p. cm.
Includes bibliographical references and index.
ISBN 1-84407-123-5 (hardback)
1. Forest management–Standards–Handbooks, manuals, etc. 2. Forest products–Certification–Handbooks, manuals, etc. 3. Sustainable forestry–Handbooks, manuals, etc. I. Simula, Markku. II. Title.

SD387.S69N87 2005
634.9'2'0218–dc22

2004022384

ISBN 13: 978-1-84407-123-4 (hbk)
ISBN 13: 978-1-138-97450-0 (pbk)

Contents

List of Illustrations

Figures

Tables

Boxes

List of Acronyms and Abbreviations

ACHIS	Chilean Safety Association
AF&PA	American Forest and Paper Association
AFS	Australian Forestry Standard
AHP	analytical hierarchy process
ANSI	American National Standards Institute
API	American Paper Institute
ATFS	American Tree Farm System
ATO	African Timber Organization
BMP	best management practice
C&I	criteria and indicators
CAR	corrective action request
CB	certification body
CBFM	sustainable community-based forest management
CBO	constituent-based organization
CCFM	Canadian Council of Forest Ministers
CCP	critical control point
CDM	Clean Development Mechanism
CEA	Canadian Environmental Auditing
CEA	certified environmental auditor
CEAA	Canadian Environmental Auditing Association
CEPI	Confederation of European Paper Industries
CIFOR	Centre for International Forestry Research
CoC	chain of custody
CODEFF	Concepción, Econativa (Chile)
CONAF	Chilean Forest Service (Corporación Nacional Forestal)
CORFO	Chilean Development Corporation (Corporación de Fomento de la Producción)
CORMA	Corporación Chilena de la Madera (Chilean Forest Owners Association)
CRC	Certification Review Council
CSA	Canadian Standards Association
DFA	defined forest area
DIS	Draft International Standard
DIY	do-it-yourself
DSM	Department of Standards Malaysia
EA	European co-operation for Accreditation
ECF	Enabling Conditions Framework
EFTA	European Free Trade Association
EMAS	The European Union's Eco-Management and Audit Scheme
EMS	environmental management system
ENGO	environmental non-governmental organization
EU	European Union
FAO	United Nations Food and Agriculture Organization
FDIS	Final Draft International Standard
FFCS	Finnish Forest Certification Scheme

FLEGT	EU *Action Plan for Forest Law Enforcement, Governance and Trade*
FLO	Fairtrade Labelling Organization
FMP	Forest Monitoring Project
FMU	forest management unit
FoE	Friends of the Earth
FSC	Forest Stewardship Council
FTC	US Federal Trade Commission
G8	Informal group of eight developed countries (Canada, France, Germany, Italy, Japan, Russia, the UK and the US), which meets annually for a Summit
GEF	Global Environment Facility
GFTN	Global Forest and Trade Network (of the WWF)
GIS	geographical information systems
GMO	genetically modified organism
GPS	geographical positioning systems
GTZ	Deutsche Gesellschaft für Technische Zusammenarbeit
ha	hectares
HCVF	high conservation value forests
IAF	International Accreditation Forum
IFF	Intergovernmental Forum on Forests
IFIR	International Forest Industries Round Table
ILO	International Labour Organization/Office
INFOR	Chilean Forestry Institute (Instituto Forestal)
INN	Instituto Nacional de Normalización (National Standards Institute of Chile)
IPF	Intergovernmental Panel on Forests
ISEAL	Social and Environmental Accreditation and Labelling Alliance
ISO	International Organization for Standardization
ITTO	International Tropical Timber Organization
IUCN	World Conservation Union
IUFRO	International Union of Forest Research Organizations
JCP	joint certification programme
LCA	life cycle analysis
LEI	Lembaga Ekolabel Indonesia (Indonesian Eco-labelling Foundation)
LTM	Legitimacy Thresholds Model
MAI	mean annual increment
MC&I	Malaysian criteria and indicators
MCPFE	Ministerial Conference on Protection of Forests in Europe
MDF	medium density fibreboard
MIV	modular implementation and verification
MTCC	Malaysian Timber Certification Council
NFPA	National Forest Products Association
NGB	national governing body
NGO	non-governmental organization
NTCC	National Timber Certification Council, Malaysia (now MTCC)
OPIC	Overseas Private Investment Corporation
OSB	oriented strand board
P&C	principles and criteria
PCF	provincial communication forum
PEFC	Programme for the Endorsement of Forest Certification Schemes (formerly the Pan-European Forest Certification scheme)
PEOLG	Pan-European Operational-level Guidelines
PRB	personnel registration body

PRF	permanent reserved forest
RAB	US Registrar Accreditation Board
RAG	Rainforest Action Group
RIIA	Royal Institute for International Affairs
RPF	registered professional forester
SAG	Chilean Agriculture and Livestock Service (Servicio Agrícola y Ganadero)
SAI	Social Accountability International
SCC	Standards Council of Canada
SCS	Scientific Certification Systems
SES	Small Enterprise Scheme
SFB	Sustainable Forestry Board
SFE	small forest enterprise
SFI	Sustainable Forestry Initiative
SFIS	Sustainable Forestry Initiative Standard
SFM	sustainable forest management
SFM TC	sustainable forest management technical committee
SFPM	sustainable forest plantation management
SGS	Société Générale de Surveillance
SIC	Sustainable Forestry Initiative implementation committee
SLIMF	small and low-intensity managed forest
SNPFM	sustainable natural production forest management
SOP	standard of performance
SSC	SvenskSkogsCertifiering (Swedish forest certification)
TBT	technical barrier to trade
TFT	Tropical Forest Trust
UFMA	Union of Forest Management Associations
UK	United Kingdom
UKAS	United Kingdom Accreditation Service
UKWAS	UK Woodland Assurance Scheme
UNCED	United Nations Conference on Environment and Development
UNDP	United Nations Development Programme
UNEP	United Nations Environment Programme
UNFF	United Nations Forum on Forests
US	United States
USDA	US Department of Agriculture
VOIT	value, objective, indicator and target
WBCSD	World Business Council for Sustainable Development
WTO	World Trade Organization
WWF	World Wide Fund for Nature

Acknowledgements

The authors would like to thank the following people for their input, comments or involvement with previous work on which we drew for this book: Steve Bass, Rebecca Butterfield, Ishmael Dodoo, Richard Donovan, Richard Eba'a, Mike Garforth, Nilofer Ghaffar, Ian Gray, James Griffiths, Sophie Higman, Steve Jennings, Neil Judd, Simon Levy, Jussi Lounasvuori, Janne Löytömäki, Pedro Moura Costa, Hanna Nikinmaa, John Palmer, Hannah Scrase, Andry Rakotovololona and Tim Rayden.

We would also like to thank the seven certification schemes that provided material for the book.

The authors of this edition of *The forest certication Handbook* would like to acknowledge the contribution of *The forest certication Handbook*, First Edition © Steve Bass and Cris Upton, 1995.

This publication is an output from a research project funded by the United Kingdom Department for International Development (DFID) for the benefit of developing countries. The views expressed are not necessarily those of DFID.

R6370E – Forestry Research Programme

1 Introduction

Forest certification is essentially the process of verifying that a forest meets the requirements of a standard. It is a straightforward concept, and certification is widely used throughout all sectors of industry to provide independent confirmation that standards are being met. So why is a handbook such as this one needed?

Firstly, the demand for certification is growing steadily, so more and more forest managers, processors and buyers are being asked to implement certification or buy certified products. Therefore, they need to understand what certification is and how to use it.

Secondly, despite the simplicity of the basic concept, forest certification has become a contentious issue within the forest products sector and remains the subject of intense debate. In order to understand and engage with this debate, anyone involved in the sector, whether they are forest owners, forest managers, processors, buyers, regulators, environmental organizations, local communities, researchers, trainers, students or other stakeholders, needs to have a reasonable understanding of what certification is, how it works and what the major issues are.

Thirdly, the range of ways in which certification is being used is growing. Therefore, it is important for new users of the tool to understand it better. It is also critical that those at the forefront of the policy debate have a sufficiently good understanding of certification to be able to assess both its potential and its limitations.

As a result, there are a number of different aspects to certification, each of which is relevant to one or more users, including:

- the practical process of getting forests and chains of custody certified;
- the basic theory of certification and design and functioning of Forest Certification schemes;
- the reality of existing Forest Certification schemes;
- policy implications and impacts;
- remaining issues and problems and potential solutions.

All of these aspects are covered in the handbook. As a result, it is not designed to be read from cover to cover, but rather to be used as a reference book that provides information on whichever aspect of certification is of interest to a particular reader at a particular time. Consequently, occasionally there is repetition of ideas or information between sections to avoid constant cross-referencing.

All of the five aspects listed above are expanded upon below to explain how they are addressed by the handbook and where the information provided can be found.

1.1 How to get certified

Practical guidance on getting certified is found in Part Two. Information is provided for both forest managers interested in certification for their forests, and processors interested in chain-of-custody certification for their products. It is aimed at providing a good understanding and strategic overview. However, this book does not provide any guidance on understanding and implementing the technical requirements of forest management standards since this information is provided in the companion volume, *The Sustainable Forestry Handbook*. The guidance that is provided covers:

- Deciding *whether or not to get certified* (Chapter 7).
- Preparing for and undergoing *forest management certification* (Chapters 8 and 9).
- Preparing for and undergoing *chain-of-custody certification* (Chapters 11 and 12).
- Setting up a *Group Certification Scheme* for forests or chain of custody (Chapters 10 and 11).

1.2 Understanding the theory of certification

Those who want a better understanding of the theory underlying certification should go to Part One. A good understanding is useful for anyone involved with discussing or implementing certification, and is particularly important for anyone involved in making judgements about the adequacy of different schemes. Part One covers:

- What is a certification scheme and how do standards, certification, accreditation and claims fit into the overall structure (Chapter 2)?
- What is the International Organization for Standardization (ISO) and why is it so important (Chapter 2)?
- Why is the process of standard setting so important and what should finished standards look like (Chapter 3)?
- What are certification and accreditation and what makes them effective and credible (Chapters 4 and 5)?
- What is chain of custody and what kind of claims can and should be made about certified forests and products (Chapter 6)?

1.3 Existing forest certification schemes

There are now many different forest certification schemes in operation. Some of these are international, others are national or regional:

- Some of the most important and well-established of the existing schemes were invited to contribute material to the book. These descriptions, provided by the schemes themselves, can be found in Appendix 1.
- A summary of the different schemes is provided in Part Three.
- The complex issue of assessing certification schemes to ascertain which are acceptable or adequate is covered in Chapter 20.

1.4 Policy implications and impacts

Certification has implications for defining and driving the policy debate, as well as influencing the governance framework. This is particularly important for governments, donors and investors. It also has a range of impacts, positive and negative, which are beginning to emerge after a decade of development and use. Discussion in the handbook includes the following:

- The *origins of forest certification* are discussed in this chapter.
- *The range of ways in which forest certification can be used* is introduced in this chapter and discussed in more detail in Chapter 21.
- *Policy and governance implications* are discussed in Chapter 14.
- *The impacts of certification* are reviewed in Chapter 16, while the Enabling Conditions Framework discussed in Chapter 15 provides an analytical tool for examining impacts.

1.5 Remaining issues and potential ways forward

Certification remains a rapidly developing and changing topic. Problems remain, but solutions are always emerging. Some of the most important are discussed, including:

- the need for, and development of, *phased approaches* to certification (Chapter 17);
- the issues and barriers faced by small forest enterprises (Chapter 19);
- the need for further *capacity-building* (Chapter 18);
- mechanisms for *assessing and choosing appropriate certification schemes* (Chapter 20).

An important issue is the use of technical language. Certification has a very long history and over the last 50 years its use has been formalized and developed through organizations such as the International Organization for Standardization (ISO) (see Chapter 2). It is now widely used in almost every sector of industry throughout the world. As a result, many technical procedures have been developed, together with associated technical language. While this is not always easy to understand at first, it is important to become familiar with the key terms and approaches since they are widely used for certification of all kinds, not only forest certification.

This introductory chapter sets the scene for the rest of the handbook, focusing, in particular, on three topics:

1 A brief overview examines where forest certification came from and how it has developed over the last decade. This overview aims to provide context for those new to the discussion and a reminder for those who have been involved for longer.
2 The various ways in which certification can be used are discussed, expanding from the original market-driven approach to encompass a range of different purposes.
3 Finally, remaining issues are identified, which are then discussed further in subsequent sections.

1.6 The emergence and growth of certification

Certification was originally developed in order to provide a mechanism for verifying that a set of required characteristics, such as technical specifications, safety or product quality, were being achieved. It has proved an extremely efficient tool for doing this and is now widely used throughout industry.

As the concerns of purchasers and consumers have moved beyond technical and safety requirements to the environmental, social or ethical credentials of a product or process, certification has followed suit. It has now been adapted for use in areas such as organic agriculture, fair trade and social accountability. The development of forest certification has mirrored this wider trend. As discussed below, increasing concern about forest destruction led to the need for a tool to verify the environmental credentials of forest management and the products that forests produce.

1.6.1 GROWING INTERNATIONAL CONCERN ABOUT FORESTS

Forests emerged in the international policy agenda during the 1980s as a result of growing international concern about the extent and rate of deforestation. Several initiatives were taken by the international community in an attempt to address the issue.

The focus of many environmental non-governmental organizations (NGOs) was on undertaking campaigns to raise awareness and advocate boycotts, particularly of tropical timber, in an attempt to reduce pressure on these forests.

Donor-supported activities, which had previously focused mainly on project-level interventions, increased assistance for capacity-building and more holistic approaches, such as national forest programmes, were adopted. The potential for an international convention on forests was also explored in various international fora as a possible solution to create more political commitment to address deforestation through international action.

1.6.1.1 United Nations Conference on Environment and Development (UNCED)

In 1992, these efforts culminated in discussions about forests at the United Nations Conference on Environment and Development (UNCED), which identified three factors indicating that action at an international level was necessary (Maini, 2001):

1 intolerable rates of deforestation and associated loss of environmental, economic and social benefits;
2 threats to the livelihoods, culture and rights of forest dwellers and indigenous people in many parts of the world who live in and around forests;
3 meeting the continuously increasing demand for forest products.

UNCED agreed upon a set of overarching Forest Principles, which are fairly general, but have nevertheless provided a framework for action for subsequent efforts. The international policy deliberations continued under the Intergovernmental Panel on Forests (IPF) during 1995–1997 and its successor, the Intergovernmental Forum on Forests (IFF) during 1997–2000. The United Nations Forum on Forests (UNFF) was established to follow up the implementation of proposals for action by its predecessors and to further explore selected issues. While these have kept the debate active and strengthened political commitment to sustainable forest management (SFM), their contribution to improved forest management on the ground has remained limited.

The Earth Summit in Johannesburg in 2002 did not consider forests as a global priority in their own right, but mapped out an agenda whereby the role of forests should be recognized and enhanced in the alleviation of poverty, biodiversity conservation and conservation of water resources.

Numerous workshops, expert meetings and conferences have been organized since UNCED that aim to develop consensus views on forest policy and practical modalities for the promotion of sustainable forest management at international, national and local levels. Perhaps the most important outcome of these has been the development of a series of sets of criteria and indicators (C&I), each with a different geographical coverage, which aim to define sustainable forest management at nationals levels to allow governments to monitor and report on the state of their forests (see Chapter 15).

1.6.1.2 The role of the International Tropical Timber Organization (ITTO)

The role of trade in forest products as a contributing factor to deforestation has long been the subject of a heated international debate. It is now recognized that the direct effect of global trade on deforestation is limited by the fact that only a proportion of industrial roundwood harvested enters international trade in different forms. Nevertheless, commercial harvesting for trade is one of the factors contributing to forest disturbance and damage, either directly through poor management or indirectly by providing access to forests for other users. Therefore, a key issue is how trade can be made to work towards promoting sustainable forest management.

This issue has been the focus of the International Tropical Timber Organization (ITTO) since its inception in 1987. ITTO Objective 2000, which was agreed upon in 1990, requires that all tropical timber traded should come from sustainably managed forests. Objective 2000, which has still not been achieved, has been the strategic focus of the organization, and a whole range of tools and actions has been developed to assist in reaching this goal. A key instrument has been the *Criteria and Indicators for Sustainable Management of Natural Tropical Forests* (ITTO, 1998), together with a series of guidelines produced by ITTO. This pioneering work played an important role in the development of the nine regional C&I processes discussed in Chapter 14.

Box 1.1 **ITTO and forest certification**

- Bans and boycotts on tropical timber imports and use were frequently debated in the International Tropical Timber Council (1988–1992).
- LEEC study on economic linkages of sustainable management of tropical forests (1992) emphasized the need to develop economic incentives; a proposal was made to develop country certification, which was, however, not pursued.
- The first global study on the status of forest certification took place, which was discussed in a working party (1994–1995). The concept of certification was clarified and issues and action needs were identified. Monitoring of the situation was considered necessary.
- The second global study, 'Certification in Transition', was prepared and focused on experiences in six countries; a study on markets for certified tropical timber was carried out (1996). Further monitoring, sharing of experience and dissemination of information were considered important.
- A third global study, 'Pending Challenges', was carried out to monitor the situation, which emphasized the need for action to build capacity in producing member countries (1998).
- A submission was made by the government of Indonesia to develop the Lembaga Ecolabel Indonesia (LEI) project. After several revisions, it was approved and implemented.
- An auditing framework, the International Tropical Timber Organization criteria and indicators (ITTO C&I), was developed to build up local auditor/assessor capacity (1999–2001).
- A study on equivalence and comparability of certification systems was initiated, and an international workshop was organized (2001).
- Training packages were developed, and auditing training on C&I through regional training courses was organized (2002–2003).

ITTO has actively debated and considered the issue of certification (see Box 1.1). Initially, it was considered a potential constraint to market access by many tropical timber-producing countries, while many consuming countries emphasized its potentially positive role in promoting good forest management. The current policy of ITTO is to be supportive of voluntary certification as a tool for improving forest management, but not to support any specific scheme or approach.

1.6.1.3 Illegal logging and forest governance

Recently illegal logging and trade in associated timber products have received increasing attention. Illegal logging is responsible for vast environmental damage in developing countries, and impoverishes rural communities who depend upon forest products for a living. According to the World Bank, it also costs governments in developing countries an estimated 10 billion to 15 billion Euros every year in lost revenue.

The issue of illegal logging was the focus of a Group of eight (G8) summit in 1998, resulting in the publication of the G8 foreign ministers' *Action Programme on Forests*. The European Union's (EU's) response was FLEGT – *Forest Law Enforcement, Governance and Trade*. The EU FLEGT action plan was adopted in 2003. It proposes measures to increase the capacity of developing and emerging-market countries to control illegal logging, while reducing trade in illegal timber products between these countries and the EU.

At the same time, the forest products industry has also recognized the importance of curbing illegal logging and trade, and an increasing number of companies, particularly large multinationals, as well as trade and industry associations, have adopted policies excluding the use of any wood or fibre from illegal sources. All of this has implications for forest certification (further information on illegal logging can be obtained from www.illegal-logging.info).

1.6.2 THE EMERGENCE OF CERTIFICATION

During the 1980s, the general public in developed countries became sensitized to the seriousness of forest loss, particularly tropical deforestation. Frustrated by lack of progress through the governmental efforts discussed above, the NGO community started a range of actions against the tropical timber trade, perceiving this as the only way in which they could influence the situation. Actions included campaigning, demonstrations at traders' and retailers' premises, and advocating total bans on tropical timber use. The NGOs took the view that banning trade would result in reduced deforestation.

Gradually, many of the NGOs involved realized that this was too simplistic since forests that do not have value for local populations are likely to be converted to other uses rather than protected in their natural state. Positive instruments were therefore needed to create such value in the marketplace and link it to responsible management.

At the same time, as a result of the campaigns a number of key retailers had realized that they had very little information about the sources of their wood and paper products, and had not taken any control over the environmental and social impacts of their purchasing decisions. Therefore, they saw the value in a mechanism that would provide a straightforward and credible way for them to source wood and paper products from forests with acceptable social and environmental management. This was the breeding ground from which the idea of certification of forest management and related product labelling emerged.

Some early attempts were made to develop certification through existing institutions. In 1989, Friends of the Earth (FoE) and several other NGOs, supported by the UK government, proposed that the ITTO carry out a project to study the possibility of labelling timber from tropical forests to indicate whether the products come from forest managed for sustainable production. However, some producer countries expressed concerns that NGOs might call for boycotts of timber that was not labelled, and the initiative was abandoned. The issue continued to be debated in the ITTO (Poore, 2003), and in 1992 a major study was undertaken. This concluded that trade in topical timber was not a major cause of deforestation (LEEC, 1992), but emphasized the need for positive economic incentives, and country certification was proposed as a means of facilitating trade in sustainably produced timber. However, this was not supported for a number of reasons, and again the idea was dropped (see Box 1.1).

Thus, it was on the US West Coast where forest certification first emerged when trade unions, NGOs and others decided to set up a scheme that could differentiate environmentally benign products in the marketplace. After a germination period, the first forest certification scheme, the Forest Stewardship Council (FSC, see Chapter 13), was established in a founding assembly in 1993 with great enthusiasm among the participants. These included a wide range of economic, social and environmental interests, including many major environmental NGOs and global retailers; however, very importantly for subsequent developments, governments and a significant part of the mainstream forest industry were not involved.

1.6.3 PROLIFERATION OF SCHEMES

Until 1997 the FSC remained practically the only operational certification system in the world. It served as a focus for policy discussions and promotion of certification. Without the FSC, certification would certainly not have made a fundamental impact on the setting of forest standards, auditing their compliance for forest management and labelling certified products in the international marketplace (Elliott, 1999). The FSC's importance and visibility in the international arena was attributed to four factors (Baharuddin and Simula, 1998):

1 strong NGO support;
2 the lack of viable alternatives;
3 availability of external funding;
4 the quality and commitment of the organization's staff.

However, as discussed above, private forest owners and important players in the global forest products industry were not involved in the FSC and saw it as an actual or potential threat. There were various reasons for this, some more explicit than others:

- concern among many tropical timber producers that certification would be a new barrier to markets, particularly in Europe and North America;
- fears in parts of the forest products industry that the FSC, an organization strongly influenced by NGOs, would gain too much influence over the industry if FSC certification was widely embraced by the market, giving the scheme a global monopoly;
- concern among small-scale private forest owners, particularly in Europe, that certification would reduce their rights to control management of their forests, was not adapted to small enterprises and would result in huge increases in costs and bureaucracy;
- resistance amongst forest owners and managers to the concept that other stakeholders had an equal right to be involved in defining what is good forest management;
- concern within some governments that the multi-stakeholder approach and international endorsement of national forest management standards required by the FSC would undermine national sovereignty over natural resources.

Initially, the reaction among the interest groups who did not support the FSC was to oppose certification completely. However, it gradually became clear that in a global economy where independent verification was widely accepted as a normal part of business, this was not a viable approach. Therefore, a number of other schemes began to emerge emphasizing the national context of certification. These initiatives were mainly promoted by interest groups who were dissatisfied with the FSC approach or even opposed to it.

One of the earliest of these initiatives was a proposal by the forest industry to create a document setting out performance requirements for forestry to be linked to the ISO 14001 Standard on Environmental Management Systems (see Chapter 3). An ISO technical committee (TC207) was set up to undertake this work. However, there was strong resistance from other industry sectors to any type of internationally agreed performance thresholds to be linked to ISO 14001 at such an early stage in its own development. Therefore, though a document was developed setting out how performance requirements for forestry should be developed in the context of ISO 14001 (ISO, 1998), it was classified as guidance only. However, one consequence was that there was initially strong opposition to ISO 14001 from many NGOs who associated it with this early attempt to develop an alternative to the FSC.

Parallel to the ISO work, national-level schemes started to emerge in a number of different countries covering a wide range of forest types, including, among many others, Indonesia, Canada, Finland, Brazil and the US . These emerging schemes were developed by a wide range of different groups using a number of different approaches. They are discussed in detail in Chapter 13.

1.6.4 MUTUAL RECOGNITION

The emerging national schemes were all faced with the problem of broader acceptance in export markets. Certification was being largely driven by retailers in Europe and North America who had three important concerns:

1 NGO *support*: certification was often being used as a means of brand protection, so it was crucially important that any scheme they used was supported by campaigning NGOs.

2 *Global coverage*: big retailers purchase globally, making it much easier to adopt one global scheme such as the FSC rather than appraising the merits of many different national schemes and getting consumers to understand different national environmental labels.
3 *Sufficient supply*: many companies made commitments to procure only certified products, so the issue of sufficient supply of all types and sources of wood and fibre was also crucial.

The FSC was already established, was internationally applicable and had support from most of the major NGOs. However, in some regions its progress on the ground had been slow, resulting in a problem of inadequate supply. Therefore, the issue of accepting other schemes gained importance among buyers. At the same time, a number of the national schemes began to look at the concept of mutual recognition as a means of providing international coverage and credibility to facilitate acceptance in the marketplace.

A number of initiatives were developed to try to develop criteria to assess the acceptability of schemes (see Chapter 20 for further discussion), thereby acting as a basis for mutual recognition for all schemes. The problem with these different initiatives, however, as with the schemes themselves, was the political nature of the issue. Interest groups tended to support their own schemes and their own set of assessment criteria while remaining critical of those developed by other interest groups. As a result, despite a series of meetings and workshops, a generally agreed framework for mutual recognition between certification schemes has not emerged.

However, in 1997, a number of national initiatives in Europe decided to set up the Pan-European Forest Certification (PEFC) scheme as a mechanism to allow mutual recognition of their national certification schemes. At the end of 2003, the PEFC decided to expand its geographical scope to become a global framework for assessing and recognizing forest certification schemes. Changing its name to the Programme for the Endorsement of Forest Certification (which allows the acronym PEFC to remain unchanged), the scheme is now effectively a global framework for mutual recognition among national schemes.

1.7 The widening remit of certification and verification

Although market demand was the original driver for the development of forest certification, it quickly became apparent that there is a range of different types of forest certification or verification (see Box 1.2) that can serve a number of purposes. Forest managers (both in the private and public sector) are increasingly called upon to make credible demonstrations of how forests are managed and for whose benefit. As a result, certification may be needed in response to the varying concerns of civil society, governments, financing institutions or other interested parties – and not just market players. Over the decade since the founding of the Forest Stewardship Council, a number of different types of forest certification have started to emerge. Although this book focuses primarily on market-oriented certification schemes, it is also important to consider the range of other approaches that exist or are being developed, and which are reviewed briefly below.

1.7.1 VOLUNTARY MARKET-ORIENTED SCHEMES

As discussed above, the original purpose of market-oriented certification was as a tool to allow the forest products industry and retailers to source their wood and fibre products from well-managed forest sources. It was widely supported by environmental and social NGOs, as well as many development organizations, because it was seen as a means of allowing market forces to contribute to the improvement of the quality of forest management around the world.

It was also supported by some producers as it was seen as a means of providing better access to environmentally aware markets or to even better prices for products from well-managed forest sources.

Box 1.2 **Forest certification and verification**

At the heart of the certification concept is the requirement for verification that each requirement of the standard has been met. Therefore, the term 'verification' is also used to describe the process. While verification is sometimes used interchangeably with certification, the two terms are more usually used slightly differently:

- Verification is used generically to cover any approach to checking that a set of requirements has been met.
- Certification is used explicitly where the verification is undertaken by an accredited certification body against the requirements of a specific standard.

This book focuses primarily on certification; but the concepts and much of the practice are equally applicable to verification.

1.7.2 CERTIFICATION OF SPECIFIC FOREST MANAGEMENT REQUIREMENTS

In both the market and the non-market context, certification or verification can be used by a wide range of actors to make sure that their specific forest management requirements are being implemented. This can be achieved either through the use of an existing certification scheme or through verification against a defined set of requirements produced by users, who include:

- *Investors and donors*: increasingly, both private-sector investors and donor organizations are under pressure to ensure that they do not invest in poor forestry. As a result, many organizations have policies requiring a defined level of forest management performance from all organizations in which they invest. Certification or verification is a credible way of confirming that these policies are being met.
- *Governments*: governments have a range of different situations in which they need to confirm that particular requirements are being met – for example, checking that concession holders are meeting licence agreements; monitoring the implementation of participatory forestry agreements; ensuring that grant aid is being spent effectively; or assessing the implementation of policies that protect the rights of indigenous or local communities. Certification provides an independent, credible mechanism for checking that specific requirements are being implemented. Another potential use of certification is to check that legal requirements are being met by forest organizations. Where this has been considered, it has generally been linked to the idea of mandatory certification, discussed below.
- *Shareholders and employees*: corporate responsibility is considered a fundamental requirement of good business practice by most responsible companies. Certification can provide a useful mechanism for demonstrating to both shareholders and employees that a forest company is implementing responsible stewardship of its resources.
- *Purchasers*: many organizations in both the public and private sectors are committed to excluding wood or paper products from illegal or other unacceptable sources. Independent certification or verification of legality is increasingly important for these organizations in meeting their commitments, particularly with the development of initiatives such as the EU FLEGT programme discussed above.

1.7.3 LEGAL VERIFICATION

As discussed above, verification of legality is likely to become an increasingly important service as large parts of the forest products industry and many retailers and governments all seek to exclude illegal timber from the supply chain.

1.7.4 MANDATORY CERTIFICATION/VERIFICATION

Governments can potentially institutionalize forest certification and make it mandatory in some situations. For example, where concessions have been awarded or where the local communities have been entrusted with the management of state forest lands, successful achievement and maintenance of certification can be a requirement for continued rights to the resource. Mandatory certification can be applied in two ways:

1. Through the development of a mandatory certification scheme owned by the government: the only country where there is already a law in place to allow this is Russia,[1] where the purpose of the mandatory scheme is to attempt to improve the enforcement of regulatory requirements. However, the country is also in the process of developing voluntary certification and the future application of the mandatory system is uncertain.
2. By making the achievement of certification through an existing voluntary scheme a mandatory requirement: such an option is being developed in some countries, such as Bolivia.

1.7.5 ENVIRONMENTAL SERVICES

Certification can also be used for verifying various compliance conditions or provision of environmental services, such as carbon sequestration, biodiversity conservation or watershed services. This type of certification is performance based, and it will often also be market oriented as certification can help to convert such services into marketable commodities (for example, carbon credits, described in Chapter 21).

1.7.6 INTERNATIONAL COMMITMENTS

Finally, certification may be applied to verify the compliance of a country or an organization with international commitments (for example, the emission reduction targets of the Kyoto Protocol – in this case, changes in carbon sinks).

Of all of these potential uses, voluntary market-based certification is still the most widely used; but other uses of certification and verification are gaining importance and are likely to become more important over the next few years.

1.8 Progress and unresolved issues

1.8.1 PROGRESS

Certification has made huge progress since it was first conceived a decade ago. Tens of millions of hectares of forest have been certified and thousands of wood and paper products have carried logos and labels. This has resulted in many real improvements in forest management and in the environmental and social contexts of forests, which are discussed in detail in Chapter 15. However, there are still a number of unresolved issues related to forest certification so that it remains an extremely contentious subject in the forestry sector. Overall, the issues can be divided into two types:

1 *Political problems*: these are issues arising from differences in both the interests and the core values of the various interest groups involved.
2 *Technical problems*: these are issues arising from the particular situation or constraints of different actors.

1.8.2 POLITICAL ISSUES

The two most significant political issues are the remaining opposition to certification per se and the ongoing conflict between different certification schemes and their supporters.

1.8.2.1 Opposition to certification

There are significant economic interests involved if certification becomes a *de facto* baseline requirement for timber and fibre being traded internationally. Individual countries are in different positions to make use of this instrument. For example, some tropical countries consider certification as an obstacle to market access as their ability to use certification is more limited than developed countries.

In addition, as discussed above, certification shifts the balance of power between different interest groups, and there is often reluctance from the groups who have had the most influence over forests to let go of this to other groups. However, as the demand for certification grows, and the range of perceived uses and benefits expands, the degree of opposition to certification per se is decreasing (see Chapter 14). Approaches such as the development of national standards not necessarily affiliated to any certification scheme seem to help this process.

1.8.2.2 Continued competition between schemes

Although some stakeholders feel that it is useful to avoid a 'monopolistic' situation in forest certification, many buyers would prefer to have one scheme and one logo on the products traded. Furthermore, financing institutions, donor agencies and other organizations who are using certification would often prefer to have a single option as they do not have the time, expertise or resources to choose between schemes.

However, it seems unlikely in the short term that there will be any agreement between the schemes to limit competition for a number of reasons. Firstly, it is increasingly recognized that part of the reason that there is still conflict about forest certification is due to real differences in the deep core values of different actors and supporters of different certification schemes (Cashore, 2004). This is particularly so with relation to forest management standards because of the fundamental role of standard-setting in defining how forests should be managed and the necessity to resolve conflicts between interest groups over how forests can be used in order to make this definition. As a result, it has been difficult to find areas of common ground from which to begin building consensus.

In addition, particularly for FSC and PEFC – the two international schemes – there is little incentive to reduce competition because:

- They are competing in the marketplace and their commercial success in building market share is linked to their future resources.
- The FSC already has a strong visibility in some market segments and is currently perceived as being more credible by most of the influential NGOs; therefore, any links between FSC and PEFC would be more likely to benefit PEFC.

As a result, there will be a continued need for users of forest certification schemes to make choices about the schemes they wish to use. This requires the development of accepted mechanisms for assessing schemes and deciding which ones are credible. This is discussed further in Chapter 20.

1.8.3 TECHNICAL ISSUES

The most pressing technical issues that remain are the barriers still faced by small forest owners and enterprises and the slow rate of progress in many developing and transition economy countries.

1.8.3.1 Barriers for small forest owners and enterprises

The difficulties faced by small forest owners and enterprises have been recognized since the early days of certification, and comprised one of the major driving forces behind the development of the PEFC. However, although mechanisms have been found that are reasonably successful in some contexts, the huge variety of small forest enterprises globally means that many are still facing barriers. This is discussed in Chapter 18.

1.8.3.2 Slow progress and the need for phased approaches

Certification was originally developed as an alternative for boycotts and bans on tropical timber use. However, about 90 per cent of the world's certified forests are located in the developed countries and the main beneficiaries of the instrument have been owners of temperate forests and related downstream industries. The markets have been slow to understand the specific problems that developing countries are facing in getting their producers certified (Eba'a Atyi and Simula, 2002).

One possible solution that has begun to emerge recently is the development of phased approaches to certification. This is discussed further in Chapter 18.

Note

1 The mandatory certification scheme in Russia is not yet operational as the respective decree has not been issued.

Part One

How Forest Certification Schemes Work

2 What is a Forest Certification Scheme?

This chapter introduces the main features of forest certification schemes and is followed by Chapters 3 to 6, which look at four critical elements – standards, certification, accreditation and claims – in more detail.

2.1 The elements of a certification scheme

Certification has been widely used in a range of sectors as a tool for providing independent verification that a defined set of requirements is being met. As a result, there is already considerable experience of what works and what does not, particularly for market-oriented schemes, which has led to generally accepted approaches that have been shown to work well in practice.

In general, all certification schemes, including those for forestry, are usually made up of three elements:

1 *Standard*: this sets out the requirements that must be met and against which certification assessments are made. Standards are developed by standard-setting bodies. Standards are discussed in detail in Chapter 3.
2 *Certification*: this is the process of establishing whether or not the standard has been met, usually carried out by a certification body or 'certifier' (also referred to as a registration body or 'registrar', especially in North America). Certification is discussed in detail in Chapter 4.
3 *Accreditation*: this is the mechanism for ensuring that the organizations which undertake certification are competent and produce credible, consistent results, sometimes described as 'certifying the certifiers'. Accreditation is discussed in detail in Chapter 5.

To emphasize the importance of each of these elements to the certification scheme as a whole, they are often shown schematically as the three sides of a triangle, which represents the certification scheme, as shown in Figure 2.1.

In addition, if the scheme is going to be used as a basis for identifying products from certified forests and for making *product claims*, then a mechanism for controlling this also needs to be in place. This requires mechanisms for:

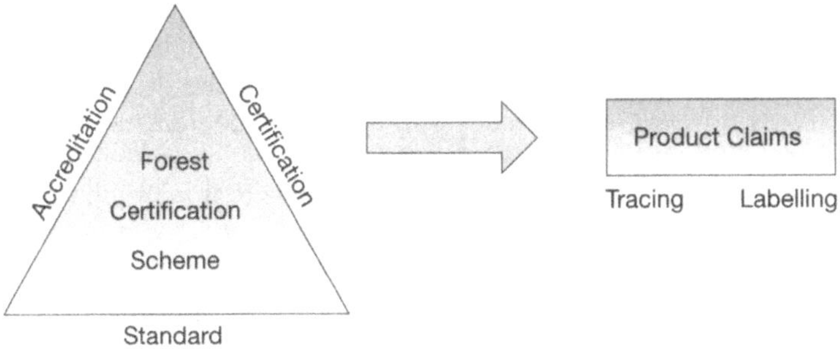

FIGURE 2.1 A schematic representation of the essential elements of a credible forest certification scheme, including product claims

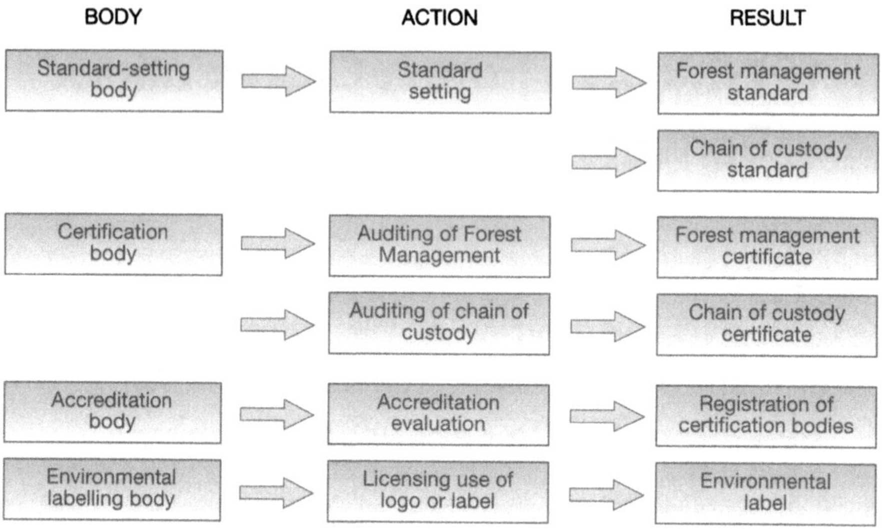

FIGURE 2.2 **Elements of a market-oriented forest certification scheme**

- *Tracing*: the material may go through many production and logistical stages between the forest and the final product and there must be a mechanism for tracing it from the certified forest through each stage. This is to provide certainty that the product or product line about which the claim is being made really does use material from a certified forest. This process is often referred to as chain of custody.
- *Claims and labelling*: it is essential to ensure that any claims made about a forest or a product or any labels are clear, credible and honest. This generally requires a set of rules to be followed by those making claims or labelling products.

Product tracing and labelling are discussed in detail in Chapter 6.

The various bodies involved in certification, the actions they take and the results can be summarized as shown in Figure 2.2.

2.2 The International Organization for Standardization (ISO)

Forest certification is a relative newcomer to the world of standards and certification; but thousands of other schemes already exist in many other sectors. One of the most important organizations in the world of certification is the International Organization for Standardization, generally known as ISO.[1]

ISO is a non-governmental organization made up of a network of the national standards institutes from more than 140 countries, large and small, industrialized and developing, in all regions of the world. Each country can have one member, with the overall coordination undertaken by a central secretariat in Geneva, Switzerland (see Box 2.1).

The exact structure of the national standards institutes that make up the membership varies. Some are part of the government structure or at least are mandated by government, while others have their roots much more strongly in the private sector, having been set up by national partnerships of industry associations.

Since it was set up in 1947, ISO has produced more than 14,300 standards, almost all of which are precise technical documents defining items such as the size of bank cards or graphical symbols for use on equipment. These standards affect many aspects of everyday life, but are generally not known outside the industry to which they apply. More recently the organization has produced more generic standards, such as the quality and environmental management system standards ISO 9000 and ISO 14001, which can be applied to any organization and which are much more widely known.

As well as coordinating and managing the development of hundreds of international standards for different industry sectors, ISO has also produced a number of guides to help with the development of certification schemes. The guides (see Box 2.2 for a list) provide guidance on:

- the development and use of standards;
- certification bodies and certification;
- accreditation; and
- claims.

These ISO guides, which are based on several decades of experience, usually provide excellent baseline requirements and are discussed in detail in the next four chapters. ISO guides are also important because ISO works with the World Trade Organization (WTO) to try to ensure that certification meets WTO requirements and does not become a technical barrier to trade.

In general, certification schemes should follow ISO guidance for each of the four elements. This guidance is extremely helpful in outlining the basic requirements for credibility and efficiency. However, it is also important to remember that the ISO guides are general guidelines, developed to be applicable in the widest possible way. They do not always go into sufficient detail to provide all necessary guidance for individual schemes, nor do they cover all of the issues that are important in a specific sector, such as forestry.

Further information about ISO, as well as copies of the ISO guides, can be found on the website at www.iso.org.

2.3 Scheme ownership and governance

It is important who owns and controls a forest certification scheme and what its governance structure is because it is likely to have an impact on technical aspects of the scheme, on what the scheme tries to deliver and on the perceptions of the scheme among different stakeholder groups. In particular, it will tend to influence:

- the way in which the different elements of the scheme are developed and implemented (the importance of this is discussed in detail in Chapters 3 to 6);
- the way in which the scheme is run both on a day-to-day basis and over the medium and long term; this, in turn, is likely to have a strong influence on what it delivers;
- the perception of the scheme among different stakeholder groups.

In fact, a significant part of the debate and controversy about forest certification over the last decade can be linked to the perceptions of key stakeholder groups about the ownership and governance of the different schemes, as many of those involved in the debate have only a limited understanding of the technical content of the schemes being debated.

There are a number of approaches to scheme ownership and governance, which are discussed below.

Box 2.1 **The structure and functions of the ISO**

The structure of the International Organization for Standardization (ISO) is shown schematically below. The roles of some of the main bodies within the structure are as follows:

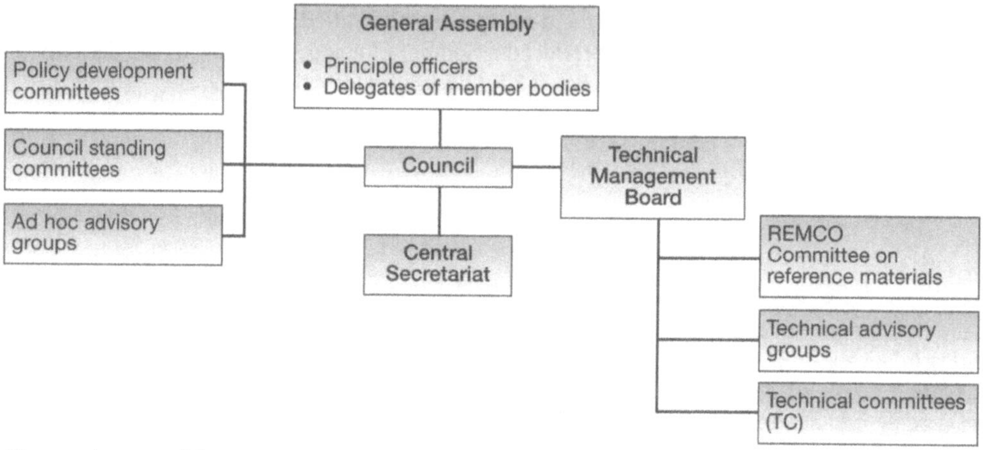

General assembly

All strategic decisions are referred to the ISO member bodies, who meet for an annual general assembly.

Council

The operations of the ISO are governed by the ISO council, consisting of the principle officers and 18 elected member bodies. The ISO council meets three times a year and its membership is rotated to ensure that it is representative of ISO membership.

Central secretariat

The ISO central secretariat is in Geneva, Switzerland, and provides administrative and technical support to ISO members, coordinates the decentralized standards' development programme and publishes the output. It is managed by a secretary-general who is a permanent member of staff. The secretary-general reports to a president who is a prominent figure in standardization or in business, and who is elected for two years.

Technical committees

ISO standards are developed by technical committees that are composed of experts from the industrial, technical and business sectors which have asked for the standards, and which subsequently put them to use. These experts may be joined by others with relevant knowledge, such as representatives of government agencies, testing laboratories, consumer associations, environmental groups and so on. The experts participate as national delegations, chosen by the ISO national member institute for the country concerned. These delegations are required to represent not just the views of the organizations in which their participating experts work, but of other stakeholders, too. According to ISO rules, the member institute is expected to take account of the views of the range of parties interested in the standard under development, and to present a consolidated, national consensus position to the technical committee.

Source: ISO website, www.iso.org, January 2004

Box 2.2 ISO guides on setting up and running certification schemes

ISO Guide 59: 1994 code of good practice for standardization

ISO Guide 61: 1996 general requirements for assessment and accreditation of certification/registration bodies. ISO Guide 61 has been replaced by new guidance published in October 2004. This is ISO 17011 'General requirements for bodies providing assessment and accreditation of conformity assessment bodies'

ISO Guide 62: 1996 general requirements for bodies operating assessment and certification/registration of quality systems

ISO Guide 65: 1996 general requirements for bodies operating product certification systems

ISO Guide 66: 1999 general requirements for bodies operating assessment and certification/registration of environmental management systems (EMS)

ISO 14011: 1996 guidelines for environmental auditing – audit procedures; auditing of EMS

ISO 14012: 1996 guidelines for environmental auditing – qualification criteria for environmental auditors

ISO 14020: 2000 environmental labels and declarations – general principles

ISO 14021: 1999 environmental labels and declarations – self-declared environmental claims (type II environmental labelling)

ISO 14024: 1999 environmental labels and declarations (type I environmental labelling) – principles and procedures

ISO/TR 14025: 2000 environmental labels and declarations (type III environmental declarations)

2.3.1 THE 'TRADITIONAL' APPROACH

Globally, the most common ownership and governance structure for certification schemes in a whole range of sectors is through control of the different elements of the scheme by different entities:

- *The standard*: the majority of international standards are developed and maintained by ISO, which has detailed guidelines for the way in which this development is undertaken (see Chapter 3).
- *Certification*: most certification bodies are commercial companies which specialize in certification. These can be either international or national.
- *Accreditation*: accreditation for most ISO standards is developed and maintained by national accreditation bodies, based on the guidance provided by ISO.
- *Control of claims*: this is generally provided as part of the accreditation process by the national accreditation body again, based on guidance provided by ISO.

This is the approach adopted for standards such as ISO 9000 and ISO 14001, as well as thousands of other ISO and national standards for product quality and safety. This model has some very strong advantages:

- The governance and ownership structure is very clear.
- There is complete separation between the organizations responsible for setting standards, accrediting certification bodies and undertaking certification assessments, which increases the level of independence and avoids conflicts.
- Each of the organizations involved is a specialist in its own field with well-developed ways of working that should provide both quality and consistency.

However, as the use of standards moves away from the traditional areas of product quality, product safety and management systems into the realms of environmental and social performance, it has become clear that there are also some weaknesses:

- Both ISO and national standards bodies are dominated by industry and government interests since traditionally these have been the groups concerned with the type of standards produced. Other groups, such as environmental and social NGOs, are slowly being included; but generally they do not yet have equal representation and influence.
- While the ISO approach is robust and valuable, it can lack the flexibility and speed needed to pioneer standards looking at social and environmental issues.

2.3.2 THE 'SPECIALIST' APPROACH

In response to the need to find development processes that are both flexible and expert enough to deal with the complexities of many social and environmental issues, another approach, which has become more common recently, is for an international organization specializing in a particular field to develop and maintain a certification scheme. For example, the international social accountability standard SA 8000 was developed by Social Accountability International (SAI) while the Fairtrade Labelling Organization (FLO) has developed international standards of fair trade for a range of commodities. In this model, there are a number of ways in which accreditation and certification are undertaken. FLO has a certification unit that undertakes certification work against its own standards, while SAI acts as an accreditation organization and accredits certification bodies to undertake certification against the SA 8000 standard.

2.3.3 GOVERNANCE OF FOREST CERTIFICATION SCHEMES

The development of forestry standards reflects this growing diversity of approaches with a range of different ownerships and governance structures.

A number of organizations have been set up specifically to develop and run forest certification schemes. Examples include the Forest Stewardship Council (FSC), the Programme for the Endorsement of Forest Certification Schemes (PEFC), the Sustainable Forestry Initiative (SFI) and Lembaga Ekolabel Indonesia (LEI).

Other schemes have utilized the existing national standards process, such as the Canadian Sustainable Forestry Standard, which was developed by the Standards Board of Canada with certification bodies accredited by the Canadian national accreditation body.

Whichever approach is used, the critical question when looking at the governance of a certification scheme is who has influence and control over the development, implementation and revision of the standard and the requirements for certification and accreditation.

2.4 The importance of technical detail

Certification is a very precise and technical process and differences in technical detail can have a very big impact on the practical outcome delivered by a particular scheme. Therefore, although there is almost universal agreement about the main elements of a certification scheme (see Section 2.1), and most schemes contain these elements, there remains a heated debate about which certification schemes are acceptable. This is because there are differences in the detailed design of each element and, therefore, in what each scheme delivers.

For example, two standards both address health and safety. One requires that all operators are adequately trained and provided with safety equipment. The other requires that safe working practices are implemented. Only the latter guarantees that a certification body must check that all operators in a certified operation are actually working safely.

Two certification processes both require consultation. One specifies that consultation must include all local interested parties. The other specifies that the certifier should publicize the forthcoming certification on its website and consult with anyone who contacts them. Only the former approach will ensure input from local indigenous people without access to a computer.

While it is clear that a forest certification scheme based on a standard that is universally thought to be inadequate is unlikely to be useful, and similarly a scheme utilizing a certification process that is unreliable and prone to error is pointless, there is, nevertheless, a wide spectrum of thought about what constitutes an 'acceptable' scheme.

In order to really understand and discuss certification, it is necessary to understand some of the detail of the design of each element of a certification scheme and the implications of the design for the way in which a scheme functions. Chapters 3 to 6 provide a detailed analysis of standards, certification, accreditation and claims. A discussion of how to assess certification schemes can be found in Chapter 20.

Note

1 Because 'International Organization for Standardization' would have different abbreviations in different languages (IOS in English and OIN in French for *Organisation Internationale de Normalisation*), it was decided to use a word derived from the Greek isos, meaning 'equal'. Therefore, whatever the country, whatever the language, the short form of the organization's name is always ISO.

3 Forest Standards

3.1 Introduction to forest management standards

As discussed in Chapter 2, standards set out the requirements that must be met by any organization wishing to be certified and against which certification assessments are made. The International Organization for Standardization (ISO) definition of a standard is:

> ... *a document, established by consensus and approved by a recognized body, that provides, for common and repeated use, rules, guidelines or characteristics for activities or their results, aimed at the achievement of the optimum degree of order in a given context* (ISO/IEC Guide 2: 1996, definition 3.2).

The content of the standard is fundamental to a forest certification scheme since the standard provides the basis for the level of forest management that will be delivered by the scheme. Only those elements that are required by the standard are guaranteed in a certified forest.

There are two types of standard that can be applied to forest enterprises – system-based standards and performance standards.

3.1.1 TYPES OF STANDARD: SYSTEM AND PERFORMANCE

3.1.1.1 System-based standards

Management system or *process* standards specify the management systems that must be in place within an organization to ensure that they are managing quality, environment or even social performance consistently. Therefore, the requirements of management systems standards relate to elements of management that must be in place, rather than requirements about the outcomes or results of management. The best-known management systems standards are the quality standard ISO 9000 and the environmental management system (EMS) standard ISO 14001. It is the latter of these, ISO 14001, which can be used as an environmental standard for forest organizations.

Systems standards have some great strengths. Firstly, they can be applied to any sector or industry. Thus, ISO 14001 can be applied equally to a forest enterprise, a pulp mill or a furniture factory. This is particularly useful for integrated companies. Secondly, systems standards can be very powerful tools for helping organizations to systematically understand their performance and ensure that it is continuously improved. Systems standards are easily adapted to organizations operating in all types and sizes of forest since they specify generic systems and not specific performance requirements. In addition, certification to a systems standard provides recognition of the organization's commitment to improve while the improvements in performance are still being achieved.

However, systems standards do not specify any minimum level of performance that must be achieved. Instead, they require forest organizations to set their own performance targets and then use the management system to ensure that they reach them. This means that two forest companies, both certified to the same system standard, can have very different levels of performance in the forest. This is highlighted in the introduction to the ISO 14001 standard, which states:

> *It should be noted that this international standard does not establish absolute requirements for environmental performance beyond commitment, in the policy, to compliance with applicable legislation and regulations and to continual improvement. Thus, two organizations carrying out similar activities but having different environmental performance may both comply with its requirements* (EN ISO 14001: 1996 environmental management systems – specification with guidance for use).

As a result, since system-based standards do not provide any 'guarantee of product quality', it is not normal to associate a product label with this type of standard.

3.1.1.2 Performance standards

Performance standards specify the level of performance or results that must be achieved, but do not necessarily specify how this should be done. Therefore, they do not require an organization to put in place any particular management system, but they clearly specify the minimum performance that must be achieved in a certified forest.

The strength of this approach is that it provides a guarantee that a certified forest meets a defined level of performance. Since performance standards provide this 'guarantee of quality', it is normal to use them as a basis for a product label.

Systems standards apply to a particular forest organization (a company, a landowner, an association of owners), while performance standards apply to a forest management unit (a defined area of forest) and the quality of management in that forest. A variety of terms are used to describe this quality of management, including 'responsible forest stewardship', 'good practice' and 'sustainable forest management'.[1]

A comparison of the two types of standard, summarized in Box 3.1, shows that they:

- deliver different benefits;
- are potentially complementary, but do not substitute for each other.

Management systems standards are very useful for providing a management framework within which improvements can be recognized and made; but they do not give any guarantee of actual performance in the forest. This is only provided by performance standards and for this reason product labelling is always associated with a performance standard.

In practice, ISO 14001 is the main management systems standard applied to forestry. All forest certification schemes have developed standards that include some degree of performance requirements, though most also include some systems requirements.

Box 3.1 Comparison of what system and performance standards deliver for forest management

	System standard	Performance standard
Guaranteed minimum level of performance in the forest	No	Yes
Recognition of ongoing improvements in management	Yes	No
Management framework	Yes	No
Application to all forest types without being adapted	Yes[a]	No
Product label	No	Yes

Note: a In practice, the bureaucratic requirements of systems standards can be a serious obstacle for small forest enterprises and for forest owners and managers who are not literate.

Box 3.2 **How ISO standards are developed**

The need for a standard is usually expressed by an industry sector, which communicates this need to a national member body. The latter proposes the new work item to the International Organization for Standardization (ISO) as a whole. Once the need for an International Standard has been recognized and formally agreed, the first phase involves defining the technical scope of the future standard. This phase is usually carried out in working groups comprising technical experts from countries interested in the subject matter.

The national delegations of experts of a technical committee meet to discuss, debate and argue until they reach consensus on a draft agreement. This is then circulated as a Draft International Standard (DIS) to ISO's membership as a whole, for comment and balloting. Many members have public review procedures for making draft standards known and available to interested parties and to the general public. ISO members then take account of any feedback they receive in formulating their position on the draft standard. If the voting is in favour, the document, with eventual modifications, is circulated to ISO members as a Final Draft International Standard (FDIS). If that vote is positive, the document is then published as an International Standard. The acceptance criteria stipulate approval by two-thirds of the ISO members who have participated actively in the standard development process, and approval by 75 per cent of all members who vote.

Every working day of the year, an average of 11 ISO meetings take place somewhere in the world. In between meetings, the experts continue the standards' development work by correspondence. Increasingly, their contacts are made by electronic means and some ISO technical bodies have already gone over entirely to electronic working, which speeds up the development of standards and reduces travel costs.

Source: ISO website. www.iso.org, January 2004

Therefore, the remainder of this chapter focuses on performance standards and their development for forestry. There are two elements to standards that must be considered:

1 the process used to develop the standard;
2 the content of the standard which is developed.

Each of these is discussed below.

3.2 Forest standard development process

The process of developing a standard is usually a relatively lengthy and complex one. Box 3.2 outlines the normal process for developing an ISO standard. Based on this experience, ISO has developed a number of guidelines for developing standards, in particular Guide 59: Code of Good Practice for Standardization (see Box 3.3). This provides a widely accepted basis for the minimum requirements expected of a certification scheme in developing and using a standard.

As the demand for environmental and social standards develops, there is also growing experience of some specific issues relating to the development of these types of standard. The International Social and Environmental Accreditation and Labelling (ISEAL) Alliance has used this experience to develop 'The Code of Good Practice for Setting Social and Environmental Standards' (ISEAL, 2004). This incorporates much of the guidance from Guide 59, but also discusses some of the additional aspects that need to be considered for the development of standards which address complex social and environmental issues.

Box 3.3

ISO guidelines for standard development

Some of the main requirements of ISO/IEC Guide 59: 1994, Code of Good Practice for Standardization are as follows.

Procedures

Written procedures based on the consensus principles should govern the methods used for standards development (clause 4.1; see Chapter 4 in this volume for a discussion of what 'consensus' means in practice).

Transparency

The procedures of the standardizing body shall be available to interested parties upon request (clause 4.1).

Complaints and appeals

The procedures of the standardizing body should contain identifiable, realistic and readily available appeal mechanisms for the impartial handling of any substantive and procedural complaints (clause 4.2).

Approval

Formal approval of standards should be based on evidence of consensus (clause 4.5).

Advancement of international trade

Standards shall not be written so as to allow them to mislead consumers and other users of a product, process or service addressed by this standard (clause 5.4).

Participation

Participation in standardization processes at all levels shall be accessible to materially and directly interested persons and organizations within a coherent process (clause 6.1).

Hierarchical framework

In addition to the above, a key requirement of the World Trade Organization (WTO) is the existence of a hierarchical framework between international, regional and national standards. The use of a hierarchical framework for forestry standards is comprehensively addressed in the 1997 Tropenbos Discussion Paper (Lammerts van Bueren, 1997).

Another important factor to consider in standard development and content are the requirements of the World Trade Organization (WTO), which establishes international rules on trade and defines what constitutes a technical barrier to trade (TBT). Guide 59 is currently being reviewed to ensure that it is compliant with WTO rules on TBTs. In the interim, anyone developing a certification scheme needs to be aware of WTO requirements (GTZ, 2000; Fern, 2003).

However, while this provides a useful starting point, there are some particular issues that make the development of performance-based standards for forests particularly complicated and, therefore, raise additional requirements for the standard-setting process.

3.2.1 THE CHALLENGE OF PERFORMANCE STANDARDS FOR FORESTS

Performance standards for forests are unusually complicated to define when compared to standards in other sectors for three reasons:

1 *Incomplete information*: most standards are based on precise factual information.

> For example, a standard specifying the minimum strength of a motorcycle helmet is based on scientific and technical data that can be used to precisely define what is strong enough to be safe in the event of an accident.

However, we do not have all the necessary information to understand and model in detail the way in which forests function, or their response to management interventions. There are many gaps where information is incomplete or absent.

Therefore, we have to base any standard on the best available information, combined with human decisions about what to do when there are uncertainties. Furthermore, forest management is an adaptive process in which knowledge is constantly being accumulated through experience that needs to be taken into account in drafting and updating standards.

2 *Conflicting requirements*: definitions of 'sustainable forest management' vary, but all agree on the basic premise that it involves a balance of economic, environmental and social requirements. However, it is often impossible to achieve all of these similtaneously and sometimes conflicts arise.

> For example, it is not possible to simultaneously achieve in the same area an economic desire to harvest trees with an environmental desire to set it aside as pristine forest. Similarly, it may not be possible to simultaneously protect wildlife for conservation purposes while meeting a social requirement to allow hunting for subsistence.

Therefore, the standard-setting process has to deal with conflicting requirements.

3 *Variability*: Forest standards have to address the very high degree of variability that exists between forests around the world. Most other standards are equally applicable anywhere.

> For example, the strength required of a motorcycle helmet to ensure that it protects anyone wearing it is the same anywhere; therefore, a standard for motorcycle helmet safety can be applied directly in any country.

Forests, however, vary enormously in their biology, climates, soils and their social and economic context, even within one country. As a result, forest certification schemes need to include mechanisms to ensure that the standard used is appropriate to the specific ecological, social and economic conditions where it is applied.

For these reasons all forest standards have to be developed using a combination of:

- best available scientific and technical information and knowledge about forests and the way in which they function and are affected by management;
- decision-making about how to address any gaps in the information available and how to balance the different demands made on forests (see Figure 3.1).

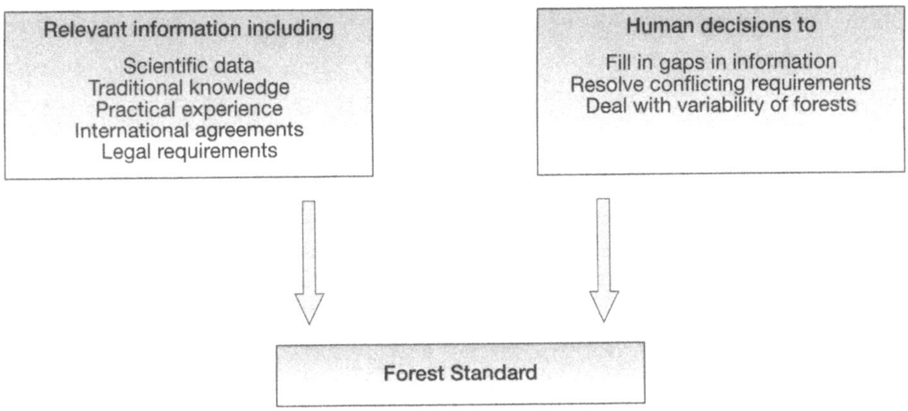

FIGURE 3.1 The two types of input needed in developing standards for forestry

The way in which fulfilling these two requirements is addressed will have a significant influence on the final standard. As a result, the standard-setting process adopted to develop a forest management standard is particularly important since this has a significant influence on the final content.

This, in turn, will depend upon two things: who is involved in the standard-setting process and how the standard-setting group works. Each of these is discussed below.

3.2.2 THE STANDARD DEVELOPMENT PROCESS

Since forest standards are very complex to develop, national standard development processes for forestry in many countries have taken several years to complete. The normal process followed is outlined in Figure 3.2.[2] There are two critical issues in the development process:

1 Who is involved in the standard-setting group and the consultation process.
2 How the standard-setting group makes decisions on the standard.

Each of these is discussed below.

3.2.2.1 Who develops the standard

Standards are usually developed by a group of people referred to as a standard-setting group or technical committee. The membership of this group can range from a few selected experts working in isolation to a group composed of all interested and affected parties.

The task of experts is to provide the input of 'best available information and knowledge' into the standard. The wider the range of experts, the greater the certainty that all relevant information will be fed into the process. Thus, for a forest management standard, relevant experts may include a wide range of specialists, comprising scientists, foresters, ecologists and wildlife managers, indigenous people, government and many others. Their input should ensure that all relevant scientific, technical and empirical information is fed into the process.

While it is relatively straightforward to identify who will be useful in proving technical input to the standard, it is more complex to decide who should contribute to the 'decision-making' element. The type of input each person or group will give will reflect their political, economic, social and environmental motivations; therefore, the precise makeup and balance of the group becomes important. The wider the range of people and interests involved, the wider the input into any decisions made.

For example, in a discussion about whether the standard should require a minimum of 5 per cent or 10 per cent of the forest area to be managed for conservation, it is likely that environmental non-governmental organizations (NGOs) will tend to favour 10 per cent, while industry prefers 5 per cent. If the standard-setting group excludes industry, there will be no one to argue the case for 5 per cent, while a group dominated by industry may not consider the 10 per cent option.

It is important to note that there is a significant difference between a process that is 'open to all interest groups' and one that actually involves all interest groups. In many situations important groups, such as indigenous people or rural communities, may be 'invited' but might not have access to information or the resources to travel to meetings. In addition, their representatives may not be able to understand the issues and provide an effective input to the process unless support is actively provided. The final standard will reflect the interests of those who were actually involved, and not those who were invited.

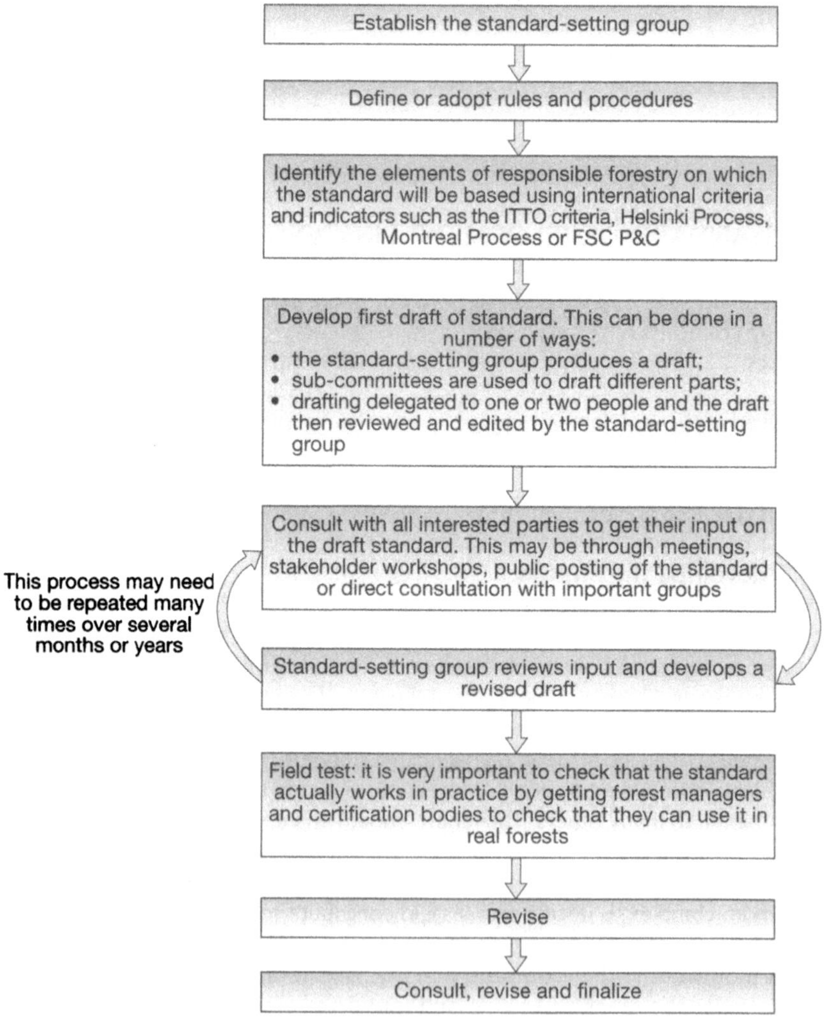

FIGURE 3.2 Outline of a typical national forest certification standard development process

3.2.2.2 How does the standard-setting group work?

It is not only the range of organizations involved in the standard-setting group or the consultation process that will affect the final content of the standard, but also the way in which the group works. Therefore, it is extremely important that the working group has clear, documented rules of operation addressing issues such as how decisions are made, how meetings are conducted and documented, how material is made available to members of the group and the wider public, and how problems or grievances are resolved.

In particular, the process by which decisions are made is of great importance in determining the content of the standard. This can range from decisions being taken by a single interest through to decision-making based on consensus.

Where decision-making is controlled or heavily influenced by a particular group or interest, the final standard is likely to reflect their views, even when other interest groups have been involved in the process. During the past, this was often the way in which government requirements for forest management were developed, with comments and input invited, but the final decision on what was incorporated (and what was not) remaining with the forest department.

It is now recognized that, for forest standards, it is important that decision-making power is shared between all of the interests who are involved in the process. This can be done in a number of ways, the most common of which are voting and consensus.

Where voting is used, it becomes very important to carefully define and control the makeup of the group who can vote in order to make sure that no group dominates in numbers and, therefore, in voting power. ISO tends to use voting and has defined procedures which aim to ensure that there is wide support for a decision and no single interest can dominate (see Box 3.2). However, for forest management standards, where there is often a clear division between social, economic and environmental interests, majority voting can create difficulties if it allows a standard to be approved in the absence of any support from one of these three broad interest groups.

This problem is avoided if the standard-setting group works on the basis of consensus. ISO recognizes that 'consensus, which requires the resolution of substantial objections, is an essential procedural principle and a necessary condition for the preparation of international standards that will be widely accepted and used' (ISO, 1995).

In a genuine consensus-based approach, a decision can only be made if there is no significant and sustained objection by a member of the standard-setting group. Where it works, it results in a standard that has the support of everyone involved in the development process. However, in practice, this way of working gives each member of the group the power of veto since a single person or organization registering disagreement (provided that it is serious and sustained) is sufficient to prevent progress.

3.3 Content of standards

It is the requirements set out in a standard that actually determine what a certification scheme delivers in practice. Therefore, the content of the standard is extremely important. There are three main elements that need to be considered when developing or assessing a forest standard:

1 *Performance requirements*: the requirements contained in the standard define what level of forest management has to be achieved in order to be certified and therefore what the certification scheme actually delivers in the forest.
2 *Wording*: standards are technical documents that should be written clearly and unambiguously to ensure that they can be consistently implemented and used for auditing.
3 *Applicability*: forests are enormously variable in type, location and size; therefore, forest standards need to be relevant to all of the forest types and local situations to which the certification scheme is intended to apply.

Each of these is discussed below.

3.3.1 PERFORMANCE REQUIREMENTS

The requirements that the standard contains are fundamental in determining what the certification scheme delivers.

For *system* standards there is broad international agreement on what these requirements should be, with ISO 14001 providing a working model.

For *performance* standards it is less clear, with no single globally accepted set of detailed requirements. However, over recent years there have been a number of international processes that have made significant progress in identifying the range of issues which must be considered in defining responsible forest management and which, therefore, need to be addressed in a performance standard.

This process can be traced back to the discussion of 'sustainable development' in the Brundtland Report (WCED, 1987) but has subsequently been greatly developed through United Nations Conference on Environment and Development (UNCED) and related processes (Grayson, 1995), together with work by the International Tropical Timber Organization (ITTO) and other regional efforts to develop sets of requirements for sustainable forest management. The most important sets of international principles, criteria or guidelines for sustainable or responsible forest management are discussed further in Chapter 14. Several analyses of these have been made (Nussbaum et al, 1996; Higman et al, 2004; CICI, 2003) and show that there is considerable agreement about what the relevant issues are.

However, although there is considerable conformity between international initiatives and definitions, there are also some significant differences. In addition, the requirements that have been established are often very general or designed for national-level monitoring, rather than for implementation at the forest management-unit level, leaving scope for widely differing interpretations. As a result, there is no single international set of detailed requirements for good forest management with universal acceptance. Nevertheless, it is possible to put together a list (see Box 3.4) that summarizes the main issues considered relevant by one or more of the international processes.

Anyone developing or assessing a forest standard must consider each of the requirements in the list and either address them or justify why any may be ignored.

In addition to the requirements set out in Box 3.4, an additional issue that forest managers are being asked to deal with in many standards is the need to ensure that there is adequate consultation with interested parties. This is discussed in Box 3.5.

3.3.2 HOW THE STANDARD IS WRITTEN

ISO guidance on structure and drafting of standards is set out in ISO/IEC Directives, Part Three: Rules for the Structure and Drafting of International Standards. The guidance is useful, but is also specific to standards that are intended to become part of the ISO family of standards; therefore, it may not fully address all of the issues related to forest standards.

There is general agreement that standards are supposed to be precise technical documents that can be unambiguously understood, implemented and audited against. In addition, for forest management standards, it is also recognized that there needs to be adequate flexibility to allow managers to achieve responsible management in the most appropriate and cost-effective way.

3.3.2.1 Precision, accuracy and clarity

As discussed in Chapter 4, the job of an auditor is to collect evidence to confirm whether or not the standard is being met. Clearly, if the standard is unclear or ambiguous it will be difficult for the auditor to carry out their job in a repeatable and objective way. It will also be difficult for a forest manager to implement the standard, potentially resulting in confusion and wasted resources.

Box 3.4

Typical requirements for forestry standards that define responsible or sustainable forest management

There is no single agreed set of detailed requirements for forest management that are universally accepted as defining responsible or sustainable forest management. There is, nevertheless, broad agreement at a general level that requirements need to address legal, technical, environmental and social issues. However, within each of these broad areas, there is less agreement about exactly what needs to be addressed. The lists below summarizes the main requirements found in one or more of the main international initiatives.

Legal requirements

Legal requirements include:

- resource rights: clear defined rights to the resource that do not threaten the rights of others;
- operating legally: full compliance with all relevant national and international laws;
- control of unauthorized activities, particularly those that could threaten the integrity of the forest.

Technical requirements

Technical requirements include:

- management planning, including both short- and long-term plans for the forest;
- forest inventory and resource assessment;
- appropriate silviculture and ensuring sustained yield;
- economic viability: forest management cannot be sustainable in the long term unless it is economically viable;
- forest operations and operational planning;
- monitoring both of operations and of the state of the forest;
- training and capacity-building to a level sufficient to ensure that the requirements of the standard are met;
- forest protection, including from pests, diseases, fire and other natural problems;
- control, minimization and proper use of chemicals and biological control;
- the proper design or restructuring of plantations.

Environmental requirements

Environmental requirements include:

- full assessment of environmental resources and impacts and adequate planning to minimize negative impacts;
- conservation and environmental protection, including the identification and good management of particularly important features and values;
- waste management, including reduction, reuse and recycling wherever possible.

Box 3.4 continued

Social requirements

Social requirements include:

- health and safety for both employees and contractors;
- workers' rights, including issues such as fair pay, the right to organize and the control of child and slave labour; many standards defer to International Labour Organization (ILO) requirements;
- assessment of social impacts and interaction with stakeholders, such as local communities and interested parties, including proper mechanisms for consultation and for dealing with complaints;
- recognition and protection of the rights and needs of forest users, including both forest-dependent people and local communities;
- encouraging and supporting employment and development for local communities.

3.3.2.2 Flexibility

Although standards need to be clear, there is also a need to build in some flexibility in order to deal with intrinsic variation in forest ecosystems and the objectives of forest owners and managers. As discussed in Chapter 1, forests vary in their ecology, climate, geography and size, while forest owners and managers differ in their approach to management and the social, cultural and economic environments within which they work. Therefore, forest management standards must allow for the range of ways in which forests can be managed while still achieving the level of performance envisaged by those developing the standard.

3.3.3 APPLICABILITY

As discussed above, forest standards need to be sufficiently precise and detailed to provide a sound basis for auditing. At the same time, there is huge variation in forest size, types and location globally, and even within many countries. This means that it is not possible to write a single global standard that is detailed and precise, as well as applicable to all forests. Even at a national level, the range of forest types in many countries is sufficiently great to make the development of a single standard very challenging. There are a number of ways of dealing with this:

- The scope of the standard should be restricted to a single forest type; this allows for a specific and detailed standard that is clear for those implementing the standard and easy to audit for the certification body. For example, a standard limited to poplar plantations in a particular country can be extremely detailed since it will be applied to only one type of forest in one cultural context. A scheme based on such a standard is only applicable to that forest type.
- The scope of the standard should cover a limited range of forest types by including guidance on interpretation. These standards are sufficiently detailed to audit against, but still include some requirements that have to be interpreted as appropriate to the specific forest being certified. This is generally the approach taken for national standards in countries with a limited range of forest types – for example, the UK Woodland Assurance Scheme standard (UKWAS Steering Group, 2000) and the Malaysian Timber Certification Council (MTCC) standard for natural forest management in Malaysia. In countries with a wider variety of forest types, more than one 'national' standard may

Box 3.5

Benefits of and requirements for participation and consultation

One requirement that is often new for forest managers, but is increasingly seen as an essential component of forest standards, is participation and consultation. Several of the international processes that have defined the scope of a forest standard have specifically addressed this issue.

Requiring mechanisms to involve local communities and other stakeholders in forest management planning and decision-making through some form of consultation or participation has a number of benefits. It:

- Decreases the likelihood that the forest management will be unacceptable to external parties or have a negative impact upon them.
- Provides a basis for managing the social impacts of forest management.
- Provides input into the process of balancing conflicting social, economic and environmental needs.
- Increases equity and empowerment, thus contributing to sustainable development.

Traditionally, the approach taken by many forestry companies has been to provide information through leaflets, brochures or (increasingly) websites with information about themselves. While this is often very useful, it is neither consultation nor participation.

Consultation involves actively soliciting feedback on information provided and plans being made. This can range from an *ad hoc* willingness to receive comments to a systematic approach to receiving and dealing with comments received. In general, the larger an organization and the greater its potential or actual impact on the forest and the communities surrounding it, the greater the need for a formal approach to consultation.

The greater the focus on responding to comments and incorporating ideas received within management planning and operations, the nearer the process moves to participation. Participation requires a process for actively involving external parties in the forest management process and, particularly, in decision-making.

The range of ways in which an organization can undertake consultation and/or encourage participation is discussed further in Chapter 4.

be needed because the variation is too great for a single standard – for example, in Brazil where forest ranges from temperate plantation in the south to tropical rainforest in the north.

- A system may be designed to produce a linked set of standards that can be applied to any forest type. There are two ways in which this can be done:
  1. The single-system approach, where a generic international standard is defined at a level that is applicable to any forest type, together with guidelines for the development and approval of more detailed national or regional interpretations. If this approach is being used, it is necessary to assess the adequacy of both the international standard and the process used to make national or regional interpretations. This is the approach adopted by the Forest Stewardship Council (FSC).
  2. The mutual recognition approach, where national or regional standards are developed independently and a system is subsequently defined for assessing whether or not the standards are compatible. In this case, the system used to carry out the mutual recognition assessments will be very important. In particular, there should be clearly defined requirements ensuring that all of the participating schemes meet an acceptable minimum standard. This is the approach adopted by the Programme for the Endorsement of Forest Certification (PEFC).

3 Either of these approaches can, in theory provide an effective way of overcoming the conflicting needs for standards that are detailed enough to implement and audit against, while at the same time remaining applicable to a wide range of forest types.

An additional point for any scheme aiming to create an international certification system is that it should ensure that WTO guidelines are met and that the resulting scheme will not create unnecessary barriers to trade (Vallejo and Hauselmann, 2000).

3.4 Small forest enterprises and standards

A number of analyses have shown that standards can be a significant barrier to small forest enterprises (SFEs) in obtaining certification (Nussbaum et al, 2001; Higman et al, 2002).

The main problems are a result of:

- *Relevance*: some requirements are not relevant in all situations, but are particularly apt for small forest owners, and managers create confusion about how to meet the standard. This may be inevitable since standards are generally developed to apply to a range of forest sizes; but it needs to be recognized that it makes the document more difficult for a small forest manager to use and that separate guidance may be needed.
- *Applicability*: some requirements are not applicable or are not feasible for an SFE to implement in a small forest area. These may be requirements that relate to the landscape-level values of the forest, which cannot be fulfilled individually at a small scale, or requirements for detailed planning and documentation by the SFE – which adds considerably to the management burden, but does little to improve forest management when applied at the small scale. Even requirements for sustained yield can be difficult to fulfil in a small area of forest.
- *Length and language*: standards are often lengthy documents. Requirements are frequently phrased in complex technical language. It may not be clear exactly what is being required and some interpretation may be needed before the requirement can be implemented. Some requirements are repeated at different points in the standards, adding to the length. The length and language of the standards create a disincentive to anyone with limited time available. They also exclude people without a formal forestry training and with low literacy levels. Owners of SFEs, who are rarely professional foresters, particularly in developing countries, and who combine forest management with other work, are more likely to fall into these categories than professional managers of medium to large enterprises. The length and language of the standard therefore affect SFEs disproportionately.

If these issues are not addressed, then a scheme may discriminate against SFEs. In order to avoid this, it is important that there are mechanisms within the scheme to ensure that any barriers are minimized – for example:

- guidance notes specifically for SFEs on how to interpret the standard;
- an appropriately simplified (but not lower) version of the standard specifically for implementation by SFEs;
- some type of group certification scheme that provides SFEs with a version of the standard designed for their forest type and size.

Notes

1 There is a problem with the use of the term 'sustainable' in the name of a forest certification standard (or any other standard) where it is planned to link the name to any claims. This is because of ISO guidance which states: 'At this time there are no definitive methods for measuring sustainability or confirming its accomplishment. Therefore, no claim of sustainability shall be made' (ISO 14021, clause 5.6). While it may be acceptable to use the term 'sustainable forest management' (SFM) in discussions about the standard, this phrase should be avoided in any claims made relating to certification against the standard.

2 This book does not provide practical guidance for those involved in a standard-setting process; but such guidance is available in *Developing Forest Stewardship Standards: A Survival Guide* (Scrase and Lindhe, 2001, available from www.taigarescue.org) or from the World Bank–WWF Alliance Pathfinder Tool for Capacity-building (available at www.piec.org/pathfinder).

4 Certification

Certification is a means of confirming that the forest and its management conform to a particular standard. The certification process involves verifying that each requirement of the standard has been met based on *objective evidence* collected from the organization and the forest being assessed. Once a certificate has been issued, provided that it is credible, it can be used by the forest organization to confirm to its customers, investors, regulators, shareholders or any other interested party that the forest is being managed in conformance with the standard.

Certification (or verification) can be separated into two components (see Figure 4.1), which are:

1 the technical process of verifying that the requirements of the standard have been met and, based on this, making a decision about whether or not to issue a certificate; and
2 the means of ensuring the credibility of the process and confidence in the result and, therefore, the certificate.

In practice, these two activities are closely linked and both form an integral part of the certification process, which is summarized in Box 4.1. Each is discussed in more detail below, while certification from an organization's viewpoint is addressed in Chapter 9.

4.1 Establishing that the standard has been met

There are two parts to the technical process of certification: firstly, collecting evidence to confirm whether or not the requirements of the standard have been met; and, secondly, based on this, making a decision about whether to issue a certificate. Each of these will be affected by the people involved. Therefore, the overall effectiveness of the process will depend upon:

- *People*: the people and organizations responsible for managing and implementing the certification process;
- *Verification process*: the methodology used to collect information in order to establish whether or not the requirements of the standard have been met.

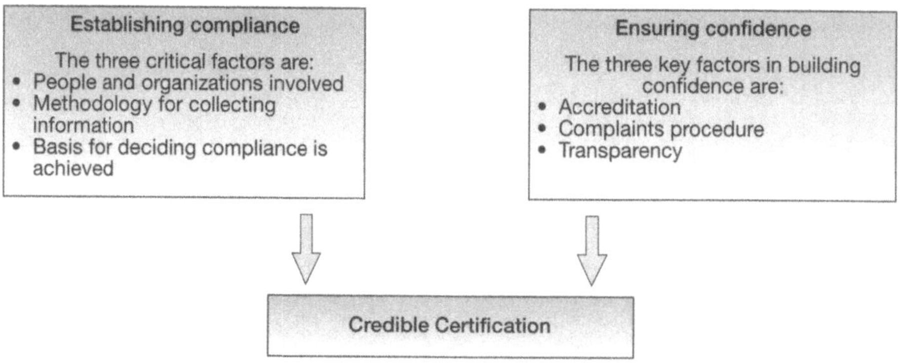

FIGURE 4.1 **Critical factors in developing and implementing a credible certification process**

Box 4.1

The certification process

The exact way in which certification is carried out varies between certification schemes. However, the outline below sets out the main steps that most schemes follow. The certification process is discussed in more detail in Chapter 9.

For forest owners who are certifying their forest through a group scheme, the experience will be somewhat different since it is the group management who goes through the certification process.

Application and proposal

For all voluntary certification the first stage is that the forest manager or management organization interested in having a forest assessed applies to a certification body, and the certification body prepares a proposal.

Pre-assessment or scoping

The certification body commonly makes a brief preliminary visit to the certification applicant with three main objectives in mind, to:

1 ensure that the applicant understands the requirements of certification;

2 plan for the main assessment; and to

3 identify any major gaps between the applicant's current management and the level required by the standard.

Closing gaps

Applicants address any gaps between current management and that required for certification until they are confident that their performance is in compliance with the requirements of the standard.

Main assessment

The main assessment is the primary opportunity to establish that the standard is (or is not) being met and is discussed in detail in Chapter 9. It is usually carried out by an assessment team whose job is to collect objective evidence that demonstrates whether or not the standard is being met. The collection of objective evidence involves a combination of document review, field visits and consultation.

When non-compliances with the standard are found, this normally results in corrective action requests, which must be addressed by the applicant in order to achieve full compliance with the standard.

Reporting and certification decision

The assessment team does not make a decision about whether or not the forest should be certified. Following the assessment, the team produces a report setting out the findings and making a certification recommendation.

The certification decision is made based on the report. This should always be done by a panel or committee who were not directly involved in the assessment in order to reduce the risk of corruption. The report can be reviewed by specialists (peer reviewers) prior to going to the final decision-making committee to get some independent feedback on the process and the results.

The report must be made available to the accreditation body. Some or all of the report can be made publicly available in order to give stakeholders access to information on the certification.

Surveillance

A critical part of the certification process is the ongoing surveillance of certified forests. Surveillance visits serve two purposes:

1 Ongoing compliance with the standard is checked in order to ensure that performance does not fall below the required level.

2 Where improvements have been required through corrective action requests, progress is monitored.

- *Decision*: the way in which a decision is made concerning whether or not there is compliance with the standard and how the decision is maintained over time.

Each of these is discussed in more detail below.

4.1.1 ORGANIZATIONS AND AUDITORS RESPONSIBLE FOR CERTIFICATION

The effectiveness of the certification process and the reliability of the certification decision will depend upon:

- the organization which is responsible overall for running the process and making the final decision: the certification body;
- the people actually carrying out the assessment to collect the information upon which the certification decision is made: the audit or assessment team.

Each of these is discussed below.

4.1.1.1 Certification bodies

Certification bodies are also known as *certifiers* or, particularly in North America, as *registration bodies* or *registrars*. Most certification bodies are commercial companies, some of them large international organizations and some smaller national companies; but there are also non-profit organizations, such as research institutes or non-governmental organizations (NGOs) which act as certification bodies. Some certification bodies certify against several, or even hundreds, of different standards, while others specialize in a particular area.

Any of these models can work well. The most important consideration for a certification body is that it must be completely independent of the organization which it is assessing in order to ensure a genuinely *third-party* assessment.

There are three ways in which an assessment or verification against a standard can be carried out:

1. First-party assessments are those carried out by an organization on itself and are often referred to as internal audits.
2. Second-party assessments are carried out by an organization which has a relationship of some sort with the organization being assessed. A common example is a supplier audit where a company audits its suppliers.
3. Third-party assessments are carried out by an organization which is completely independent of the organization being assessed.

First- and second-party audits are very useful for internal or company-to-company communication, but they are *not* certification processes. Many organizations have an internal audit programme and, indeed, it is a requirement of systems standards such as ISO 9000 and ISO 14001. However, clearly neither first- nor second-party assessments are independent; therefore, credible certification schemes are based on third-party assessments.

There are a number of International Organization for Standardization (ISO) documents that set out the way in which a certification body must be set up and run (see Box 4.2). As with standard development, these rules have been developed based on many years of experience and provide the baseline for any certification body and certification process.

The quality and independence of the certification body are so critical to both the technical success and the credibility of the whole process that most schemes also require 'certification of the certifiers' through a process called accreditation. This is discussed further on in this chapter and in more detail in Chapter 5.

One exception to the approach described above is where the certification scheme takes on the

Box 4.2 ISO requirements for certification bodies (ISO Guides 62 and 65)

There are two International Organization for Standardization (ISO) guides that set out the requirements for certification bodies which operate certification schemes. Guide 62 focuses on requirements for those working with system-based standards and Guide 65 focuses on requirements for those working with product (or performance) standards. Both are relevant to forestry and some of the main requirements are summarized below.

Organization

General requirements include non-discrimination towards certification applicants; no impediments or inhibitions to access to certification; services to be made available to all applicants; and no undue financial or other conditions (Guide 62, clause 2.1.1; Guide 65, clause 4.1).

Detailed requirements include impartiality; separation of responsibility for certification decision and certification evaluation; freedom from commercial or financial pressure that may influence decisions; ensuring that the activities of related bodies do not affect confidentiality; objectivity and impartiality; no advice given or consultancy services provided to the applicant regarding the methods of dealing with matters that are barriers to the certification requested (Guide 62, clause 2.1.2; Guide 65, clause 4.2).

Quality system

This involves a requirement to document and operate an effective quality system appropriate for the type, range and volume of the work performed. The quality management system must include, among other things, the procedures for the recruitment, selection and training of certification body personnel and the monitoring of their performances; procedures for handling non-compliances and for assuring the effectiveness of any corrective and preventive actions taken; procedures for implementing the certification/registration process, including conditions for issue, retention and withdrawal of certification documents; surveillance and reassessment procedures; and procedures for dealing with appeals, complaints and disputes (Guide 62, clause 2.1.4; Guide 65, clause 4.5).

Conditions for certification

Conditions for certification include requirements to specify conditions for granting, maintaining and extending certification and the conditions under which certification may be withdrawn or suspended; to document and make available upon request procedures for certification assessments, surveillance and reassessment; and to identify non-compliances and the need for corrective action (Guide 62, clause 2.1.5; Guide 65, clause 4.8.1).

Personnel competence

This category includes requirements to define minimum relevant criteria for the competence of certification body personnel; maintain information on the relevant qualifications, training and experience of accreditation body personnel; to define minimum relevant criteria for the competence of auditors and technical experts; to have a procedure for selecting auditors and technical experts on the basis of their competence, training, qualifications and experience; and to ensure that the skills of the audit team are relevant and appropriate (Guide 62, clause 2.2; Guide 65, clause 6).

Source: ISO Guide 62: General Requirements for Bodies Operating Assessment and Certification/Registration of Quality Systems and ISO Guide 65: General Requirements for Bodies Operating Product Certification Systems

job of carrying out certification assessments itself. Certification schemes do not usually carry out certification activities directly, as this almost always proves to be inefficient, as well reducing the independence of the scheme overall. However, this is sometimes used when schemes are first set up and need to gain experience and clients. In this case, all of the same requirements and constraints must be applied as for certification bodies in general.

4.1.1.2 The assessment team

The assessment team is responsible for most of the technical process of collecting and analysing information in order to establish compliance with the standard. Therefore, the competence of the team is fundamental to the effectiveness of the certification process. This importance is reflected by the International Organization for Standardization (ISO), which provides specific guidelines for members of assessment teams (see Box 4.3). These guidelines are very general since they must apply to all types of certification, but they include the requirement for further elaboration of sector-specific guidance.

The assessment team has a number of key functions (each is discussed further in subsequent sections) and its makeup must ensure that it performs all of these functions adequately:

- *Interpretation of the standard*: there will almost always be some degree of interpretation of the standard for the specific situation of the forest management unit being assessed.
- *Collecting objective evidence*: the team must have sufficient expertise in order to adequately seek out and collect objective evidence, including making decisions about how much is enough.
- *Identifying and weighing non-compliance*: the team must be able to identify non-compliance with the standard and differentiate between that which is major and that which is minor.

In order to perform these different functions, at least two types of personnel are required:

1. *Auditor*: the team must include at least one competent auditor who understands how to audit and how to interpret the standard. Most certification schemes have specific requirements for training auditors. This usually consists of a combination of theoretical and practical training:
   - theoretical training on an auditor training course;
   - practical training through observation of assessments and carrying out assessments while being monitored by an experienced auditor.
2. *Sector specialists*: the team must include adequate technical expertise in order to ensure that it is able to:
   - judge the appropriate interpretation of the standard's requirements;
   - assess whether the requirements are being met.

This expertise can be provided by auditors who are also sector specialists or by a combination of an auditor and sector specialists. In either case, for assessments against forest management standards, the makeup of the team needs to ensure adequate knowledge of legal, technical, environmental and social issues.

4.1.2 DETERMINING WHETHER THE REQUIREMENTS OF STANDARDS HAVE BEEN MET: AUDIT METHODOLOGY

A certification decision must be justified by clear, rigorous, objective evidence that the standard is complied with. There are a number of factors that will influence whether or not the methodology used to collect and use this objective evidence is adequate:

- the types of objective evidence that are sought;
- the sampling method used to select what is actually examined;

Box 4.3

ISO requirements for assessment team personnel

With regard to International Organization for Standardization (ISO) requirements related to certification body personnel (ISO Guide 62), the certification body is required to:

- Define the minimum relevant criteria for the competence of personnel in order to ensure that evaluation and certification are carried out effectively (clause 2.2.1.1).
- Maintain information on the relevant qualifications, training and experience of certification body personnel (clause 2.2.1.2).
- Provide clearly documented instructions to certification personnel, describing their duties and responsibilities (clause 2.2.1.3).
- Define minimum relevant criteria for competence of auditors and technical experts (clause 2.2.2.1).
- Have a procedure for selecting auditors and technical experts on the basis of their competence, training, qualifications and experience (clause 2.2.3.1).
- Ensure that the skills of the audit team are appropriate (clause 2.2.3.2), including:
    - being familiar with applicable legal regulations, certification procedures and certification requirements;
    - having a thorough knowledge of the assessment method and assessment documents;
    - having appropriate technical knowledge of the specific activities for which certification is sought;
    - having a degree of understanding sufficient to make a reliable assessment;
    - being able to communicate effectively;
    - being free from any interest that may cause partiality or discrimination.

- the balance of evidence for system implementation versus outcomes;
- the way in which the standard is interpreted;
- the identification and treatment of non-compliances with the standard.

Each of these is discussed below.

4.1.2.1 Types of objective evidence

The primary job of an assessment team carrying out a certification assessment is to collect evidence (which must be as clear and objective as possible) that confirms that each requirement of the standard is, or is not, being met. There are four possible types of objective evidence: documentation; observations in the forest; interviews with management and workers; and evidence from external parties.

Documentation

Documents provide three sorts of information:

1 Where the standard specifically requires particular documents (for example, a written management plan), then the objective evidence of compliance comes directly from checking the document.

2 Where the standard requires management actions to be implemented, documents showing that this is planned (for example, operating procedures) contribute towards the objective evidence required, though further verification in the field or through interviews is needed to ensure that there is compliance in practice.
3 Where the standard requires collection and analysis of information (for example, a forest inventory and growth-and yield data to form the basis for the calculation of sustained yield; incident records to confirm that safety measures are working; or records of bird numbers to monitor impacts on biodiversity), documents in the form of records provide information about what has been done.

Observations in the forest

An essential source of information is the forest itself. Field visits provide information on current activities and operations. In addition, a somewhat unique feature of forest assessments is that forests often show clearly the effects of biotic or abiotic disturbances long after the events occur. Thus, if a harvest crew fails to leave a riparian reserve, anyone visiting the site for the next few years will be able to observe this. Field visits to the forest can be used to:

- Collect evidence that documented plans and procedures are (or are not) being followed in practice.
- Check that data which has been collected is accurate by examining the accuracy of a sample of the data.[1]
- Check the actual performance of operators in the field to collect objective evidence on compliance with areas such as health and safety, environmental protection and training.
- Collect objective evidence that the state of the forest does (or does not) meet the requirements of the standard – for example, that riparian areas are intact, that thinning has been carried out, and that measures to control erosion on roads are in place and working.

Management and workers

Interviews comprise the third important source of evidence, though they are complex since people do not always tell the truth, and information collected may not be consistent. However, there are several types of information that can provide objective evidence, including the following:

- If there is information people must know in order to meet the standard, then establishing that they *do not* know it provides evidence of non-compliance, while establishing that they *do* know it is a step towards demonstrating compliance.
- If several people on separate occasions all give the same information in response to a question (for example, what to do in the event of an accident), then this cumulative information provides objective evidence.

External parties

Objective evidence can also be collected from people outside the forest organization. This type of information has two main uses:

1 If the standard requires interaction with the community (for example, consultation, participation in forest activities or a 'good neighbour' policy), then information on whether this is implemented needs to be collected from the community in question.
2 Assessments are always conducted over a short period (days or, at most, weeks); therefore, if there are any issues that are seasonal or short lived, these may not be identified by the team during an assessment. However, if they are serious, they are likely to be mentioned by local people or organizations consulted.

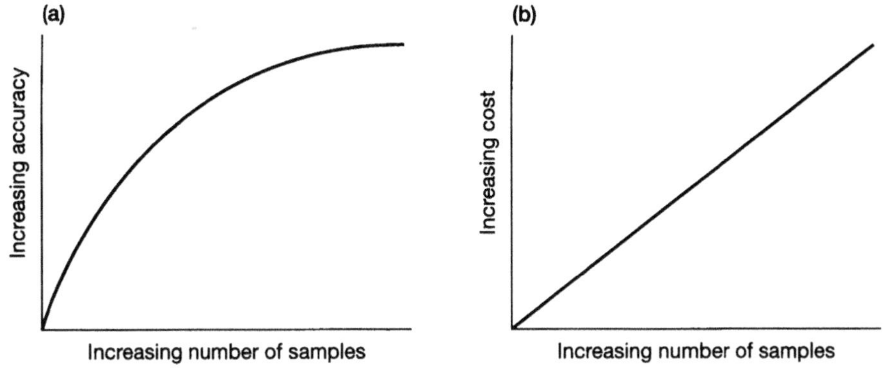

FIGURE 4.2 The effect of increasing numbers of samples on (a) the accuracy of the assessment result and (b) the cost of the assessment

4.1.2.2 Sampling

Whatever source objective evidence is collected from, it is impossible (except in the smallest forests) for the assessment team to examine everything. Therefore, the team has to use a methodology that will sample a proportion of the 'population' (or forest) at large, which is standard practice in the world of certification.

Sampling is a valid approach because a randomly selected sample of a population, if sufficiently large, will give a reasonably good picture of the population as a whole. For example, if the 'population' is harvesting sites, then visiting a sufficient number of randomly selected sites will give a good indication of harvesting performance over all sites. There are two issues to consider in achieving representative, reliable sampling: the size or proportion of the sample relative to the total population and the technique used to select the sample.

Size of the sample

The greater the number of samples taken, the closer the picture obtained will be to the population as a whole. However, as can be seen in Figure 4.2a, above a certain point each additional sample gives progressively less improvement in the overall accuracy. At the same time, as shown in Figure 4.2b, the cost of each additional sample is approximately linear, so that excess sampling will result in rapidly decreasing returns on investment.

Therefore, the sample size needs to balance the need for statistical validity with cost. In general, the greater the variation in the whole population, the greater the number of samples needed to provide a reliable picture, while, conversely, a population with very limited variation will be adequately described by a smaller number of samples.

Selecting samples

Statistical theory is generally based on the assumption that the sample will be selected in an unbiased way. Any bias in the selection will decrease the likelihood that the results are an accurate reflection of the population as a whole and therefore make it less acceptable to make general conclusions based on the results from the sample. One way of avoiding any bias is to use completely random sampling. In practice, sampling in audits is rarely entirely random, but the importance of unbiased sampling should not be underestimated. Very subjective or directed sampling is likely to result in a bias so marked that it destroys the statistical validity of the results of an audit.

For example, if the organization being assessed is allowed to select the samples (for example, the documents to be reviewed or the harvesting sites to be visited), it will almost certainly select the best cases and exclude anything that does not conform with the standard, giving a false impression of compliance.

Even if the independent certification body selects the samples, to use a subjective approach is very risky.

For example, an auditor might decide to choose the nearest field sites, thus allowing more visits in one day. However, it is likely that these nearby sites are also the ones regularly visited by management and therefore are the best managed. If this were the case, then the information obtained by the auditor would be inaccurate. In any case, if only nearby sites are selected, the auditor does not have an adequate statistical basis for extending the results of the audit to the whole forest.

On the other hand, if totally random sampling is used, this may result in the first two sites chosen to visit both being so difficult to reach that there would be no time left for any other visits, or all of the samples randomly falling into only half of the management areas being assessed, or none of the samples falling in an area where there are a number of stakeholder issues to check.

Therefore, it is generally considered appropriate to use some form of stratified or directed sampling in order to ensure adequate coverage, combined with random selection. Indeed, in some cases it is good auditing practice to do this in order, for example, to follow up on complaints or information suggesting that a problem exists. However, this approach should always be clearly justified and the statistical implications for the reliability of the results understood.

4.1.2.3 Assessment of evidence

One of the main differences between assessments of system-based and performance-based standard criteria is in the balance of objective evidence collected from each of the four possible sources identified in Section 4.1.2.1. The first part of the assessment process is very similar for both types of standard:

- The assessment team must assess what the forest organization being audited plans to do and how it plans to do it. This is done mainly from checking documents and from interviewing staff. Based on the information collected, the assessment team then needs to decide whether what is planned is adequate to meet the requirements of the standard.
- It is then necessary to collect objective evidence to verify that the plans are actually implemented. This is done based on a combination of visits to the forest in order to observe operations and locations, checking of documented records and discussions both internally and with external stakeholders, as appropriate.

At this point, the assessment of a system standard is complete; but for performance standards there is still a final step. It is necessary to establish the adequacy of the actions – that is, the level of performance required by the standard actually being delivered as a result of the actions being taken. This means collecting sufficient objective evidence on the actual state of the forest, and from documents and people, to confirm that the performance levels in the standard are being met.

For example, the requirement of the standard is that all operators work safely:

- The plan is to provide all operators with safety equipment and attendance at a half-day training course on safe working.
- The mechanism for achieving this establishes where equipment will be purchased and who should run the course and the topics it should cover.
- Implementation can be checked by establishing that the equipment has been bought and the operators trained (for example, from training records, signed attendance sheets and discussions with operators).

So far, this seems adequate. However, for a performance standard the assessment team also needs to establish whether, as a result, all operators are now operating safely – in other words, the adequacy of the actions. If the equipment is not being used or the training course was not a very good one, and operators were still operating unsafely, then the performance requirement has not been met.

As a result, it is very important for the assessment of performance standards to ensure that sufficient objective evidence is collected on the actual state of the forest, together with documentation and people so that it is possible to establish that they are all in conformance with the standard. Therefore, verifying compliance with performance criteria takes more time in the field (including consultation) than is necessary for a system criterion.

4.1.2.4 Interpretation of the standard

There are three types of interpretation that have to be performed:

Requirements that are not precise

With the exception of standards produced exclusively for a single forest type, all forest standards include some requirements that have to be interpreted at the level of an individual certification assessment.

For example, the standard requires 'appropriate' levels of documentation. A decision must be taken on what 'appropriate' means for each forest, depending upon the size and complexity of both the forest management unit and the organization managing it.

Method of meeting the standard

Performance standards tend to specify what has to be achieved, rather than defining how this must be done. Therefore, a decision must be made as to whether the particular approach taken by the organization to meet a particular requirement is appropriate. Ideally, this would be based on information demonstrating that the approach was working in practice; but, in some cases, this can take years and it would not be realistic to delay certification for such a long period.

For example, the standard requires that rare species are protected. The audit team need to decide whether the rare species identified are the right ones and the plans to protect them are appropriate. This will often be happening before there has been time for the plans to impact on species numbers which will ultimately show whether or not they are working.

Balancing conflicting requirements

Almost all forest organizations will be in a situation of having to balance the various requirements of the standard. A decision must be taken on whether the compromises made between different requirements are appropriate.

For example, the standard requires chemical use to be minimized and requires conservation areas to be protected. Is a medium-term increase in the use of herbicides appropriate in order to control exotic species from threatening a conservation area? The standard requires the maintenance of customary indigenous activities and the protection of wildlife. Should local indigenous people be allowed to continue to hunt, although this has adverse impacts on wildlife?

The quality of the interpretation made will depend upon the complexity of the issues being faced, the knowledge and expertise of the audit team and the degree to which additional expertise or inputs are sought through consultation. In general, the more complex the issues, the more important the input of a range of expertise and views.

4.1.2.5 Non-compliances and corrective action requests (CARs)

In the real world, perfection is seldom achieved and assessment teams almost always find evidence of incomplete compliance with one or more requirements of the standard. The normal way to deal with this is through the issue of corrective action requests (CARs), which set out details of the non-compliance and the requirement for action to be taken.

Experience shows that the seriousness of non-compliance varies widely. Some are crucial and clearly need to be addressed before a certificate can be issued.

> For example, if an audit team visited 20 harvesting sites and found that marking of riparian reserves was incomplete at 18 sites, this would be a serious problem that would need to be addressed before certification could proceed. However, if the audit team visited the 20 sites and found incomplete marking at just one, while this would still be an issue of non-compliance, it is much less serious as it does not point to a systematic problem.

To require every minor non-compliance to be completely addressed prior to the certificate being awarded would add significantly to the time and cost of an audit. Therefore, it is accepted practice that minor non-compliances should not preclude certification, provided that problems are addressed within a reasonable (and agreed) time frame.

To implement this in practice, it is necessary to have a mechanism for differentiating between non-compliances that are serious and those that are not. This is usually achieved through the classification of major and minor non-compliances:

- Major non-compliances result from a complete failure to address a requirement of a standard or from a systematic failure in implementing plans or procedures.
- Minor non-compliances arise where a requirement is inadequately addressed or where there is isolated non-systematic failure to implement plans or procedures.

> Examples of major non-compliances include:
>
> - a company operating in natural tropical forest that does not have any written operating procedure for harvesting and is not implementing low-impact harvesting techniques;
> - a community organization operating a plantation where chemical herbicides and pesticides are routinely used adjacent to water courses in contravention of the national Plantation Code of Practice;
> - a group scheme where none of the group members take any measures to protect or enhance biodiversity.
>
> Examples of minor non-compliances include:
>
> - a company operating in natural tropical forest where one contractor out of several observed is found to be inadequately implementing the company's operating procedure for low-impact harvesting;
> - a community organization operating a plantation where two operators applying herbicides were unaware of the requirement to maintain a minimum distance from water courses;
> - a group scheme where one member out of ten visited is failing to implement the biodiversity protection measures planned in the member's management plan.

Clearly, there are will be some borderline cases between major and minor non-compliance; therefore, it is important that auditors are well trained and have clear guidance on how to deal with these situations.

If non-compliances are classified as major, they must be addressed prior to a certificate being awarded, whereas if they are classified as minor, then a certificate can be awarded, conditional upon the minor non-compliances being addressed within an agreed time. This has to be checked by the certification body as part of their ongoing surveillance (see 'Complaints procedures and resolution').

4.1.3 MAKING AND MAINTAINING A CERTIFICATION DECISION

A decision to certify has to be based on a decision that there is compliance with the standard and may be maintained only as long as there continues to be compliance with the standard. There are two main components to achieving this:

1 *Decision-making*: there needs to be a clear mechanism for making a final decision to issue a certificate.
2 *Surveillance*: there needs to be a mechanism for the certification body to monitor ongoing compliance with the standard after the forest has been certified in order to ensure that the standard continues to be met, a process usually known as surveillance.

Each of these issues is discussed below.

4.1.3.1 The decision-making process

The guidelines set out by ISO for certification decision-making provide a very sound basis. They specify that the decision must be based on objective evidence and that the final certification decision should always be made by a person or group who is independent from the assessment itself (ISO Guide 62, clause 2.1.2). This requirement has been established to minimize the risk of auditors being threatened, bribed or otherwise unduly influenced.

It can also be useful to incorporate other requirements within the decision making process, such as the requirement for peer review of the final report or for a special panel to make the final decision. Either of these can help to add another level of independent confirmation that the:

- objective evidence collected is sufficient;
- interpretation of the evidence is reasonable;
- standard has been met.

Peer review is the process of engaging one or more independent specialists to review the certification report and recommendations produced by the assessment team. It is particularly useful when the panel making the final certification decision does not have experience of the particular forest type and location under assessment. In this case, the peer reviewers act as 'sector specialists' for the final decision-making panel.

4.1.3.2 Surveillance

Certification is not complete at the moment when a certificate is awarded; it is an ongoing process. The certification body which awards the certificate needs to be sure that during the whole period for which the certificate lasts (three or five years are common) the organization continues to comply with the standard.

This process is usually referred to as surveillance. If surveillances were not carried out, organizations could become certified and then immediately start operating in contravention of regulations, while claiming to be sustainable for the next five years. However, if regular surveillances are conducted, then if organizations stop complying with standards, the certificate can be withdrawn.

In addition to a general check on continuing compliance, there are a number of other issues that require continued monitoring during surveillance visits:

TABLE 4.1 Mechanisms for creating confidence in the certificate decision

Accreditation	Complaints procedures and resolution	Transparency
Certification process and decisions are systematically checked by an independent body, which confirms that certification decisions are appropriate.	Interested parties can get a response if they are concerned about specific issues related to the process and results.	Interested parties have access to sufficient information to decide for themselves whether the certification process and decision were appropriate.

- Any outstanding minor corrective action requests must be monitored to ensure that they are addressed within the agreed time frame (otherwise they are subject to major CARs; if requests are still not addressed, the certificate is withdrawn).
- Any new problems must be identified and raised as new corrective action requests.
- Any changes to the law or the certification standard (which is usually reviewed at least every five years) must be adequately implemented.
- Any comments or complaints from stakeholders related to the certified organization need to be addressed.
- Any changes in the forest organization or forest area must be monitored.

Therefore, it is essential that the certification process includes the requirement for regular monitoring or surveillance visits, including guidance on the frequency of such visits, the personnel involved and the methodology to be used for surveillance audits. There must also be a mechanism for suspending and withdrawing the certificate if a certified organization ceases to comply with the standard.

4.2 Ensuring confidence in the process and decision

Clearly, confidence in the process and the decision will depend upon all of the components described in Section 4.1 being properly implemented. One of the most important mechanisms for delivering this is accreditation. In addition to accreditation, an adequate complaints procedures and a reasonable degree of transparency are ways of ensuring confidence and credibility (see Table 4.1).

4.2.1 ACCREDITATION

Accreditation is the internationally accepted basis for confirming that certification bodies are credible, independent and operating properly. Accreditation prevents a situation where any organization can simply decide to become a certification body and carry out certification, whatever their experience or ability. Accreditation aims to ensure that all certification bodies operate above a certain level and that there is consistency between the approaches and, most importantly, the results, of different certification bodies.

Thus, accreditation is the process of 'certifying the certifiers'. It is so fundamental to a credible certification process that it is considered one of the three essential elements of a certification scheme and is discussed in detail in Chapter 5.

Box 4.4

Dealing with complaints about the certification process or its outcome

1 A designated person within the certification body should be responsible for dealing with complaints and guidance on making complaints, which is available to all clients and stakeholders.

2 When a complaint is received, it should be passed to the person responsible for complaints and recorded in a register or log. It should then be passed to an appropriate person for investigation.

3 The complainant should be informed that an investigation is underway, what the process is (including timing) and who is in charge. Where the complainant is a stakeholder, it is also important to clarify that while they will be kept as fully informed as possible about the outcome, they will not have access to confidential information about a client. Where necessary, additional information should be obtained from the complainant.

4 When the investigation is complete, the outcome should be documented together with recommendations for any necessary actions. This should be communicated both internally within the certification body and externally to the complainant. Where the complainant is a stakeholder, confidential information should not be included.

5 If the complainant is not satisfied with the outcome, there needs to be a mechanism for appeal. There must be a clear procedure for this that is documented and available to anyone who makes a complaint. The appeal may be dealt with by an appeals panel within the certification body, an independent appeals panel appointed by the certification body or the certification system itself.

6 Certification bodies should regularly review complaints and appeals to ensure that they are all being adequately resolved within the prescribed timeframe and to make sure that any problems within the organization that are leading to complaints are being identified and addressed.

4.2.2 COMPLAINTS PROCEDURES AND RESOLUTION

While accreditation is very important, it is recognized that it is still necessary to provide a mechanism to deal with situations where there is objection to a certification decision, either by the organization being assessed or by a third party.

It is not generally considered sufficient to merely document complaints. ISO guidelines specify that certification bodies 'should have policies and procedures for the resolution of complaints, appeals and disputes' (ISO Guide 62, clause 2.1.2) and provide some guidance as to how these should be developed.

Experience from the first decade of forest certification has shown that complaints about forest certification are quite common relative to certification in other sectors and that inadequate responses by certification bodies (and, ultimately, accreditation bodies) can lead to a reduction in the credibility of the scheme as a whole. Therefore, it is particularly important to have an effective mechanism for dealing with and resolving complaints and disputes. Box 4.4 sets out an outline of a complaints procedure.

4.2.3 TRANSPARENCY OF THE CERTIFICATION PROCESS

With regard to technical standards, such as those for product safety or quality, it has been usual to depend upon accreditation, together with a complaints mechanism, in order to provide credibility for

Box 4.5 ISO guidelines for provision of information by certification bodies

ISO Guide 62, clause 2.1.7.1, sets out a requirement for certification bodies to provide, update and make available on request the following information:

- **a documented statement of its product certification system, including rules for granting, maintaining, extending, suspending and withdrawing certificates;**
- **information about the evaluation procedures and certification process;**
- **a description of the means by which the organization obtains financial support and general information on the fees charged to applicants and to suppliers of certified products;**
- **a description of the rights and duties of applicants and suppliers of certified products;**
- **information about the procedures for handling disputes, complaints and appeals; and**
- **a directory of certified products and their suppliers.**

certification. However, with the advent of increasingly complicated standards covering social and ecological requirements, as well as technical ones, it has been recognized that to provide credibility it may be necessary to allow interested parties direct access to information about the process and the results of a certification assessment. This allows the interested parties to make their own decision about whether or not the result is acceptable. However, this raises a number of issues, the most important of which are:

- *Confidentiality*: forest managers and enterprises need to be able to maintain the confidentiality of sensitive information.
- *Cost*: transparency costs money since it results in more time and resources being spent on the audit.

Both of these must be balanced against benefits of greater transparency. Transparency can be provided by a number of mechanisms, including provision of information on the certification body, consultation and participation, and provision of publicly available information. Each of these is discussed below.

4.2.3.1 Information on the certification body

Information about the certification body is important since it allows interested parties to check whether the organization is genuinely competent and independent. ISO Guide 62 provides guidelines on the provision of information (see Box 4.5).

4.2.3.2 Consultation and participation

Consultation and participation are relevant to each element of a certification scheme, which sometimes leads to confusion about exactly what is appropriate at each stage. Box 4.6 provides a summary of the role of consultation and participation within the whole scheme. This section examines the role of consultation and participation in the certification process.

Certification is a technical process, so consultation is usually considered more appropriate than participation since the latter might jeopardize the independence of the process. Consultation, however, can play a number of important roles, each of which has been discussed in previous sections. These are:

- provision of objective evidence on compliance or non-compliance with requirements relating to interaction with consultees (see 'Types of objective evidence');
- general information on the organization being assessed (see 'Types of objective evidence');
- identification of issues that may otherwise not be apparent to the auditors (see 'Types of objective evidence').
- input into the interpretation of the standard for the specific organization being certified (see 'Types of objective evidence');
- contribution to the credibility of the final decision (see 'Making and maintaining a certification decision').

The effectiveness of consultation can vary substantially. Two important factors in its effectiveness are:

1 *The diversity and representativeness of those consulted*: there is a wide range of potential consultees for most certification processes, including government, industry, local communities, indigenous people, environmental and social NGOs, trade unions and workers. The more representative the sample, the more effective the consultation process is likely to be. However, this also has implications for cost; therefore, there needs to be a clear justification of the strategy adopted.
2 *Consultation methods*: there are numerous ways of consulting, including letters, emails, phone calls, and private and public meetings. There is no single correct way to carry out consultation since different situations will require different approaches. But it is important that the appropriateness of the approach is considered.

> For example, consultation by letter will not be effective in eliciting input from illiterate local communities, whereas face-to-face discussions work well. Local people may not feel comfortable discussing issues at a public meeting where management from the forest organization is present and would prefer meetings alone with the audit team.

Since there are so many possible ways of consulting and the choice of the right method depends upon the situation, it is very important that certification bodies have clear procedures that set out:

- the possible options for consultation;
- the way in which the most appropriate method or methods for a particular situation will be chosen.

It is also important that auditors are properly trained.

4.2.3.3 Publicly available information

Finally, transparency can be increased by the provision of publicly available information. The level of transparency provided depends upon the amount and quality of information. The extent of information provided on an assessment can range from a bare minimum, stating the name, size and location of the forest certified, to a report setting out the results of the assessment in full. There are disadvantages to both of these extremes.

Providing only factual information on size and location does little to increase transparency, although associated costs are low. Setting out the results in full provides complete transparency; but it is costly, time-consuming and may result in disclosure of confidential information.

In general, providing some information on the results, while excluding confidential information and removing unnecessary detail, can greatly increase transparency without considerably augmenting the cost.

There are two key considerations to make when deciding how much information needs to be available:

Box 4.6 **Consultation and participation in certification schemes**

The need for, and importance of, participation and consultation has been widely discussed in relation to forest certification (see, for example, Ervin, 1996; World Commission on Forests and Sustainable Development, 1999; Prabhu and Colfer, 1999; Higman et al, 2004). Despite this, there is often a lack of clarity about where it is needed within certification schemes. This box summarizes the main components of certification schemes where consultation and participation are relevant. Further detail can be found in the sections cited.

During the development of the standard (see Chapter 3, Section 3.1), the certification scheme:	• Provides input of technical information. • Provides input into the decision on how to deal with gaps in information. • Provides input into the decision on how to balance conflicting requirements. • Ensures that the standard has support.
As a requirement of the standard to be carried out by the forest organization (see Chapter 3, Section 3.2), the certification scheme:	• Provides the basis for interaction with local communities and other stakeholders. • Promotes equity and empowerment, thus contributing to sustainable development. • Contributes to the management of social impacts. • Provides input into the process of balancing conflicting social, economic and environmental needs, which the forest managers may need to undertake.
As part of the certification process (see Chapter 4, Section 4.1), the certification scheme:	• Provides input into the interpretation of the standard for the specific organization being certified. • Provides the assessment team with information on the organization being assessed. • Provides objective evidence on compliance or non-compliance with requirements relating to interaction with consultees. • Contributes to the credibility of the final decision.
As part of the accreditation process (see Chapter 5), the certification scheme:	• Provides the accreditation body with information and objective evidence relating to the compliance certification body. • Contributes to the credibility of the accreditation process.

1 *Confidentiality*: most organizations have some information that they need to keep confidential. This is usually either because it relates to personnel or because it is commercially sensitive. It is not usually considered appropriate to force organizations to make this information public if they wish to be certified, although it will be made available to the assessment team and should be noted in the full report. As a result, full results are not usually published.
2 *Costs*: the more information that has to be included in a public report, the higher the cost of writing and circulating it. In addition, public reports have to be more accurately worded in order to ensure that they can be fully understood, thereby excluding abbreviations and local or company jargon. Cost will also be a particular issue if the forest is small, but extensive information must be presented.

Therefore, it is necessary to decide how much information can be made public without breaching confidentiality or entailing excessive costs. It is important that there are clear guidelines setting out what information must be publicly available, and how interested parties can access this information.

4.3 Certification of small forest enterprises

An analysis of the barriers to certification for small forest enterprises (SFEs) suggested that the complexity and resulting cost of the certification process can act as a significant barrier to certification (Nussbaum et al, 2001).

However, it also suggested that there are a number of mechanisms that can reduce this barrier, the most important of which was identified as the various forms of group certification. If designed and run effectively, these can provide a credible and cost-effective mechanism for allowing SFEs to access certification. Group certification is discussed in detail in Chapter 10.

Note

1 The certification process should not include the collection of primary monitoring data, such as growth and yield, water flows or numbers of birds or animals. This type of data may be required in order to demonstrate compliance with the standard, but must be collected by, or collected on behalf of, the forest manager. The job of the assessment team is to check and verify the data, but not to collect it. This is very important since the assessment period is almost always too short to allow any primary data collection. Therefore, if the data does not exist, this should be treated as a failure to demonstrate compliance with the standard and the forest manager should be required to collect the data.

5 Accreditation

Accreditation is the process that provides assurance that a certification body is competent, that it meets all of the requirements of the scheme, and that its assessments and decisions are sound and consistent with other certification bodies assessing against the same standard. It is, in effect, the 'certification of the certification body' and is often confused with certification itself.

More formally, the International Organization for Standardization (ISO) definition of accreditation is a 'procedure by which an authoritative body gives formal recognition that a body or person is competent to carry out specific tasks' (ISO 1996a).

Accreditation is generally accepted as an essential component of credible certification. Without accreditation, any organization could claim to be a certification body and issue certificates. Accreditation stops this from happening and, if it is done properly, ensures a uniformly high standard of performance from all of the accredited certification bodies. This, in turn, gives value to the certificates awarded by accredited certification bodies. If the standard of accreditation is not high, this undermines the value of any certificates.

A striking example of the importance of a high standard of accreditation is provided by an analysis of the application of the international quality standard ISO 9000. In common with most ISO standards, accreditation is the responsibility of national accreditation bodies, and almost all such bodies offer ISO 9000 accreditation. However, the perceived competence of the different national accreditation bodies varies considerably. As a result, ISO 9000 certificates issued under the accreditation of some national bodies are not accepted by large parts of the market. Certification in these countries is often carried out under the accreditation of a foreign accreditation body with a greater perceived competence.

As a result of the importance attached to the accreditation body, this is usually shown on a certificate, together with the name and logo of the certification body.

Traditionally, accreditation of certification services for most international standards has been carried out by national accreditation bodies. However, with the growth of international trade and increasing globalization, many certification bodies offer certification services internationally and they need accreditation that is recognized in every country in which they operate. Companies that are buying or supplying from more than one country need to be able to rely on the accreditation services available in those countries. There are three ways in which this problem can be addressed:

1 *International standards for accreditation services*: ISO has developed an international standard for the assessment and accreditation of certification bodies: ISO/IEC Guide 61 (ISO, 1996b). Compliance with this standard ensures that accreditation services meet the same basic requirements. Guide 61 has recently been superceded by revised guidance published in October 2004 as ISO 17011.
2 *Mutual recognition between national accreditation bodies*: there are international arrangements that provide for mutual recognition between national accreditation bodies – for example, at the global level, the International Accreditation Forum, and, at the regional level, the European Accreditation. These arrangements allow a certification body to provide services under a single accreditation in more than one country, as well as providing assurance that participating national accreditation bodies are operating to the same standards.

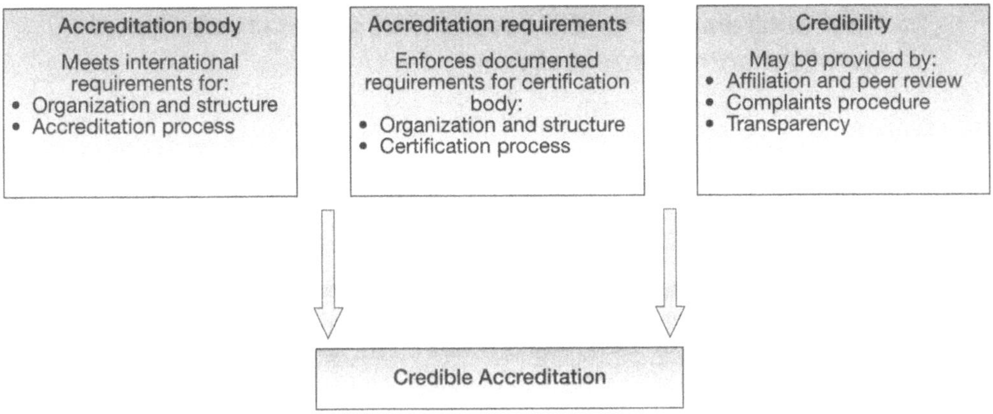

FIGURE 5.1 Components of an accreditation system

3 *Accreditation with an international scope offered by international bodies*: accreditation may also be offered by international accreditation bodies. These are usually sector-specific international bodies, such as the International Federation of Organic Agriculture Movement's Accreditation Programme for organic agriculture (IFOAM, 2001) and Social Accountability International for SA 8000 (SAI, 2004). Many of these international accreditation bodies are members of the International Social and Environmental Accreditation and Labelling (ISEAL) Alliance.

To elaborate upon the accreditation element of certification schemes, three components of the accreditation element are identified (see Figure 5.1), each of which is discussed in detail in the following sections:

1 *Requirements for accreditation bodies*: these include guidelines governing the internal organization of accreditation bodies, as well as the accreditation procedure, including ongoing monitoring of certification body performance.
2 *Rules for certification bodies*: the accreditation body must lay down adequate requirements for the internal organization of certification bodies, as well as the way in which the certification bodies conduct the certification process. These requirements become critically important when the scope of accreditation includes activities that are especially complex, as is the case with forest management certification.
3 *Credibility*: because the purpose of accreditation is to provide credibility to certification, the credibility of the accreditation body itself is vital. An accreditation body can attain credibility by fulfilling the rules for accreditation bodies and ensuring adequate rules for certification bodies (above), but may wish to enhance its credibility further by affiliating with other bodies, having adequate complaints procedure mechanisms and making its activities transparent.

5.1 Requirements for accreditation bodies

Requirements for accreditation bodies fall into two categories, each of which is discussed below. The first concerns requirements for the internal organization of the accreditation body. These should follow internationally recognized standards, which are designed to ensure that the accreditation body is competent to conduct accreditation assessments. The second category involves the way in which accreditation is carried out – the accreditation process. These are necessary to ensure that accreditation decisions are consistent and reliable.

Box 5.1

Key requirements for the assessment and accreditation of certification/registration bodies

Below is a summary of the key International Organization for Standardization (ISO) requirements relevant to the organization of accreditation bodies, taken from ISO Guide 61.

Non-discrimination and accessibility (clauses 2.1.1.1, 2.1.1.2)

Accreditation bodies' policies and procedures must be non-discriminatory and their services must be accessible to all applicants whose activities fall within their declared field of operation, regardless of the size of the applicant body or the number of bodies already accredited.

Impartiality (clauses 2.1.2.a, 2.1.2.e, 2.1.2.f)

Accreditation bodies are required to act impartially and to have a documented structure that safeguards their impartiality. The structure must enable the participation of all parties significantly concerned in developing policies and principles regarding the content and functioning of the accreditation system. A person or persons different from those who carried out the assessment must make the accreditation decision.

Conflict of interest (clauses 2.1.2.l, 2.1.2.m, 2.1.2.o)

Accreditation bodies are required to have policies and procedures that distinguish between accreditation and any other activities in which they are engaged. Accreditation bodies and their staff must be free from any commercial, financial and other pressures that might influence the result of the accreditation process. Activities of related bodies must not affect the confidentiality, objectivity or impartiality of accreditation decisions. In particular, accreditation bodies must not offer or provide directly or indirectly those services that it accredits others to perform; consulting services to obtain or maintain accreditation; or services to design, implement or maintain a certification scheme.

Source: reference should be made to the original text for the complete set of requirements (ISO, 1996b)

5.1.1 ACCREDITATION BODY ORGANIZATION

The internationally recognized standard for general requirements of accreditation bodies is ISO/IEC Guide 61 (ISO, 1996b). The requirements related to accreditation body organization and how it operates are summarized in Box 5.1. These requirements are widely accepted as providing an appropriate basis for operating an accreditation scheme in any sector, including forestry.

5.1.2 ACCREDITATION PROCEDURES

Just as the certification process is based on established rules for assessment and decision-making, accreditation follows a defined procedure. In practice, the steps in the accreditation process are similar to those for certification, as set out in Box 5.2.

In common with certification, and discussed in detail in Chapter 4, the effectiveness of the accreditation process will depend upon the people involved, the way in which information is collected to ensure compliance with accreditation requirements and the final decision made. An additional issue of some importance to accreditation is the scope of the service. Each of these issues is discussed below.

Box 5.2 The accreditation process

Application	The certification body applies to the accreditation body. A contract is signed that specifies the scope of the accreditation applied for and the terms and conditions under which the applicant is evaluated and accreditation is granted and maintained.
Evaluation	The accreditation body carries out an evaluation of the certification body's organization, systems, procedures and certification assessments and decisions. The evaluation team collects objective evidence that demonstrates whether the requirements of accreditation are met. At the end of the evaluation, the evaluation team holds a closing meeting with the applicant certification body to present its findings.
Reporting	The accreditation body prepares a report of the evaluation. A copy of the report is given to the certification body applicant, who is invited to comment upon it. The report describes any non-compliances identified by the evaluation team and the corrective action requests raised by the team.
Addressing non-compliances	The applicant certification body may be required to close out corrective action requests before accreditation is granted. Alternatively, if the non-compliances are minor, accreditation may be granted subject to corrective action requests being closed out within a specified time.
Accreditation decision	The accreditation decision is made on the basis of the report and the outcome of corrective action requests (if appropriate). Accreditation decisions must be made by a person or persons different from those who carried out the assessment.
Surveillance	Following accreditation, the accreditation body maintains surveillance over the certification body in order to ensure that any corrective action requests raised before accreditation have been closed out, and to ensure continued compliance with the requirements of accreditation and the close of subsequent corrective action requests.

5.1.2.1 Competence of accreditation body personnel

The key objective of an accreditation assessment is to determine whether a certification body is competent to undertake assessments against a particular certification standard. The accreditation assessment team must be able to determine whether the certification body is operating an effective organizational structure and an adequate management system, and is deploying competent assessment teams, which will result in certification decisions that are sound and repeatable.

As is discussed in Chapters 3 and 4, forest certification standards and forest certification procedures are relatively complex. Accreditation auditors and the staff of accreditation bodies who make the accreditation decision need to have sufficient understanding in order to make reliable and consistent assessments of certification bodies. In order to address the requirements of ISO Guide 61, clause 2.2 (ISO 1996b), a body offering accreditation for forest certification would be expected to define the minimum forest management and forest certification competencies of its assessment and accreditation staff, and to describe the procedures that it uses to ensure that staff have these competencies, including recruitment, training and continuing professional development.

Box 5.3

Criteria for selecting UKAS technical experts for the assessment of certification bodies wishing to certify against UKWAS

The requirements below were developed by the United Kingdom Accreditation Service (UKAS) to describe the skills needed by staff responsible for assessing certification bodies wishing to offer certification services against the UK forest standard UKWAS (UK Woodland Assurance Scheme). Only those criteria that address forest management and forest certification knowledge, skills and experience are listed in this box. General criteria are not listed and can be obtained from the original text (UKAS, 2001).

Knowledge of forest certification systems

The individual should be able to:

- Explain what third-party verification means and why it has become important in the international forest products market.
- Describe the main components of, and the differences between, the Forest Stewardship Council (FSC) and the Programme for the Endorsement of Forest Certification (PEFC) and describe the relationship of UKWAS to the FSC and PEFC.
- Describe the relationship between certification bodies and accreditation bodies and the role of UKAS in relation to PEFC.

Familiarity with the development of sustainable forest management principles internationally

The individual should be able to:

- Outline the main events and processes from 1990 onwards that have driven the development of sustainable forest management principles.
- Outline the Pan-European criteria and indicators (C&I).

Familiarity with the UK policy and regulatory environment within which UKWAS operates

The individual should be able to:

- Describe the institutional arrangements for forestry in the United Kingdom.
- Describe the main areas of law and regulation that are relevant to forestry.
- Outline the UK Forestry Standard and describe the supporting guidelines and forest practice guides.

Familiarity with the application of the UK Forestry Standard

The individual should have recent experience of forest management in a position that has required judgement regarding how to interpret the silvicultural, environmental and social elements of the UK Forestry Standard.

Box 5.3 continued

Understanding and interpretation of the requirements of the UKWAS certification standard

The individual can outline the main requirements in each section of the standard, identify problems of interpretation and suggest practical solutions, and provide evidence of:

- significant involvement in the development of the UKWAS standard; or
- evidence of significant involvement in redesigning forestry systems to comply with the requirements of the UKWAS standard; or
- having led a third-party assessment of forest management in the UK; or
- having participated in at least three third-party assessments of forest management.

Professional standing

The individual should have corporate membership of the Institute of Chartered Foresters and at least five years' experience in positions that have required judgement regarding how to interpret the silvicultural, environmental and social elements of the UK Forestry Standard, or a degree or diploma in forestry and at least ten years' experience in positions that have required judgement regarding how to interpret the silvicultural, environmental and social elements of the UK Forestry Standard.

Auditors from the accreditation body should meet the requirements of ISO *Guidelines for Auditing Quality Systems Part* 1 (ISO 10011-1: ISO, 1990) and ISO *Guidelines for Auditing Quality Systems Part* 2 (ISO 10011-2: ISO 1991), and they should have a thorough technical knowledge of the specific activities for which accreditation is sought and a degree of understanding sufficient to make a reliable assessment (ISO/IEC Guide 61, clause 2.2).

The minimum competencies for accreditation body personnel will depend upon the nature of the standard. Assessments of certification bodies offering forest certification against a systems standard will require a different set of minimum competencies compared to accreditation for certification against performance-based standards. Box 5.3 lists the criteria used by the UK accreditation body UKAS (United Kingdom Accreditation Service) in order to select technical experts for the assessment of bodies offering certification against the UK forest standard UKWAS (UK Woodland Assurance Standard).

5.1.2.2 Accreditation assessment methodology

Just as for certification assessments, the way in which an accreditation body collects objective evidence during an accreditation assessment can have a significant bearing on the outcome. As noted in Box 5.1, ISO Guide 61 requires accreditation bodies to specify their assessment and accreditation procedures (clause 2.1.7.1). This should include the means by which the assessment team obtains objective evidence.

The assessment team can collect objective evidence from a variety of sources. Accreditation assessments usually begin with an assessment of the certification body documentation to confirm that all of the necessary systems and documents are in place. This, by itself, however, does not tell the accreditation body whether or not the system is actually working, in practice, and whether the certification body is carrying out assessments properly.

Therefore, accreditation bodies usually include assessments of organizations that have been evaluated by the applicant certification body to see whether the process has been carried out properly. This provides objective evidence of whether the certification body's systems work in practice.

Some accreditation bodies go beyond this and invite external comments on applicant certification bodies. As with the use of consultation during the certification process, this increases the probability that the accreditation body will be aware of any specific issues likely to compromise the accreditation.

5.1.2.3 Geographical scope of accreditation

One final issue is the geographical scope of the accreditation. As discussed in Chapter 1, accreditation has often been carried out by national accreditation bodies that provide the accreditation for certification bodies to work in their country.

However, with the increasing globalization of trade, there is a corresponding demand for some form of global approach. This applies not only to standards, but also to the accreditation under which the certification is carried out. There are a number of reasons for this:

- Many companies which trade internationally do not want the additional bureaucracy and inconvenience associated with evaluating each national accreditation service in order to decide whether it is acceptable (necessary because of perceived differences in competence). They prefer to chose one or two national or international services and then demand that all certificates are issued under these accreditations (often by the same certification body).
- Many certification bodies work internationally, and it is a significant increase in cost to seek and obtain accreditation for the same standard in many countries. Since this cost is passed on to the clients, ultimately it makes certification more expensive for the organization being certified.
- As the number of standards increase, national accreditation services in many small and developing countries are unable to keep up with the need to develop new accreditation services and many do not offer them. Certification bodies and organizations who wish to be certified in these countries would be at a disadvantage unless they can seek accreditation and certification through an accreditation service that already exists.

There are three approaches to providing a wide geographical scope to accreditation services:

1 National accreditation bodies can recognize each others' accreditation so that a certification body accredited by one accreditation body will be recognized as accredited by the others (see Section 5.3.1).
2 A certification body accredited by a national body is able to work in a wider geographical framework. This must be agreed with the national accreditation body who undertook the accreditation since it must ensure that its procedures are adequate to cover the wider scope. This approach is now common practice for international certification bodies with accreditation bodies from countries such as the US, Switzerland, the UK and Japan, who support international use of their accreditation.
3 The accreditation body can be international. In this case, it must have systems and procedures in place which ensure that its accreditation is appropriate to all countries where the accredited certification body may work.

5.2 Requirements for certification bodies

The main task of the accreditation body is to establish that both the certification body organization and the certification process are adequate. To do this properly the accreditation body must have clearly defined requirements for:

- the organization and structure of the certification body;
- the certification process used.

These are both discussed in detail in Chapter 4. All of the requirements and issues discussed need to be documented by the accreditation body as the basis for accreditation. This can either be in the form of internal documents developed by the accreditation body, or external documents developed by a certification scheme, but used by the accreditation body.

5.3 Confidence in accreditation

As discussed at the start of this chapter, the issue of credibility for accreditation is a real and serious issue. If there is a perception that a particular accreditation body is weak, this undermines and devalues all certificates issued under that accreditation.

Therefore, just as for certification, it is important that accreditation assessments and decisions should have the confidence of interested and affected parties. There are a number of ways of achieving confidence, in addition to the requirements of competence, independence and honesty that are set out in ISO/IEC Guide 61 (ISO, 1996b). They are affiliation to international bodies, peer review, complaints procedures and transparency, each of which is discussed below.

5.3.1 AFFILIATION TO INTERNATIONAL BODIES AND PEER REVIEW

One way of providing credibility for accreditation bodies is through affiliation to an international organization which is itself credible. Two such organizations (see Box 5.4) are the European Accreditation organization (EA, 2001) and the International Accreditation Forum (IAF, 2001). Both organizations offer membership to nationally recognized accreditation bodies. The EA is open only to those from the member countries or the candidate countries of the European Union (EU) and the European Free Trade Association (EFTA),[1] while IAF is international. IAF also offers membership to regional groupings of accreditation bodies whose aims include the maintenance of regional multilateral recognition agreements that recognize the equivalence of their members' accreditations (IAF, 2001).

The aim of organizations such as EA and IAF is to ensure a consistent level of quality in all members through multilateral agreements. Evaluation of applicants is based on peer assessment by other accreditation bodies who are already members and therefore have both the technical competence to undertake such as assessment and a strong vested interest in ensuring that standards are maintained.

EA and IAF do not admit international accreditation bodies. Therefore, a number of international accreditation bodies working with social and environmental standards have formed the Social and Environmental Accreditation and Labelling (ISEAL) Alliance. The ISEAL Alliance aims to gain international recognition and legitimacy for their programmes, to improve the quality and professionalism of the respective member organizations, and to defend the common interests of international accreditation organizations (ISEAL Alliance 2000).

Alternatively, international accreditation bodies can set up their own system of regular peer review by either another international accreditation body, or by a credible national accreditation body.

5.3.2 COMPLAINTS PROCEDURES

It is important to provide a mechanism for dealing with situations where there is an objection or a complaint related to:

- the accreditation decision, either from the applicant certification body or a third party;
- the performance of an accredited certification body, particularly any complaints about its certification decisions.

Box 5.4 International accreditation organizations

European Accreditation's objectives

The European Accreditation's (EA's) objectives include achieving a uniform approach to accreditation throughout Europe; achieving universal acceptance of accredited certificates; building and maintaining confidence among nationally recognized accreditation systems; and supporting the harmonized implementation of accreditation standards (EA, 2001).

Charter of the International Accreditation Forum

The charter of the International Accreditation Forum (IAF) states:

> *The International Accreditation Forum, Inc, is an international association of organizations that have agreed to work together on a worldwide basis to achieve common trade facilitation objectives. We are a major world forum for developing the principles and practices for the conduct of conformity assessment that will deliver the confidence needed for market acceptance. We act through the accreditation of those bodies that certify or register management systems, products, personnel and/or inspection.*
>
> *We promote the worldwide acceptance of certificates of conformity issued by inspection, certification and/or registration bodies accredited by an accreditation body member of IAF, and seek to add value for all stakeholders through what we do and through our programmes.*
>
> *We bring together, on a worldwide basis, partner accreditation bodies and representatives of stakeholder groups that seek to facilitate global trade through the acceptance of accredited certificates of conformity.*
>
> *We develop and/or recognize appropriate processes and practices for the conduct of conformity assessment worldwide, and ensure their universal application by IAF accreditation body members and their accredited certification, registration and/or inspection bodies.*
>
> *We consult widely with stakeholders in developing our programmes, and we work to deliver the best possible standard of conformity assessment in order to provide our stakeholders with a value-added outcome.*
>
> *We influence world trade through linking, and working, with other key international organizations and industry groups* (IAF, 2001).

ISO/IEC Guide 61 requires accreditation bodies to have policies and procedures for the resolution of complaints, appeals and disputes received from applicant and accredited certification bodies or from other parties about the handling of accreditation or any related matters (clause 2.1.2.p). As with certification bodies, it is crucial that the mechanism used aims to resolve complaints or disputes, rather than simply being a system for documenting and responding.

5.3.3 TRANSPARENCY

ISO Guide 61, clause 2.1.7.1, requires that accreditation bodies make available various types of information upon request. They should include information about:

- the authority under which the accreditation body operates;

- a documented statement of their accreditation system, including their rules and procedures for granting, maintaining, extending, reducing, suspending and withdrawing accreditation;
- information about the assessment and accreditation process;
- a description of the means by which the accreditation body obtains financial support;
- a description of the rights and duties of applicants and accredited bodies; and
- information on procedures for handling complaints, appeals and disputes.

These requirements provide a basic degree of transparency in accreditation and should form the basis for any accreditation system.

Transparency of accreditation can be enhanced in the same ways as for certification. Information can be made publicly available about bodies who are seeking or have been granted accreditation. Adding a summary of non-compliances allows interested and affected parties to see whether there are any key weaknesses, while providing a summary of results allows interested and affected parties to make their own judgement on the accreditation decision.

However, the same considerations apply to transparency of accreditation as to certification. Confidential information needs to be respected and the costs of the accreditation process must be kept reasonable since they will, ultimately, be passed on to the certification clients, thus increasing the cost of certification.

Note

1 European Accreditation offers associate membership to nationally recognized accreditation bodies in the European geographical area who can demonstrate that they operate an accreditation.

6 Product Tracing and Claims

For some forest managers, the aim of certification is to allow them to make immediate claims about the quality of their forest management. This is particularly important where the demand for certification comes from investors, governments, shareholders or local communities. However, the biggest driver behind forest certification remains the market demand for products that come from well-managed forests. It is therefore necessary to have a mechanism which links products to the forest where the original tree (or non-timber forest product) was grown. This is known as product tracing, supply-chain management or, most commonly, chain of custody. This chapter examines both chain of custody and claims.

6.1 Chain of custody

Linking a claim directly to a certified forest is fairly straightforward. Making claims about products made with raw material from certified forests is more complex. Manufacturing processes in the forest products sector are often very complicated. Once a log leaves the forest it may go through a range of manufacturing processes before it becomes a final product. The wood may be cut, peeled, chipped or broken down into fibre, divided into separate loads, will probably change ownership more than once, and will generally be processed and reprocessed. At any one of these stages there is the risk that it could be mixed with similar material from uncertified forests.

Looking at the same process from the viewpoint of the manufacturer making a final product, the raw material may be sourced from a number of suppliers, each of whom has, in turn, sourced from several suppliers, and so on. In practice, many processors have material that originated in tens or even hundreds of sources, as illustrated in Figure 6.1.

Yet, if a credible product claim is to be made, it is necessary to have sufficient control over the entire production chain to be able to make clear and accurate claims about the source of the material in the final product. This requires some form of product tracing or *chain of custody*.

A chain of custody is a *verifiable system of traceability* for certified timber or other material at each stage through which it passes from the forest to the final product (see Box 6.1 for definitions of chain of custody). Each time ownership of the material changes or processing is undertaken, another link is added to the chain. It is important to demonstrate at each stage that the material being transported,

Box 6.1 Definitions of chain of custody

Definitions of chain of custody include the following:

> *To provide clear separation or demarcation of certified and uncertified forest products, at all stages, including forest sites, processing, shipping, manufacturing, and wholesale and distribution stages (FSC Statutes, Forest Stewardship Council, 1994).*
>
> *All the changes of custodianship of forest products, and products thereof, during the harvesting, transportation, processing and distribution chain from the forest to the end use (PEFC Council Technical Document Annex 1: PEFC Terms and Definitions, adopted 22 November 2002).*

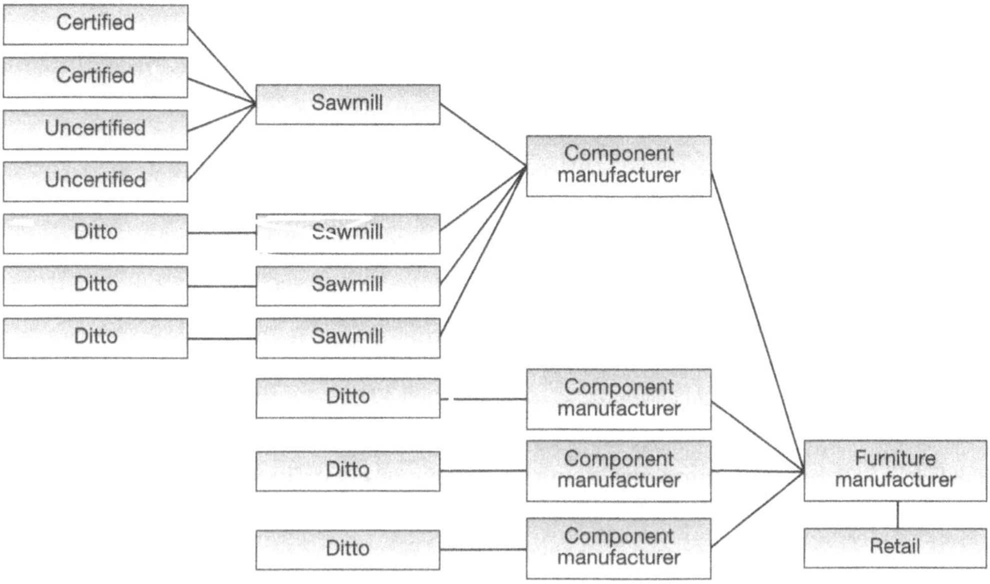

FIGURE 6.1 Schematic representation of a supply chain for a furniture manufacturer, showing how complicated such supply chains can be

processed or sold is certified, and that it has not been mixed with or 'contaminated' by material from other sources.

Chain of custody is usually implemented and controlled separately at each stage of the manufacturing process, as shown in Figure 6.2. At each point it is necessary to ensure that purchasing, processing and sales are all managed to guarantee that certified and uncertified materials and products are not mixed. This is usually achieved through a combination of:

- *Segregation*: by keeping certified materials physically separate, the chances of mixing are removed.

> For example, there should be separate storage areas, separate manufacturing lines, separate drying facilities or separate areas for completed products.

- *Identification*: making sure that certified material and products are clearly labelled reduces the risk of any accidental mixing.

> For example, identification includes bar-coding of certified logs, using different coloured labels for certified and uncertified material, and using different packaging for certified and uncertified products.

- *Documentation*: in order to ensure that there is no uncontrolled mixing, it is also very important to have detailed documentation, covering procedures, operating information and records.

> For example, records of all certified raw material received, records of all material processed, procedures setting out the rules for segregation in the storage areas, and information on certification status on both orders and invoices should all be documented.

The process of implementing chain-of-custody controls is discussed in detail in Chapter 12.

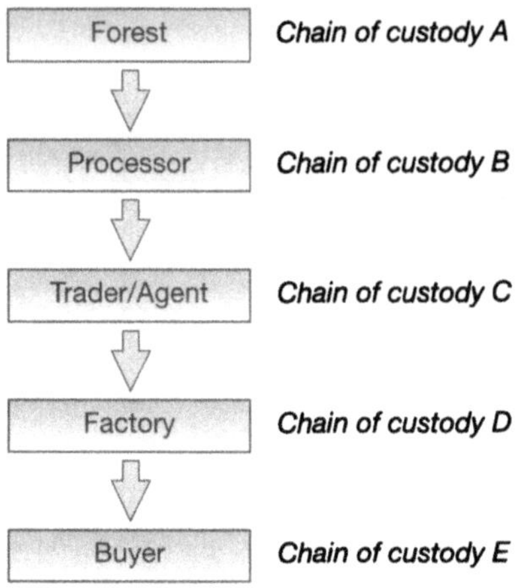

FIGURE 6.2 Schematic representation of a supply chain indicating how each organization must have its own chain-of-custody system and certification

6.2 Claims

One of the main purposes of forest certification is to provide a verifiable and credible basis for making claims about the quality of forest management. For forest owners, a claim that their forests are managed to a defined standard can help them to sell timber to their immediate customers, to win public recognition and support, and, in some cases, to secure loans and investments and aid funding for management. For forest products manufacturers and retailers of wood products, a claim that their products are made from trees grown in such forests can give them a marketing edge or help in their public relations activities.

It is vital to the effectiveness and long-term credibility of a forest certification scheme that any claims which are made about certified forests or products are accurate and not misleading. Inaccurate claims very quickly undermine the entire value of the scheme since stakeholders no longer believe the claims that are made.

6.2.1 WHAT IS AN ENVIRONMENTAL CLAIM?

The International Organization for Standardization (ISO) considers the overall goal of environmental labels and declarations to be:

> *... through communication of verifiable, accurate information, that is not misleading, on environmental aspects of products, to encourage demand for and supply of those products that cause less stress on the environment, thereby stimulating the potential for market-driven continual environmental improvement* (ISO 1999a).

Claims related to certified forests fall within the ISO definition of an environmental label or declaration (see Box 6.2). These can take different forms. For a business-to-business transaction, the purchaser may require only a verifiable statement from the supplier. A company selling to the public may require a label on the product for simple and effective communication to its customers.

ISO recognizes a number of different types of environmental claims as described in Box 6.2. A claim about the quality of forest management, or that a product (or part of it) originates in certified forests, does not fall completely within any of ISO's three categories of environmental claim. This is because although they fulfil most of the criteria for type I claims, forest certification schemes address only one aspect of the product life cycle – namely, production of the raw material (Vallejo and

Box 6.2 **ISO definitions of environmental labels and claims**

An environmental label or environmental declaration is a:

... claim which indicates the environmental aspects of a product or service.

Note: an environmental label or declaration may take the form of a statement, symbol or graphic on a product or package label, in product literature, in technical bulletins, in advertising or in publicity, amongst other things (ISO, 2000b: ISO 14020, clause 2.1).

Types of environmental claims include the following:

- *Type I – Environmental labelling programme (ISO 14024)*: voluntary, multiple criteria-based third-party programme that awards a licence which authorizes the use of environmental labels on products indicating overall environmental preference of a product within a particular product category based on life cycle considerations (ISO, 1999b).
- *Type II – Self-declared environmental claim (ISO 14021)*: environmental claim[1] that is made, without independent third-party certification, by manufacturers, importers, distributors, retailers or anyone else likely to benefit from such a claim (ISO, 1999a).
- *Type III – Environmental declaration (ISO/TR 14025)*: quantified environmental data for a product with preset categories of parameters (ISO, 2000a). The parameters are based on independently verified[2] systematic data presented as a set of parameter categories. The information is presented in a format that facilitates comparison between products.

Notes: 1 Statement, symbol or graphic that indicates an environmental aspect of a product, a component or packaging (ISO 1999b).
2 Independent verification for the purpose of type III environmental labelling need not necessarily involve certification.

Hauselman, 2000). Such single-issue claims are different from life cycle analysis (LCA)-based claims that consider the environmental impacts over the entire life cycle of the product.

Claims need to be credible, otherwise they will not be effective in the long term and may have damaging impacts. Environmental claims do not have a particularly good track record in this regard, with several cases of 'greenwash' (unsubstantiated claims of the environmental credentials of a product or company made for commercial advantage). This has led to the production of various guidelines that deal with environmental claims, such as the UK Green Claims Code (DETR, 2000) and ISO guidance.

6.2.2 GENERAL PRINCIPLES GOVERNING CLAIMS

As has been noted already, claims can serve many different purposes; but whatever the motivation for making a claim, there are a number of principles that have been developed by ISO and that need be considered.[1] ISO's general principles for environmental labels and declarations (ISO, 2000b) are set out in Box 6.3.

6.2.3 SPECIFIC ISSUES FOR CLAIMS ABOUT CERTIFIED FORESTS AND PRODUCTS FROM CERTIFIED FORESTS

The principles derived from ISO 14020 in Box 6.3 are relevant to claims about certified forests and products from certified forests. There are, however, two issues that are of particular importance to forest certification schemes: accuracy of claims and product life cycle.

Box 6.3

ISO 14020 Environmental labels and declarations: general principles

Principle 1

Environmental labels and declarations shall be accurate, verifiable, relevant and not misleading.

Principle 2

Procedures and requirements for environmental labels and declarations shall not be prepared, adopted or applied with a view to, or with the effect of, creating unnecessary obstacles to international trade.

Principle 3

Environmental labels and declarations shall be based on scientific methodology that is sufficiently thorough and comprehensive to support the claim and that produces results that are accurate and reproducible.

Principle 4

Information concerning the procedure, methodology and any criteria used to support environmental labels and declarations shall be available and provided upon request to all interested parties.

Principle 5

The development of environmental labels and declarations shall take into consideration all relevant aspects of the life cycle of the product.

Principle 6

Environmental labels and declarations shall not inhibit innovation that maintains or has the potential to improve environmental performance.

Principle 7

Any administrative requirements or information demands related to environmental labels and declarations shall be limited to those necessary to establish conformance with applicable criteria and standards of the labels and declarations.

Principle 8

The process of developing environmental labels and declarations should include an open, participatory consultation with interested parties. Reasonable efforts should be made to achieve a consensus throughout the process.

Principle 9

Information on the environmental aspects of products and services relevant to an environmental label or declaration shall be available to purchasers and potential purchasers from the party making the environmental label or declaration.

6.2.3.1 Accuracy of claims

In addition to principle 1 of ISO 14020, ISO 14021, clause 5.3, states that:

> *... an environmental claim that is vague or non-specific or which broadly implies that a product is environmentally beneficial or environmentally benign shall not be used. Therefore, environmental claims such as 'environmentally safe', 'environmentally friendly', 'earth friendly', 'non-polluting', 'green', 'nature's friend' and 'ozone friendly' shall not be used.*

On claims of sustainability, ISO 14021, clause 5.5, states that:

> *... the concepts involved in sustainability are highly complex and still under study. At this time there are no definitive methods for measuring sustainability or confirming its accomplishment. Therefore, no claim of achieving sustainability shall be made.*

This guidance would appear to be inconsistent with a claim that a forest is 'sustainably managed' or that the wood in a product comes 'from sustainably managed forests'.

6.2.3.2 Product life cycle

As noted in Section 6.1, current forest certification schemes cover only the production of timber, and make no environmental claims about other parts of the manufacturing and transportation process. This does not meet principle 5 of ISO 14020, which encourages all environmental aspects to be considered. Therefore, it is particularly important that the claims made about products containing raw material from certified forests must pertain only to the management of the forests and do not imply reference to other environmental aspects of the production process.

6.3 Percentage-based claims

A further issue that is extremely important for forest certification schemes comprises percentage-based claims.

For some supply chains, it is relatively straightforward to ensure that all of the material used at each stage of the process originated in a certified forest.

> For example, this is the case with an integrated pulp and paper mill supplied entirely from certified plantations owned by the same company, or a furniture manufacturer supplied by a single sawmill that buys all of its logs from two or three certified local forests.

However, a significant proportion of the wood and fibre industry works with very complex raw material sourcing. This is particularly relevant to chip and fibre products such as MDF, pulp, paper, chipboard and OSB, all of which utilize large volumes of wood that is often derived from multiple sources. In many cases, it has proved extremely difficult to restrict sourcing only to certified forests. There are various reasons for this depending upon the location, but the most important include the following:

- The total area of certified forest in some regions is not yet large enough to supply all of the material needed.
- The distribution of certified forest in some regions is such that using only certified material would involve much longer transport distances for raw material, with negative impacts both on the economics of production and the environment.
- The uncertified sources in some regions tend to be small- and medium-sized private forests with limited ability to access other markets, which would therefore be seriously disadvantaged if they were excluded from being suppliers.

Manufactures of solid wood products have also experienced problems – for example, where some components of a product are already available certified, but others are not.

Clearly, the long-term solution is to get more forest certified; but it is widely recognized that this can take many years. As a result, most certification schemes have developed mechanisms to allow certification and claims for products containing only a proportion of certified material, with the remainder of the content coming from uncertified sources.

If percentage-based claims are allowed, there are four main issues that need to be considered:

1 How is the percentage content controlled and calculated?
2 Is there a minimum threshold certified content before claims are allowed?
3 How is the uncertified content dealt with?
4 What claims and labels are appropriate?

Each of these is discussed below.

6.3.1 CONTROLLING PERCENTAGE CONTENT

If percentage-based claims are allowed, the first issue that arises is what the percentage refers to. In practice, certification schemes use different approaches, including control of percentages at the level of individual products, of the manufacturing process or product line, and of the processor as a whole. Each of these approaches, and the implications for the product and any associated claims, is discussed below.

6.3.1.1 Percentage in a product

Where percentage-based claims relate to an individual product, then it is necessary to have controls that ensure that every product made has the minimum percentage of certified material claimed. In practice, this requires the same approach to chain of custody as for products containing only certified material, with separation of certified and uncertified material throughout the production process. The only difference is that, at some point, there is a controlled combination of certified and uncertified material or components into the final product.

A very common example is where a product is made from more than one type or species of wood. Some components are certified, but others not.

For example, a kitchen cabinet may have an MDF carcass made from certified material, together with a veneer covering, also certified, and a solid wood door that is not certified.

However, it is also possible to have a percentage claim related to a product where only one species of wood is used.

For example, a chair may be made of solid pine. All of the components for the legs and seat are made from certified pine, while the components for the back are made from uncertified material.

Where percentage content is controlled at the level of individual products, the link between the certified forest and the certified product is maintained, with all certified products containing raw material from a certified forest.[2] Claims can reflect this. However, they must also make clear what the percentage of certified raw material is.

6.3.1.2 Percentages in a process or product line

It is not always possible to control the content of an individual product because many manufacturing processes can only control raw material input to the process and not to products. This is particularly the case with chip and fibre products such as paper, MDF or chipboard. In this case, percentage-based claims have to be associated with a product line or process, rather than an individual product.

Percentages in batch production

Where the process uses batch production, control is straightforward and can be achieved by managing the proportion of certified and uncertified material used for each batch.

Percentages in continuous production

A problem for this approach is products that are made using continuous processes, which is very common. Since the process is a continuous one, it is not possible to control raw material input on a normal batch basis.

One solution is to ensure a consistent feed of the same percentage of certified and uncertified material all of the time; but, in practice, this is seldom practical as most of these types of operations have a very quick turnover of raw material and limited space for storage and therefore depend upon what gets delivered to feed the process.

As a result, the most common approach is to work on the basis of a nominal batch. A nominal batch is a period of time such as a day, a week, a month or a year which is designated as representing one batch. During this period, the amount of certified and uncertified raw material used has to meet the requirements for the percentage proportions in the product line.

For example, if the nominal batch length is one week and the proportion of certified material required is 50 per cent, then half the raw material used during the week must be from a certified source.

The relative proportions of certified and uncertified material entering the process at any one time can fluctuate from entirely certified to entirely uncertified, provided that the average input over the period meets the requirements. In practice, this means that the process can run for a considerable period with no certified raw material being used, resulting in products which – although certified and able to carry a label – in fact, contain no material from a certified forest.

For example, in the case above, it would be possible to use exclusively uncertified material for three and a half days, and then exclusively certified material for the other three and a half days. Half of the output will then contain no certified material, although it will be certified.

In order to try to reduce the extent to which the proportion of certified and uncertified material fluctuates, some schemes require the input to be calculated using a rolling mean or rolling average approach. In this case, instead of calculating the average only at the end of each nominal batch period, the average is continually recalculated (for example, daily or weekly) and must always exceed the minimum threshold. However, even with this approach, it is possible to have extended periods when no certified raw material is used.

An example of this is a system that requires a minimum average of 30 per cent certified raw material calculated based on a nominal batch length of 30 days and a rolling average recalculated daily. As can be seen from Figure 6.3, this can be achieved while using 0 per cent raw material for 20 days out of every 30, so that up to two-thirds of the product contains no certified material.

If this approach is used, the physical link between the certified forest and the content of a product is broken. The chain of custody is no longer based on the tracing of material from a certified forest. Instead, it is based on the tracing of certified material that is produced following the rules for percentage production.

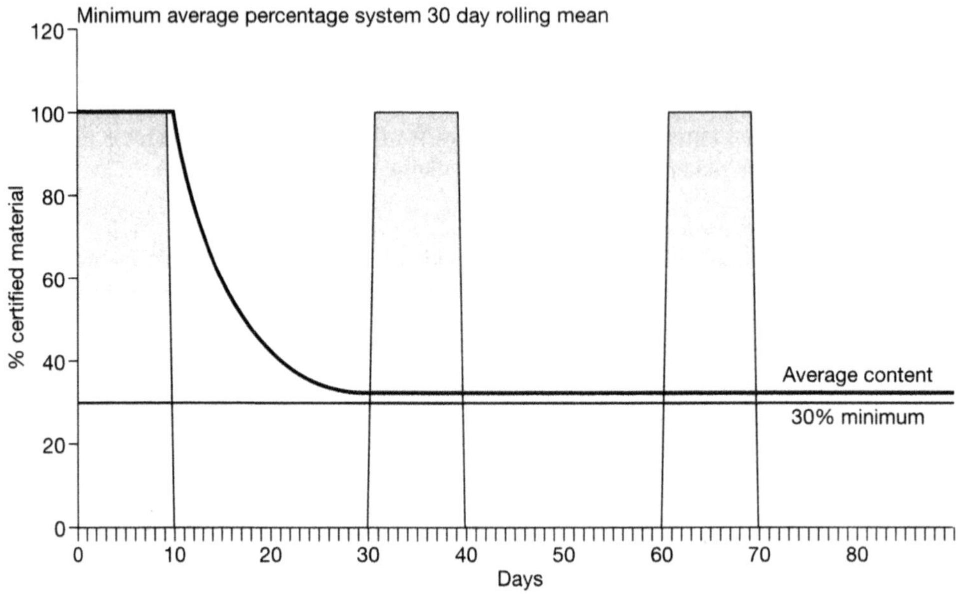

FIGURE 6.3 Graphic representation of the amount of certified raw material used to produce certified products in a 30-day rolling mean approach to percentage labelling

It is therefore very important that any claims or labels make it clear that the percentage relates to the *product line* or *process* and not the product itself. The origin of the uncertified raw material is also extremely important. as discussed below.

6.3.1.3 Percentage in–percentage out, or volume accounting

In a percentage in–percentage out system, also called *volume* or *inventory accounting* or *input–output*, the proportion of certified raw material entering a process is monitored, and an equivalent proportion of product is deemed to be certified. Thus, if 50 per cent of the raw material is certified, then 50 per cent of the output can also be certified without any need to undertake internal tracing or control.

It is very important if this approach is used to maintain the link between the *type* of raw material and the *type* of product. Thus, it should not be possible for a manufacturer to buy certified pine and sell 'certified' mahogany, or to buy certified MDF but to sell 'certified' solid oak. If the certified raw material is pine-sawn timber, then only products made from pine-sawn timber can be sold as certified.

There are many who support this approach on the basis that the aim of forest certification is to improve forest management, not to impose complicated raw material management regimes on manufacturers. Using percentage in–percentage out recognizes and promotes the purchase of material from certified forests because the greater the proportion of certified raw material that is purchased, the larger the amount of certified product that can be sold. But it does this without the additional expense of keeping certified raw material segregated and identified throughout the manufacturing process.

In fact, it could be argued that percentage in–percentage out provides more support for certified forests than other percentage approaches. This is because the quantity of certified products made is directly proportional to the quantity of certified raw material purchased, and so any increase in demand for certified products will require increases in the purchase of certified raw material. If the process-based approach is used, 100 per cent of the production is certified, allowing the same manufacturer to meet growing demand with no change in the amount of certified material purchased.

Those who do not support the system argue that consumers, who are, ultimately, the driver behind market-driven certification, will be confused by an approach which means that the 'certified' product they buy does not actually come from a certified forest at all. It should be noted, however, that this is also the case with percentage content claims for processes, as discussed in Section 6.3.1.2.

Again, the most critical element is that any claim which is made is clear, accurate and does not mislead consumers. In addition, as with any other percentage content claims, the origin of the uncertified raw material is extremely important.

6.3.1.4 Processor certification

An alternative to product tracing that has been adopted by some schemes is processor certification. When this approach is used, rules are laid out for the proportion of products from certified and uncertified origin that a certified producer must utilize. If this is complied with, then all products sold by the certified processor are certified.

As with the percentage in–percentage out approach, processor certification completely severs the link between a certified forest and the content of a certified product.

Where this approach is used, it is important that any claim or label relates to the involvement of the processor in the certification scheme, rather than actual content of the product, and that it is clear that this allows uncertified as well as certified material to be used.

Again, as with any percentage-based approach, the control of the origin of the uncertified material used is very important.

6.3.2 MINIMUM THRESHOLDS FOR CERTIFIED MATERIAL

An important issue for any percentage-based approach is what the minimum amount of certified material used in the product or the process must be in order to allow a claim to be made or a label to be used. Clearly, it will be less than 100 per cent and more than 0 per cent. However, there is no particular precedent or guidance to show where it should lie between these two extremes.

Some groups feel that the minimum should be relatively high (50 to 70 per cent) in order to ensure that when people buy a certified product it is predominantly certified. Others argue that by reducing the minimum it allows more certified product to enter the market, boosting demand and thereby increasing the demand for certified forests. A compromise that is sometimes adopted is to allow a relatively low minimum threshold, but to require a commitment from the producer to continually improve the proportion of certified material in the process or product.

The only approach that avoids this issue is percentage in–percentage out, where the amount of certified material used does not need to be controlled other than to ensure that it forms the basis for the proportion of product deemed to be certified.

In general, the main issues to consider are that:

- the proportion of certified material is clearly and accurately reported;
- there is adequate control over the uncertified material.

An additional complication is that many composite products such as chipboard, MDF, pulp and OSB contain a variety of other components, such as waste and recycled material, as well as virgin fibre. Therefore, most certification schemes recognize three categories of raw material for the purposes of calculating percentages:

1 *Certified raw material*: this material is derived from certified forests or is the certified output of a chain of custody. If the raw material used was certified using a percentage-based approach, then only the certified proportion counts.

For example, if a major component of a cupboard is MDF, which accounts for 70 per cent by weight of the raw material used, and the MDF is a 50 per cent product, then the cupboard has a total of 35 per cent certified raw material (50 per cent of 70 per cent).

2 *Uncertified raw material*: this comprises any material that is not certified and is not neutral, including the uncertified component of percentage-certified raw material.
3 *Neutral material*: this can include material such as recycled fibre (it is important whether this is post-consumer only or also includes pre-consumer; some schemes include the pre-consumer material in the uncertified category), waste wood (for example, old pallets and wood from demolition sites) and urban wood (waste from the maintenance of trees on roads and in parks).
   Generally, in the calculation of percentage content, neutral sources are not included and the calculation is based only on the proportion of certified and uncertified material.

For example, if an MDF process uses 50 per cent neutral recycled material, 25 per cent certified chips and 25 per cent uncertified chips, then the certified percentage would be 50 per cent.

The precise definition for each of these categories and the way in which the calculation is made are extremely important to the accuracy and credibility of any claim or label.

6.3.3 DEALING WITH THE UNCERTIFIED COMPONENT

As soon as percentage claims of any sort are allowed, the question of the uncertified component arises. If a percentage claim is being made, it is because a percentage of the product is definitely *not* from a certified forest. The obvious question this raises is: if it is not from a certified forest, where is it from?

Anyone buying certified products is likely to be doing so because they want to support good forest management. The credibility of a scheme is likely to be quickly undermined if it becomes known that the uncertified material in labelled products originated from illegal logging operations or other unacceptable sources. However, unless there is some control over the uncertified portion of a percentage-labelled product, then there is a very significant risk that it will be just such a source from which it derives.

This has become a particularly important issue recently as governments and companies in many countries make commitments to stop purchasing any wood or paper products from illegal sources. Purchasing certified products has been seen as a straightforward way of meeting these commitments; but it only works if there is certainty that the uncertified component of a percentage-labelled product does not contain illegal or other unacceptable wood.

6.3.3.1 Unacceptable sources

Most certification schemes have recognized this and define a number of sources that are not permitted in certified products. This can include both illegal and other controversial sources.

Illegal timber

Illegal logging is a massive problem in the forest products industry. Some estimates suggest that up to 10 per cent of timber traded globally is illegal, with the proportion much higher for certain countries. If the source of uncertified timber is not known, then there could be a serious risk that some of the timber is from illegal sources.

Other unwanted sources

There are a number of sources of wood that are particularly controversial and, as with illegal timber, could undermine the credibility of percentage-certified products if they are also one of the sources of raw material. Different certification schemes and different interest groups have different definitions of what constitutes an unwanted or controversial source; those that are often discussed comprise the following:

- protected areas or forests that have been proposed for national parks but have not yet been formally protected;
- forests that have particularly high value for conservation (high conservation value forests) or are biodiversity hotspots and are not certified or demonstrably well managed;
- forests where there are serious tenure disputes, particularly where these involve the failure to respect the customary rights of indigenous or local people;
- forests that are inappropriately converted to other uses.

As with illegal wood, many governments and companies are seeking to exclude some or all of these controversial sources from their purchasing. If they are to use certified products to help them meet their purchase policy objectives, it is essential that the products do not contain material from these sources.

It is therefore very important that the rules for all types of percentage claims include adequate provisions for controlling the uncertified component of the raw material being used.

6.3.3.2 Controlling the source of uncertified material

There are a number of ways in which the source of uncertified material can be controlled.

Independent verification

The most secure way of dealing with the uncertified portion is to require the same type of independent verification of origin for uncertified material as is required for certified material.

In practice, this type of verification would be often be extremely expensive and complex to implement. For example, many pulp and chip mills buy wood from hundreds or even thousands of small forests. To require independent verification of source for each purchase would be very costly.

However, for some high-value sources such as tropical hardwoods, this is an option that is increasingly being considered, particularly in countries or regions where illegal logging is a major problem. A number of certification bodies and other organizations carry out generic assessments of origin for raw material.

For lower value wood where there is a risk of illegality or other unwanted sources, even if it is too expensive and complex to implement independent verification of source for all raw material supplied, it may be useful to verify a sample of the total supply base. This is discussed below.

Other certification schemes

Many schemes recognize other schemes as providing adequate confirmation that the material is not from an unacceptable source.

Internal control of sources

An alternative approach is to require an internal system to check the origin of all uncertified raw material used. This is the approach used by most certification schemes. The system needs to ensure that every batch of incoming raw material is checked based on the documentation provided, and that the level of internal verification is appropriate to the level of risk of the source of the raw material.

> For example, if wood is being supplied entirely from small local producers in a country with a long history of good forest management and a strong legislative framework, then a system that requires a check on the transport documentation for each load may be adequate.

If the wood is being purchased from a country with a known problem with illegal logging, then a system based on checking an export declaration is probably not adequate because it is well documented that there are many ways in which illegal timber can be 'laundered' to make it appear legal at the point of export. In this case, it would be necessary to have a system that could trace the timber back to the original forest source.

Internal control with verification of a sample

An approach that combines the two approaches outlined can be effective. A processor is responsible for implementing and maintaining a system to check the origin of uncertified raw material. A sample of the raw material received is then randomly selected and the adequacy of the source independently verified. If problems are found, this indicates a problem with the system overall, and it will need to be revised.

The issue of implementing internal controls for uncertified material in certified products is discussed in more detail in Chapter 12.

6.3.4 PERCENTAGE-BASED CLAIMS AND LABELS

As discussed in Section 6.2, the most important thing about any claim or label is that it is clear and not misleading. Therefore, for percentage-based production, any claims or labelling needs to reflect the fact that only a percentage of the material is certified.

Principle 1 of ISO 14020 requires that claims are accurate, and although ISO has no specific guidance relating to percentage-based claims for certified material, the guidelines for claims about recycled content (ISO 14021) provide a useful parallel:

- Where a claim of recycled content is made, the percentage of recycled material shall be stated (ISO 14021, clause 7.8.2.1).
- If a symbol is used for a recycled content claim, it shall be ... accompanied by a percentage value stated as 'X per cent', where X is the recycled content expressed as a whole number calculated in accordance with [clause] 7.8.4. The percentage value shall be located either inside [the symbol] or outside and immediately adjacent to [the symbol] (ISO 14021, clause 7.8.3.2).
- If the percentage recycled content is variable, it may be expressed with statements such as 'at least X per cent', or 'greater than X per cent' (ISO 14021, clause 7.8.3.3).

When applying this guidance to product labels associated with forest certification schemes, it seems reasonable to require such labels to state the percentage of the wood fibre in a product or product line that comes from certified forests. It is also important that it is clear whether it is the product or the product line that is being referred to.

Where percentage in–percentage out or processor certification is adopted, then claims and labels must make it clear exactly what is being referred to. The simple use of a logo without any further explanation would be unlikely to meet ISO guidelines in this circumstance, particularly if the same logo is used to promote products that contain certified material.

6.4 Certification of chain of custody

The basic components of chain-of-custody certification are the same as for forest management certification: a standard, accreditation and certification. The ISO guidance on accreditation and certification referred to earlier in relation to forest management certification is equally relevant to chain-of-custody certification. The standard sets out the procedures that a company must implement in order to ensure that a claim is truthful and verifiable. The implementation and certification of chain of custody are discussed in detail in Chapters 12 and 13.

Notes

1 ISO is not the only organization to have developed principles or guidelines related to environmental claims. National governments have done this – for example, the UK Green Claims Code (DETR, 2000) – and the European Commission is currently considering a proposal for European Union guidelines.

2 The exception to this is when the certified component is itself percentage labelled as a result of a production process that does not maintain the link between the certified forest and the product, as discussed in Chapter 6.

Part Two

Forest Certification in Practice

7 Getting Started

7.1 To certify or not

7.1.1 WHY ASK THE QUESTION?

With all of the discussion that forest certification has generated, it is sometimes difficult to remember that, just like any other type of certification, it is a *means* to an end and not an end in itself. Certification is an efficient mechanism for providing reliable, independent verification that a particular standard has been met. However, it also costs both time and money. Therefore, it is important to consider whether or not independent verification is really needed.

7.1.1.1 Certification for forests

Certification in the forest can be a long and expensive business; therefore, it is particularly important for forest managers to be sure that it is the right decision before starting. This is not the same as considering whether or not to manage the forest responsibly or sustainably. Good management should be the goal of every responsible manager. Certification is the process of *confirming* the level of management, and it is this which needs to be considered because it is only useful to confirm that you are implementing the requirements of a responsible forestry standard if there is a reason for doing so.

There are a number of reasons why a forest manager may decide to be certified. The most common comprise the following scenarios:

- Customers demand certified products.
- There is the potential to use certification a means of accessing new markets.
- An investor or donor demands certification as a condition on a loan or grant.
- An insurer demands certification as a condition of insurance.
- The owners, shareholders or management see certification as a useful tool to achieve management goals.

All of these potential drivers and what they mean for a forest manager making a decision about certification are discussed below.

7.1.1.2 Certification of products

For those involved in the processing chain, the decision about certification is usually more straightforward since it is driven by market demand or potential market access, rather than by any other factors. For either of these drivers, the questions that processors need to ask about moving into certification are broadly the same as for the forest manager.

7.1.2 REASONS TO GET CERTIFIED

Each of the different drivers that might make a forest manager or processor consider certification is discussed below, with some comments on the process for making a decision about how to proceed.

7.1.2.1 Demand for certification from customers

This is an important driver of certification for both forest managers and processors. In either case, the process for making a decision about whether to proceed is relatively similar.

Firstly, consider how many customers are looking for certified products, what proportion of sales they currently represent and whether the proportion is likely to grow in the future.

It is important to be as accurate and realistic as possible in assessing the demand for certification both currently and in the future. Experience shows that if an organization is reluctant to become certified, then there is a tendency to underestimate demand. On the other hand, if there is enthusiasm for certification, then it is easy to overestimate how many customers are interested in certified products.

One way of assessing demand is to ask customers a series of simple questions, such as:

- Do you currently purchase any certified material from other suppliers? If so, what is it?
- If we supplied certified material, would you purchase it now or in the future?

It may be possible to obtain information by recording requests for certified products, through discussions with the sales team or by analysing the use of certified material in the sector supplied. It is also important to look at the current and developing policies of important customer groups, such as trade associations or government purchasing departments, many of which are now making commitments to purchase timber from verifiably legal or sustainable sources.

Once the information has been collected, it is necessary to decide whether there is sufficient demand to justify seeking certification either immediately or in the future.

An additional question that may arise is which certification scheme(s) customers are interested in. It may be useful to ask this question when collecting other information. It is important to avoid making prejudgements based on individual or company preferences for a particular scheme because, ultimately, there is little point getting certified against a standard in which customers are not interested. Where certification is market driven, it is the market that decides upon the best scheme to use.

7.1.2.2 Certification as a way of accessing new markets

Careful thought is needed before seeking certification as a means of getting into new markets. The prospects need to be properly researched in order to ensure that there really is a demand. If there is, then it is absolutely essential to be sure that it is also possible to deliver the other demands of the new market, such as quality, species, reliability of supply and competitive prices.

It is also important to be sure which certification scheme the potential market prefers. Access to a market will only be provided by the certificate that the market is looking for.

Only after all of this research is complete is an organization in a position to decide whether or not it is worthwhile to seek certification.

7.1.2.3 Pressure for certification from investors, donors or insurers

If there is strong pressure to certify from an investor or donor, particularly if they make certification a condition of a loan, then there is little option other than to seek certification, and the question is how to become certified.

In this case, it may be useful to assess whether a phased approach to certification (see Chapter 17) is the most appropriate approach. This is allowed, for example, by the World Bank forest policy.

7.1.2.4 Certification as a tool for meeting management goals

Where certification is being used as an internal tool, then it is important to decide which certification scheme best meets the needs of the organization and the time frame during which it wants to achieve certification.

It is also important to consider whether any of the drivers discussed above are likely to become important in the future, in which case this should be factored into the decision. It is particularly important to consider the standard that will be implemented. Identifying gaps and implementing the requirements of the standard (see Chapter 8) is usually the major part of preparing for certification. If an organization is going to initiate the process, it is worth including all of the requirements that might be needed in the medium term.

For example, it may suit management goals in the short term to be certified against a national standard. However, if it is likely that there will be growing market demand for products certified to a regional scheme in the medium term, then it may be worth ensuring that the requirements of both the national and the regional schemes are implemented. Certification can then proceed against the national scheme; but if demand for the regional scheme grows, it will be straightforward to seek the additional certification since all the work of implementation is already complete.

7.2 Choosing a certification scheme

As discussed above, the other issue that forest managers or processors have to consider when deciding whether to become certified is which certification scheme is the best one to get certified with. This is not always an easy question to answer and it remains the subject of much debate and disagreement. As the discussion on assessing certification schemes in Chapter 20 makes clear, there is no single answer to the question 'which scheme' because it depends upon the reason for becoming certified. The manager needs to go through the process of considering why certification is needed and, based on that, which certification scheme is best.

Sometimes this will be very clear because customers or investors will specify a particular scheme. At other times, the reasons will be less obvious; in these situations, the best advice to managers is to ignore the politics and decide which scheme will give them the greatest benefits.

For example, if the reason for seeking certification is the potential to access markets that demand certificate X, there is little point in seeking certification under scheme Y. Chose the scheme that the market wants.

If the reason for certification is to boost internal morale and improve relationships with local stakeholders, then it may be better to use a popular local scheme rather than an unpopular international one, even if the local scheme provides little market benefit.

In some cases, the best option is to seek certification against two schemes simultaneously. This option is increasingly offered by certification bodies in countries where two or more schemes are operating.

7.3 Routes to certification: Individual or group

Once a forest manager or a forest products processor has decided that certification is needed and has chosen the scheme or schemes to certify against, the next decision is the most appropriate route by which to achieve certification. There are two main possibilities: individual certification or some form of group scheme.

7.3.1 INDIVIDUAL CERTIFICATION

Individual certification involves a specific forest area (forest management unit) or processing operation being certified directly by a certification body. This is the most common approach to certification for medium- and large-sized enterprises. It is the responsibility of the organization seeking certification, whether it is a forest enterprise or a processor, to:

- implement the standard in the forest or the factory (see Chapters 8 and 11);
- engage a certification body and undergo the certification assessment (see Chapters 9 and 12).

If the assessment is successful, a certificate is issued to the organization that was assessed, covering the forest management unit or process which it manages.

7.3.2 GROUP CERTIFICATION

While individual certification works well for most medium- and large-sized enterprises, it can be a major challenge for small enterprises, whether these are small forest owners or small-scale producers of wood products. They do not have the economies of scale that their larger competitors have; therefore, the cost and complexity of understanding and implementing the standard and engaging a certification body can be a major barrier to certification.

As a result, most certification schemes provide a mechanism that allows certification through a group scheme. A group scheme is managed by a group manager who is responsible for ensuring that all the group members, whether they are forest owners or small-scale producers, understand and implement the requirements of the standard. The group manager then engages the certification body and manages the certification process on behalf of the members.

Group certification for forests is discussed in detail in Chapter 10 and for chain of custody in Chapter 11; in summary, there are two major advantages for small enterprises in seeking certification through some form of group scheme:

1. The group manager takes on the challenge of understanding and interpreting the requirements of the standard and can help group members to understand and implement them in practice.
2. By undergoing the certification assessment as a group, economies of scale are regained so that the cost per small enterprise is significantly reduced.

Therefore, any small- or medium-sized enterprise wishing to become certified should consider the advantages of obtaining certification through a group scheme.

7.4 Starting the process

As discussed above, when making a decision about certification, it is important to consider carefully whether you need to be certified or not. If you decide that certification is required, then it is important to consider:

- the certification scheme or schemes that best suit your needs;
- whether to seek certification directly or to join a group scheme.

Once these decisions are made, it is time to begin thinking about the practical issues of implementing the standard and undergoing the certification process. This is discussed in Chapters 8 to 12, starting with forest certification and then moving on to chain of custody.

8 Forest Certification: Implementing the Standard

Whatever the certification scheme or approach to certification that is chosen (see Chapter 7), there are always two parts to the process of achieving certification for the forest manager:

1 implementing the requirements of the standard in the forest;
2 undergoing the certification audit or assessment to confirm that the requirements are in place.

In many cases, forest managers who discuss and contemplate certification become very focused on the second of these requirements; but it is important to remember throughout that certification is predominantly about managing a forest in compliance with a standard. The 'certification' part is simply a confirmation that this has been achieved. Therefore, this chapter discusses the process of implementing the requirements of the standard in the forest, while Chapter 9 deals with the assessment process.

This book does not deal with the technical basis for meeting the requirements of forest standards since it is dealt with in the companion volume, *The Sustainable Forestry Handbook* (Higman et al, 2005). However, it is useful to consider the process of implementing and ensuring compliance with a standard since, as discussed above, for most forest managers this is by far the biggest part of 'getting certified'. The different suggestions are not obligatory requirements for meeting a standard, but aim to provide an idea of processes that may help the forest manager to progress towards certification.

8.1 Understanding the standard

All certification schemes have their own forest standards that must be implemented by any forest organization wanting to get certified. While these vary somewhat between schemes, they are likely to cover some or all of the issues summarized in Box 3.4.

The first step in the process of certification is to obtain the standard and to understand its requirements. However, there are a number of reasons why this is not always straightforward for forest managers:

- Forest management standards are usually written in technical language that is unfamiliar to foresters and can be relatively complex.
- The order and presentation of the requirements may not follow the same logic as forest management activities so that the link between the requirement in the standard and an activity in the forest is not always immediately obvious.
- The requirements are often not very specific about what the forest manager needs to do in practice; therefore, they may need a considerable amount of interpretation. Even national standards or national interpretations of global standards may not clearly specify what is required in every situation. This is partly to allow flexibility in the application of the standard; but it does mean that interpretation is required.

As a result, it is worthwhile spending some time at the beginning of the process in developing a good understanding of what the standard actually requires. In addition to reading and discussing the standard internally within an organization, there may be other sources of information or support:

- *Certification schemes*: some certification schemes may be able to provide information or guidance on interpreting and implementing the standard. It is certainly worth checking. Even if they do not provide this information directly, they may be able to suggest other organizations that do.
- *Certified forest organizations*: one of the best sources of information can be other certified forest organizations. Since they have been through the process of implementing the standard and being audited against it, they should have a very thorough understanding of what is required. Where certified organizations are prepared to share this information, it can be very useful.
- *Certification bodies*: certification bodies must have a very thorough understanding of what the standard requires in order to undertake audits. While they are not supposed to provide direct consultancy, they can often provide considerable help in understanding the standard. This is often part of what happens during a pre-assessment (see Chapter 10, Section 10.2).
- *Consultants and specialists*: there are a growing number of specialists and consultants, including many academics, who work with forest certification and have developed a good understanding of what the standard requires and the different ways in which it can be implemented in practice.
- *Guides and handbooks*: the companion volume to this book, *The Sustainable Forestry Handbook* (Higman et al, 1999), provides guidance on implementing standards requirements, particularly those of the International Tropical Timber Organization (ITTO) and the Forest Stewardship Council (FSC).
- *Training*: there are numerous public and private training courses available that are run by certification bodies, consultants, professional associations, universities and donor organizations.
- *Group schemes*: for small forest owners or any other manager joining a group scheme, the best way to get information about what the standard requires is through the group who provides this type of information to its members (see Chapter 10).
- *Formal programmes*: there are a number of different initiatives available to help companies understand and implement certification standards. These include producer groups established by the World Wide Fund for Nature (WWF), Global Forest and Trade Network (GFTN), specialist organizations such as the Tropical Forest Trust and the Tropical Forest Foundation, and certification body schemes such as the SGS Certification Support Programme (see Appendix 2 for further information).
- *Involvement in standard development*: another way of gaining a better understanding of the requirements in countries where a national standard or a national interpretation of a global standard is being developed is by participating in the process. This has proved a very valuable way of building capacity, while contributing to the content of the standard for many forest organizations.

The most appropriate source of information will depend upon the certification scheme and the type and size of the forest organization.

8.2 Identifying gaps

The next stage in implementing a standard is to assess which of the requirements are already being met and which still need to be implemented. In other words, the gaps between current performance and the requirements of the standard must be identified. This can be done in a variety of ways and is referred to by various names, including *gap analysis*, *baseline review* or *initial assessment of performance*. Some of the possible approaches are discussed below.

8.2.1 INTERNAL REVIEW

For many organizations the best way of identifying gaps is through an internal process. It is a good idea to identify a particular person or small team to be responsible for leading this process; but it is also important in larger organizations to involve a wide range of people. This is for two reasons.

Firstly, increasing the number of people who contribute to the review provides more information, making it more accurate and comprehensive. Secondly, it is a good way of getting people within the organization to think about the standard and its implementation.

For organizations which have already implemented ISO 14001, there will be parallels with the initial environmental review process, although for a forest certification scheme the review will be relative to the defined requirements of a performance standard.

8.2.2 EXTERNAL REVIEW

While an internal process works well for some organizations, others find it more effective to bring in external support in order to help with the process. There are a number of reasons why this might be a useful approach:

- If there is still uncertainty about what the standard actually means, an experienced consultant can simultaneously help with interpreting the standard and identifying the gaps in current management practice.
- If there is uncertainty about how to deal with any gaps that are identified (see 'Planning and implementation'), a competent consultant can help to provide ideas and guidance.
- If the forest organization lacks the internal resources to commit much staff time to the process, an external specialist can provide the additional short-term resource needed to get the process moving.

It is very important that if external specialists are used, they work closely with the staff of the organization since it is usually these staff, and not the external specialist, who will be implementing any changes or improvements that are needed.

It is also imperative that any external specialists are competent. Being a consultant does not automatically give someone a good understanding of a standard and its implementation. Therefore, it is important to be cautious in selecting external support and to make sure that the individuals involved really do have a good understanding of implementing the requirements of forest standards. Certification bodies or certification schemes may be able to recommend competent specialists.

8.2.3 BASELINE AUDIT

An alternative, more formal approach is to undertake a baseline audit against the standard and identify any gaps or non-compliances in this way. This can be done either internally by staff of the forest organization, or externally by consultants or by a certification body. Each approach has advantages and disadvantages:

- *Internal audit*: as with an internal review, if the baseline audit is undertaken internally, it ensures that there is ownership of the process and a thorough understanding of the issues identified within the forest organization. This option is also likely to be considerably cheaper. However, the disadvantage is that if there is uncertainty about the requirements of the standard, or staff are inexperienced, then the assessment may not identify all of the existing gaps.
- *External audit by consultants*: there are two advantages to this approach. Firstly, if the consultant is really familiar with the standard, they should ensure that all of the possible gaps are identified. Secondly, if the consultant undertakes the baseline review, they can then assist with the development and implementation of the action plan (see below), ensuring that the information from the review is efficiently utilized. The disadvantage is that if the consultant is not good, then the results of the work will be correspondingly poor. This approach is also likely to be more costly than undertaking an internal audit.

- *External audit by a certification body*: there are a number of advantages to this approach. Firstly, the certification body should certainly understand what is required by the standard and should be able to identify all non-compliances effectively. Secondly, if the forest organization wishes to communicate its activities externally, particularly if it is involved (or likely to become involved) in some type of formal phased approach (see Chapter 17), then results produced by a certification body are likely to be the most credible. Thirdly, where the forest organization is already meeting many of the requirements of the standard, the baseline audit can become the first stage in a certification process, acting as a pre-assessment (see Chapter 9). The disadvantage of this approach is that certification bodies are not supposed to provide consultancy to clients and so will be unable to help in developing a programme to address the gaps identified. In addition, it is likely to be the most expensive of the three approaches.

Whichever approach is adopted, the most important thing is that the staff of the forest organization end up with a thorough understanding of where the gaps are between current performance and the level required by the standard.

8.3 Planning and implementation

8.3.1 DEVELOPING A PLAN TO ADDRESS THE GAPS

Once the gaps have been identified, the next stage is to plan how they are going to be addressed. For each requirement where a non-compliance or gap has been identified, the following elements will need to be focused upon:

- *Action*: decide exactly what needs to be done in order to achieve compliance with the requirement of the standard. The actions required will depend upon the standard and the forest organization. They could range from relatively routine activities, such as revision of operating procedures or provision of training, to completely new actions, such as undertaking a social impact assessment or setting aside an area of forest for conservation.
- *Responsibility*: define clearly who is responsible for undertaking the action. Where a range of people or departments need to be involved, then make sure it is clear who is taking a lead or coordinating activities.
- *Resources*: identify the resources that will be required. This may include capital expenditure (for example, new safety equipment, ongoing expenditure on training or consultants) or staff time (for example, if a member of staff will need to be seconded to a particular activity for a few months).

If many changes are required within an organization, it may be impossible to address them all at the same time with limited staff and resources. In this case, the actions to be taken need to be prioritized and some form of action plan developed that includes a timetable (see also Chapter 17, which discusses phased approaches in detail). There are various factors that should be considered when developing a timetable for implementation:

- Some things need to be done in order to provide a foundation for other activities. Make sure that the timetable specifies that the foundation activities should occur first.
- Different actions will involve different individuals and departments. Try to devise a timetable that spreads the work of particular individuals or departments over a reasonable period, rather than organizing all of the actions that they are involved in together.
- Some tasks might be completed early on because they are relatively easy or need little extra effort; these will provide a sense of progress in their completion. More daunting tasks might be left until the organization has gained some experience of implementing the certification

requirements and has a clearer understanding of how they have been addressed in other similar forests. However, it is important not to leave all of the more challenging tasks until the end of the process as this is likely to result in delays.

In many organizations senior managers set a deadline for achieving certification before the process of identifying gaps and action planning is undertaken. If this deadline is not compatible with the timetable for implementing the requirements of the standard, this needs to be highlighted early on, discussed with senior management and the deadline revised. Pursuing an unrealistic timetable is likely to result in failure to meet the requirements at the certification assessment.

The action plan needs to bring together the plans for addressing each gap and the overall timetable for implementation. The result may be a specific document developed especially for this purpose or, alternatively, may be integrated within existing documents such as:

- *Annual plan of operation*: many forest organizations have annual operating plans that are based on the longer-term management plan, but set out the detailed planning for the next 12 months. This is a good place to integrate the activities that will be needed to address the identified gaps. This ensures that implementing the standard becomes part of the mainstream activities of the company rather than the responsibility of the environment or certification departments.
- *Environmental programme*: organizations that have implemented an environmental management system standard (for example, ISO 14001 or the EU's Eco-Management and Audit Scheme (EMAS)) will have an environmental programme that is used to implement the environmental policy and targets. It may be possible to expand this existing approach to include addressing the identified gaps.
- *Objectives and targets*: some organizations set annual objectives and targets for both the organization and individual members of staff. It may be possible to link the actions required to meet the standard to these objectives and targets.

Whichever approach is utilized, the action plan needs to set out in detail how each of the identified non-compliances or gaps is going to be addressed. The people who will have to implement it should be involved in its development in order to make sure that it is realistic and achievable. Other assistance may also be required from consultants, specialists in particular fields, non-governmental organizations (NGOs) or certification bodies; but it is critical that the planned actions are developed and therefore understood and owned by the forest organization.

8.3.2 IMPLEMENTATION

Once the plans have been made, they need to be implemented. For small operations, this may be relatively straightforward as there are usually only one or two people involved in implementation. For larger organizations, where one person or department is responsible for coordinating the implementation, but other individuals and departments are responsible for actually undertaking the activities, it can be more challenging to ensure that plans are implemented. In forest organizations which already operate a quality or environmental management system, this should be relatively routine as systems for ensuring that plans are implemented in practice are central to management system approaches.

Where this is not available, it is important to ensure the following key elements:

- *Commitment*: there needs to be clear and unequivocal support for implementation from senior management. If possible, it is useful to find ways of ensuring that individuals and departments are judged on their success in implementing the plans, along with other responsibilities. Implementation is unlikely to be prioritized in practice if achievement targets and related recognition and other benefits are not linked to implementing the standard.

- *Responsibilities*: it is critical to ensure that all of the individuals and departments identified as being responsible for implementing the actions in the action plan are aware of their responsibility, understand what they need to do and are committed to doing it.
- *Resources*: it is crucial that the resources identified in the action plan are available at the time when they will be needed. For example, if training is required to begin the process of improvement, then the money to pay for the training must also be available at the beginning of the process.

8.4 Monitoring progress with implementation

Once the action plan has begun to be implemented, it is important to have a programme that periodically monitors progress and makes sure that:

- the planned actions are being implemented in line with the agreed timetable;
- the actions are effective in delivering compliance with the standard.

This type of monitoring can range from a quick, informal check every now and then to a formal internal audit and review programme.

8.4.1 INFORMAL MONITORING OF PROGRESS

For small forest organizations or owners, or even larger ones with relatively little to do in order to meet the standard, an informal approach to monitoring progress may well be sufficient. In this case, the person with overall responsibility for ensuring that the standard is met checks periodically (perhaps every two or three months) how things are progressing relative to the action plan developed.

8.4.2 FORMAL MONITORING OF PROGRESS

In larger organizations, or where there are many things that need to be done in order to meet the standard, it may be useful to have a more formal approach to monitoring progress with implementation. There are two ways in which this monitoring can be conducted:

1. Information on progress is regularly reported by each person or department responsible for implementing an action.
2. Internal audits are undertaken periodically in order to assess progress with implementation of each action.

The advantage of the first approach is that it is relatively simple and can be integrated within regular reporting schedules. However, there are two disadvantages:

1. It depends upon accurate reporting, and experience shows that while this can work well in some organizations, in others, particularly large organizations, people are reluctant to report failure to meet targets.
2. If the person responsible for implementing the action has not understood what to do, or has not noticed that the outcomes are not what is expected, this will not be reflected in any report that they produce.

The advantage of using internal audits is that it provides more objective information on progress, and may also pick up problems which the people implementing the actions have not noticed themselves. The disadvantage of internal audits for small organizations is the cost of setting up such a programme. However, many larger organizations, particularly those which already operate a management system, will have an internal audit programme in place and can simply extend it to cover the monitoring of the action plan.

8.4.3 MANAGEMENT REVIEW

Information gathered through monitoring, as well as the experiences of staff implementing the certification standard, need to be fed back into management practices in order to learn from experience. In small organizations, this can often be done informally since only a small number of people are involved. For larger organizations, it is useful to have a more formal programme of regular management reviews every three or six months. Where a management system standard has been implemented, this should already be in place.

8.5 Bringing in the certifiers

Once the forest manager is satisfied that sufficient progress has been made in meeting the requirements of the standard, the certification process can begin. As discussed in Chapter 9, Section 9.2, certification usually begins with a pre-assessment (scoping) visit where the certification body checks whether there are any remaining gaps between performance and requirements.

It is therefore possible for forest organizations to decide to begin the certification process and undergo a pre-assessment before they have completely implemented all of the requirements in the standard. This provides an opportunity to confirm that all of the gaps really were identified during the initial review and that the actions being taken to address them are considered adequate by the certification body. If any additional gaps or inadequate actions are identified, then the final phase of the action plan can be amended to take account of this.

However, as discussed in Chapter 9, Section 9.5, it is very important that the actions are all fully implemented and are in place before the main assessment begins since the certification body has to assess actual compliance with the standard rather than plans for compliance in the future.

9 Forest Certification: Getting Certified

As discussed in Chapter 7, Section 7.3, there are two ways in which forest managers can become certified:

1 *Individual certification*: here, the forest area managed by a single forest organization is certified.
2 *Group certification*: several forest areas that are owned or managed by different organizations or people join together in a group, which is then certified. Group certification is discussed in detail in Chapter 10.

All forest organizations or groups who want to become certified will have to go through the certification process. This chapter looks in detail at how certification is undertaken. It is based on the process for the Forest Stewardship Council (FSC) scheme, which has been the most widely used to date; but certification against most other certification schemes will be similar.

Forest certification comprises a number of stages, which are outlined in Figure 9.1. Each of the stages is discussed in further detail below. The certification process is also discussed in detail in Chapter 4.

The experience of forest owners or managers who are being certified as part of a group will be slightly different as their group manager will be the main interface with the certifier. This is discussed further in Box 9.1.

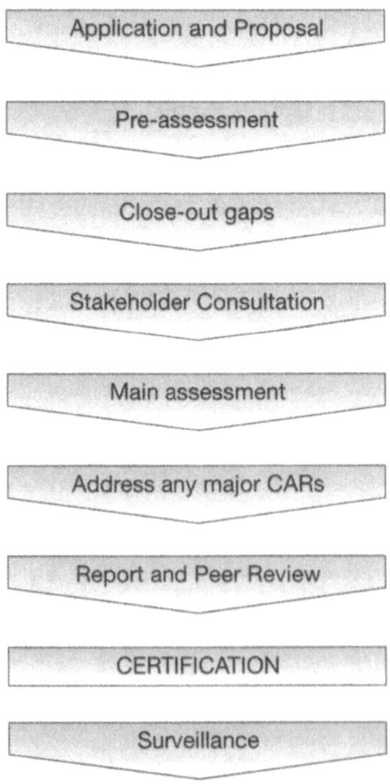

FIGURE 9.1 The stages in a typical forest certification process

Box 9.1 **The certification process for members of group schemes**

Assessment of members of a group scheme is carried out jointly by the group manager and the certification body.

It is the responsibility of the group manager to check that every member of the group scheme complies with the group requirements, which includes compliance with the standard. To do this, most group schemes have a programme in place to monitor their members, generally through an annual visit. Monitoring may be performed by the group manager, by staff of the management organization, by consultants hired to carry out the visit or by other members.

During the main assessment, the certification body will then check that this monitoring programme is in place and is working. Unless the group is small, it is unusual for an audit team to visit each member. Instead, they will choose a sample of members, check their monitoring records and then confirm that the real situation corresponds with the report from the monitoring visit.

As a result, only this sample of group members will actually meet with the audit team or undergo an audit visit. However, if any non-compliances are found, they will apply to every member of the group since the basis for this approach is that the sample of members visited is representative of the group as a whole; therefore, a problem found in the sample is also assumed to exist in the rest of the group.

It is important for group managers to explain this to group members so that they understand the need to let audit teams visit, if requested, but also understand why they may not be visited and, in particular, that they may still have to respond to non-compliances even if they were not visited.

The need to visit only a sample of members, rather than every forest, is one of the reasons why group certification is so much cheaper per forest owner than individual certification, since field visits are time-consuming and, therefore, expensive.

9.1 Application and proposal: Choosing your certifier

Once an organization has decided that it wants to become certified, and has chosen the scheme that it wants to be certified with, it will need to choose a certifier (also referred to as a *certification body*, a *registration body* or a *registrar* – they are all the same thing). Information on accredited certifiers should be available from either the accreditation body for the scheme or from the scheme management. Nearly all forest certification schemes now have a website, which is a good place to start looking (see Appendix 2 for details).

The certifier will probably send an application form that needs to be filled in and, based on the information provided, a proposal outlining both the certification process and the costs. This part of the process is normally free and does not involve any obligation on the part of the organization requesting the proposal.

If there are a few certification bodies working in the country or region, then it is probably sensible to get a proposal from several or all of them in order to see which one gives the best deal. The key factors to consider when choosing which certifier to work with are:

- *Costs*: how much do they charge? Look at all parts of the costs. Most certifiers charge fees for staff time and then charge all of the costs of flights, hotels, food and other expenses. Check what these expenses are likely to be – for example, will they include an international airfare; will auditors stay in the cheaper local hotel or the expensive hotel in town; and so on.
- *Efficiency*: how quickly can they start and continue with the job? Forest certification is always quite a long process; but sometimes if certifiers are very busy it can take even longer. Things to check include:

  - How soon can they carry out the pre-assessment?
  - How soon after the visit will the report be available?
  - After a decision has been made to proceed with the main assessment, how quickly will action be taken?
  - How soon after the main assessment is complete will a draft report be produced and how long after that will a certification decision be made?

  If the certifier finds non-compliances and has to raise corrective action requests during the main assessment, then the timings will be delayed and the new timetable will depend upon how quickly the organization being assessed is able to respond. However, certification bodies should be able to provide some idea of timings in the absence of major non-compliances.
- *Local service*: for regional or international schemes, check whether the main contact is local and what language will be used for the process.

Where possible, it is also useful to have discussions with the managers of organizations that have already been certified to find out about their experience of their certifier.

9.2 Pre-assessment: The initial visit

Once a certifier has been chosen and a contract is signed with them, the next stage for most certification schemes is a pre-assessment (occasionally called an initial visit or scoping visit). This is a short visit made by one or two people from the certification body and it has three main purposes:

1 It gives the certifier an opportunity to get to know the forest enterprise or group scheme and to plan for the main assessment.
2 It gives the organization seeking certification a chance to meet the certifier and to ask any questions about how the certification process will proceed.
3 It allows the certifier to go through the requirements of the standard (including the requirements for group schemes if it is a group certification) and to see if there are any areas in which the organization is clearly not complying. This is often referred to as 'identifying gaps'.

Of the three purposes, the third is the most important and time-consuming. The certifier will probably want to spend time in the office, discussing forest management or the group system, and will look at documents and records, and visit the forest or, for a group, some of the group members in order to talk to them and study their forests.

However, this is not an audit. The pre-assessment team will ask many questions about how things are done; but they will not confirm if the answers are accurate. If problems are hidden, then gaps will not be identified. As a result, it is very important that the organization being assessed is as open and honest as possible so that all of the gaps will be identified and they will know what they have to do before the main assessment. It can be tempting not to give accurate answers and to try to impress the auditors by only providing favourable details. However, it is important to resist this temptation because, in the end, it makes the pre-assessment much less useful. Even if a problem is hidden during the pre-assessment, it will almost certainly be identified during the main assessment, resulting in a non-compliance.

Following the pre-assessment, a report will be prepared that summarizes the findings and highlights the gaps that need to be addressed before the main assessment.

As mentioned above, pre-assessments are not required by all certification schemes. Where pre-assessments are not compulsory, it may well seem that opting out of one is an easy way to save time and money. However, experience shows that, in general, they are very important in ensuring that forest organizations are properly prepared for the main assessment, and it is much more efficient to identify gaps during a pre-assessment than to find that there are major non-compliances during the main assessment.

9.3 Closing out gaps and deciding to proceed

If gaps have been identified in the current forest management or group scheme, the next stage is to get them sorted out. This is, in fact, a continuation of implementing the requirements of the standard discussed in Chapter 8.

There is no set time for how long this takes. For some forest organizations or groups it is just a few days or weeks; for others it can take months. The time depends upon what issues have to be addressed and what resources are available to address them.

However, once all of the problems identified in the pre-assessment have been addressed, together with any others identified internally, it is possible to proceed with the main assessment.

9.4 Stakeholder consultation

Generally, forest certification schemes allow certification bodies to undertake a pre-assessment confidentially. This is important for many forest organizations and groups because they want to check that they are close to meeting the standard before it becomes public information that they are seeking certification.

However, once a decision is made to undertake a full main assessment, many schemes require the certification body to undertake consultation with a range of stakeholders as part of the process.

Usually consultation will already have been undertaken by the forest organization or group as part of implementing the certification standard. Therefore, the forest organization or group manager and members should already have contacts with local communities, local government and other key organizations such as non-governmental organizations (NGOs). Where this is the case, it is a good idea to explain in advance about the pending certification assessment so that contacts know what it is and why they are being contacted by a certification body. This does not mean telling people what to say, and certainly not coercing them into only saying good things which, if identified by an auditor, would certainly be a major non-compliance with the standard. But it is useful to explain what certification is, why the forest organization or group scheme is seeking certification and why auditors may want to talk to different stakeholders.

The purpose of stakeholder consultation as part of the certification process is to allow any person or group with an interest in the management of the forest or forests to provide information to the certifier that they think may be relevant to the certification.

Stakeholders include such groups as local communities, the forest department, local government, environment and conservation groups, social organizations, workers and employees and anyone else who is affected by management. The larger the forest organization or group scheme, the more national or even international the scope of stakeholder consultation is likely to be.

In order to compile a list of stakeholders to be consulted, the certification body will probably ask the organization being assessed for key contacts. It will then add to this list based on the experience and knowledge of the audit team and on advice from the certification scheme, and, in some cases, by checking who has been contacted in other assessments.

For some certification schemes, such as the FSC, the requirement for consultation is very strong and includes a public announcement on the certification body website and contact with key consultees at least four weeks before the main assessment begins. Other schemes have a lower profile approach to consultation.

Certification bodies may consult by letter, on the telephone, through private meetings, by holding public meetings, by advertising in the local paper or any other way that seems appropriate. The number of stakeholders contacted and the method used for contacting them will depend upon the size and location of the forest or group scheme and the type of stakeholders being contacted. In general, the larger the forest or group scheme, the wider the consultation process. Conversely, for a single small forest, consultation might be as simple as three or four phone calls to key stakeholder representatives.

Many forest managers find the idea of stakeholder consultation quite intimidating. In particular, there is a concern that some people may hold grudges or have other personal reasons for saying negative things about the forest organization and its staff, or the group scheme and its members. However, the certification body should have a very systematic way of dealing with stakeholder comments to ensure that real information is treated seriously while unfounded accusations are not.

When stakeholders provide information, the assessment team must look at each issue and decide what to do about it. In general, there are three possibilities:

1 The issue raised is not covered by the standard, or it is covered by the standard but is not the responsibility of the forest organization or group being assessed.

For example, a stakeholder complains that there is no training for local people in improved agricultural methods. While this is genuinely a problem, there is no requirement in the standard that certified forest operations provide agricultural training; therefore, it is not relevant to the certification.

In another example, a stakeholder says that the forest on the steep slopes of hills is being overexploited and causing soil erosion. This is a real issue and is relevant to the standard; but the forest in question is managed by the state forest department and is not part of the forest being assessed. Therefore, the issue is not relevant to the certification.

2 The issue is relevant to the standard and the forest or group being assessed, but there is no evidence to show that it is a current problem.

For example, a stakeholder says that group members are planting trees adjacent to watercourses, which is against the law and affects water flow. This would definitely contravene the standard; but the stakeholder is unable to give any specific location of where this occurs. Although the assessment team check for non-compliance at every site visited during the main assessment, there are no examples of planting by watercourses. Therefore, the claim cannot be substantiated.

3 The issue is relevant to the standard and the forest or group being assessed, and there is evidence to show that it is a current problem.

In the example above, if the stakeholder could give an indication of where planting adjacent to watercourses occurs and, on visiting these sites, the information was found to be accurate, or the audit team found examples of such sites themselves, this would be classified as a non-compliance with the standard.

Quite often, the information that arises from the consultation exercise is very useful for the forest organization or the group management and its members, as well as for the certifier. Frequently, an issue is raised that can easily be addressed, but of which the manager was previously unaware.

For example, in one case it emerged that the growing antagonism toward a forest company was the result of logging truck traffic along the main town road, which coincided with children making their way to school. The forest company stopped trucks from moving during this short, but critical, period and relations immediately improved. In another case, it emerged that contractors who were planting seedlings were leaving polythene bags on site, which subsequently blew into gardens in neighbouring communities. Again, the company was able to take quick action to resolve the problem, thereby improving community relations.

In conclusion, forest managers should not see the consultation phase of the certification process as a threat, but as a continuation of their own consultation process and as an integral and useful part of the main assessment.

9.5 Main assessment

The purpose of the main assessment is for the audit team to collect objective evidence in order to demonstrate that the forest management or the group and its members do (or do not) meet the requirements of the standard. The number of people and the number of days that it will take to do this will depend upon the size and type of the forest operation or group being assessed. In general, the larger the operation or the group, the more people and time will be required.

9.5.1 THE ASSESSMENT TEAM

The assessment team will have a team leader and usually (unless the forest or group is very small) one or more team members. Usually each member (including the leader) will have particular areas of expertise, such as forest management, conservation and ecology, community forestry or legal requirements. The makeup of the team will be a result of the information collected in the pre-assessment, and the forest organization or group being assessed should be informed of the name of each team member in advance.

If the organization or group has a problem with a particular team member, then they can inform the certifier and ask to have them replaced. However, a certification body is only likely to agree to this if there is a specific reason for disqualifying the person, such as:

- The person has worked for the organization being assessed during the last three to five years.
- There is a relationship between the person and a member of staff.
- There has been a specific problem between the organization being assessed and the proposed team member that could potentially make them biased in their assessment of performance.

9.5.2 OPENING MEETING

An assessment usually starts with an opening meeting. This is run by the audit team leader who explains what will be happening over the next few days. This is a useful opportunity for people in the organization being assessed to make sure that they understand what is happening and to ask questions. Therefore, it is a good idea to involve as many personnel as possible in the opening meeting so that they can meet the assessment team and have some idea of what to expect.

The exact content of an opening meeting varies depending upon both the certification body and the individual team leader; but it is likely to cover some background, an introduction to the team, the aims of the audit and the way in which the work will be undertaken. Auditors are taught that they should be in control of their own audits from the start, including the opening meeting, so the team leader will generally want to run the meeting.

The team leader may ask for a short presentation about the forest organization or group scheme. This is useful in order to provide an overview to the team and to set the context for work. The assessment team will also finalize the planning of whom its members want to see and where they want to go. Usually, the team will provide the organization being assessed with some idea of its plans before arriving; but it is not good auditing practice to give too much warning of which offices or sites have been selected for visits since this allows them to be preferentially improved prior to the audit. Therefore, much of the selection is likely to be made at the opening meeting or even, particularly for visits to operational sites, during the course of the audit itself.

9.5.3 THE AUDIT

The audit consists of three main activities:

1 *Reviewing documents*: members of the team will want to see a range of documents and records and to check their content. This sometimes occurs prior to the start of the main audit in a separate 'document review' phase. Team members will need to check a number of things. Firstly, do the documents meet the requirements of the standard?

For example, if the standard requires a management plan that includes a particular set of topics, then members of the team will need to check that the management plan exists and covers the right issues. If the standard requires written operating procedures, these will need to be identified and checked

Secondly, are appropriate records are being kept? Sometimes these are checked at the start; but they may also be required as the audit progresses.

For example, the auditors may ask to see training records to confirm that training has occurred, records of accidents to establish whether health and safety is being properly managed, or records of wildlife monitoring.

Thirdly, do all actions conducted by the organization or group staff, the contractors or the members comply with the documentation? This, in turn, should result in compliance with the standard.

2 *Visiting the forest*: a random sample of sites within the forest will be studied; the team may also visit with certain members of the group scheme. The audit team will check whether plans, procedures and rules are actually being implemented in practice.

At least one staff member of the forest organization or the group manager should always accompany the audit team (or each part of the team, if they split into groups). From a practical point of view, it will be necessary to arrange adequate transport and food and drink for the assessment team and anyone accompanying them.

3 *Discussions and interviews*: finally, the team will spend considerable time talking to staff, contractors or members about what they do and how they do it. Everyone should be encouraged to respond openly and honestly. It is likely that when talking to field staff or contractors, the team will not want management or even the group member involved in the discussion, but will ask them to wait some distance away. This is normal auditing practice since experience shows that field staff find it difficult to talk in front of management and that management find it hard not to answer questions for field staff! Therefore, it is easier for the auditor to converse with field staff alone.

9.5.4 FINDINGS, NON-COMPLIANCES AND CORRECTIVE ACTION REQUESTS (CARS)

During the course of the assessment, the team will be looking for objective evidence that the standard is being complied with. This evidence will come from the documents that they review, their visits to forest sites and the people to whom they talk, both within and outside the organization being assessed.

If they find evidence that the forest organization or group is not meeting a requirement of the standard, this is defined as a *non-compliance*. Whenever a non-compliance is identified, a corresponding *corrective action request* (CAR) is raised. A CAR sets out the details of the non-compliance and requires that the organization or group scheme undertake corrective action to resolve the problem. There are two types of CAR (see also Chapter 4, Section 4.1.2.5):

1 Major CARs (also called *pre-conditions*) are raised when there is complete failure to comply with a requirement of the standard or a systematic failure to implement plans and procedures. If a major CAR is raised, it must be adequately addressed *before* certification (this is why it is sometimes called a pre-condition as it has to be addressed pre-certification).
2 Minor CARs (also called *conditions*) are raised when there is partial compliance with a requirement or a non-systematic failure to implement plans or procedures. If a minor CAR is raised, certification can proceed but only on the condition that the issue is adequately addressed within an agreed time frame (hence, the alternative name of condition). It is important to remember that if a minor CAR is not addressed within the time agreed, it will automatically be raised to a major CAR and must then be addressed within a short period (for example, one month) or the certificate will be suspended.

Most assessments result in one or more CARs. It is almost unheard of not to receive any, and it is quite usual to be issued ten or more. Forestry standards are tough and complex, so even the best forest management is likely to miss full compliance in one or two areas.

9.5.5 CLOSING MEETING

Assessments nearly always end with a closing meeting where the team reports back to the organization on its findings, though a good audit team should keep the organization being assessed informed of progress throughout the audit.

One problem of many closing meetings is that, because there are usually time constraints, it is rare for the assessment team to be able to go through all of the findings in detail. As a result, the focus is on non-compliances and resulting CARs, rather than on the areas of compliance. This can appear somewhat negative, so it is good for organizations to be prepared.

It is important that the organization which has been assessed asks questions and feels confident about discussing the findings. In particular, if it is possible that the team misunderstood something, or raised a CAR wrongly because it did not see a critical document or field site, then it should be informed.

A very important point is that the assessment team does not make the final certification decision. The team can only summarize its findings and confirm what its recommendations will be. The final decision is made after a further process of review, and for almost all certification schemes, the decision is independent of the assessment team's proposals. This reduces the potential for pressure to be put on assessment teams to ignore evidence or change their findings.

9.6 Closing out major CARs

If the main assessment uncovered any major non-compliances with the standard then one or more major CARs will be raised. These have to be addressed or 'closed out' before a certificate can be awarded.

This will be discussed at the final meeting and probably confirmed in writing soon after the assessment team leaves. It is then up to the forest organization or group management and members to address the non-compliance and, when this has been done, to inform the certification body.

The certification body will then need to check that the action taken is adequate. Sometimes this can be done without another visit, particularly if the non-compliance relates to documentation; but often 'closing out' a major CAR requires a visit from some or all of the assessment team.

The costs of a visit to close out a major CAR are usually additional to the cost of the certification process originally quoted. This highlights the importance of making sure that all issues are identified in the pre-assessment and addressed *prior* to the main assessment.

9.7 Report and peer review

Following the completion of the main assessment the team leader, assisted by the team members, will write a report. The FSC and several other schemes require that a summary of the report – including background information and the results of the assessment – is made public, so the report will include both the public summary and the more detailed assessment report. Other certification schemes do not require this type of public summary – only information on the forest or group (for example, size, location and type) has to be made public.

Some schemes, including the FSC, require the report to be peer reviewed before any final certification decision is made. When the report is complete, it is sent to two or three independent specialists who have been selected as peer reviewers. As with the team members, the peer reviewers will be selected by the certification body; but the organization being assessed should be informed of who they are and can tell the certification body if there is any problem.

The peer reviewers are asked to comment on whether, based on the information provided in the report, and their expert knowledge of the issues in the region and the type of forest being assessed, the assessment seems to be adequate and the findings reasonable.

Any issues raised by peer reviewers must be responded to by the certifier and, if appropriate, can lead to new or revised CARs.

9.8 Certification and surveillance

Once all major CARs or preconditions have been adequately addressed, and a response has been made to any peer review comments, the certification body will finally make a certification decision. This is normally done by a certification panel that is independent of the assessment team (though the team leader may join the panel to answer questions and provide information).

Certification is also referred to as 'registration' by some schemes, particularly in North America. Usually this means exactly the same thing so that reference to a 'registered forest organization' means the same as a 'certified forest organization'.

Certificates are usually valid for five years but are conditional on the findings from annual surveillance visits, which the certification body will carry out in order to verify that the forest management in the forest organization or group scheme continues to comply with all of the standard's requirements.

Surveillance visits are usually like a shorter version of the main assessment, and are undertaken by a team which follows the same process. The main elements that the surveillance team will focus on are:

- checking that any current minor CARs have been adequately addressed and can be 'closed out';
- checking that the actions taken to address previous major and minor CARs are implemented and are successful in addressing the non-compliance identified;
- visiting new sites or members where new forest areas have been acquired, where work has started in a new part of a concession or where new members have joined a group scheme;
- following up on any complaints or stakeholder comments received since the last visit;
- checking that any changes to the standard have been implemented – standards are usually revised at least once every five years and certificate holders usually then have 12 months to comply with any changes; the surveillance team will need to check that this is being done;
- assessing continued compliance against the standard.

As with the main assessment, any evidence that there is a non-compliance with the standard will result in corrective action requests.

10 Forest Certification: Setting Up a Group Scheme[1]

Owners and managers of small forest enterprises often find it difficult to gain access to forest certification. They may be located in remote locations where it is difficult to get information about certification requirements and procedures. It is often much harder for them to understand and interpret the requirements of certification standards, and even to know which certification scheme would be appropriate for them. In addition, the costs of certification are relatively high for small forests compared to larger forest organizations because they do not benefit from economies of scale both in implementing certification requirements and in the cost of the certification assessment itself.

Group certification can help to overcome these barriers. Group certification allows a number of small forests to work together under a single 'group manager' who can both provide information to the forest managers and organize a single certification assessment for the group, allowing them to benefit from the economies of scale of belonging to a larger entity. The group manager might be an individual person, an organization or association, a company or other legal entity. The group manager is responsible for ensuring that group members meet the requirements of the standard through their support and monitoring of members' activities. Most forest certification schemes make a provision for some sort of group scheme (see Box 10.1).

The requirements for group certification schemes outlined below are largely based on the Forest Stewardship Council (FSC) system where group certification has been most widely implemented.

10.1 Group and resource managers

There are two main approaches to group certification:

1 *Group manager schemes*: the group manager runs a group which forest owners or managers can join. The forest owner or manager is responsible for managing the forest. The group manager is responsible for ensuring that all group members comply with the certification standard. Group

Box 10.1 Certification scheme provisions for group certification

Forest Stewardship Council (FSC)	Group certification programme and streamlined procedures for groups of small (or low-intensity managed) forests.
Programme for the Endorsement of Forest Certification (PEFC)	Group certification and regional certification schemes can be developed by national schemes endorsed by the PEFC – for example, the Finnish Forest Certification Scheme (FFCS) regional certification approach.
Sustainable Forestry Initiative (SFI)	Group certification provided through a mutual recognition agreement with the American Tree Farm System.
Canadian Standards Association (CSA)	Certification operates at the level of defined forest area (DFA). No specific provision is made for group certification; but a DFA may effectively be a group of small forest management units with common implementation of the CSA requirements.

managers can be individuals, companies, forest owners organizations or non-governmental organizations (NGOs).

2 *Resource manager schemes*: the resource manager sets up and manages the group, just as in group certification. The difference is that a resource manager is also contracted by the forest owner to manage the forest. Thus, the resource manager takes on the roles of both group manager and forest manager.

Both of these approaches are widely used with great success and, in practice, there are many groups which are a hybrid, with group managers taking on some forest management tasks for members, while others are still undertaken by members themselves.

10.2 Requirements for a group scheme

The requirements for group and resource manager schemes are similar and are described together here. Where there are slight differences this is noted. The requirements set out below are generally applicable to group schemes; however, since there are differences between different certification schemes' requirements for the group scheme, it is important to check the precise requirements of the scheme being implemented.

10.2.1 LEGAL ENTITY

In group certification, the contract for certification is made between the group manager and the certification body. The certificate is held by the group manager on behalf of the group. In order to make such a contract, the group manager needs to be some sort of legal entity, which could comprise a company, an association, a co-operative, an NGO, a government organization or a community group.

The most appropriate type of legal entity will depend upon the group manager's circumstances and the types of organization considered as legal bodies in the country. If the group manager is not already a legal entity, it will be necessary to establish one.

10.2.2 GROUP MANAGEMENT STRUCTURE

Groups can vary in size from two or three members to several hundred, or even more. Depending on the size of the group, the group management may consist of a single person, or may involve a group management team, with several management levels working from different offices. In all circumstances, a single person should be designated as the 'group manager', with overall responsibility for ensuring that the group functions correctly.

10.2.3 WHO CAN JOIN?

Some certification schemes set limits on the type, size or number of members which a group can have, or require the group manager to define who will be allowed to join the group. It is important to make sure that the group manager is capable of running the group effectively: the larger and more complex the group, the more complicated it will be to run.

Defining who can join the group may involve decisions about:

- a maximum or minimum size of forest;
- the total number of members;
- forest types (such as plantation or semi-natural forests) that the group manager is able to work with;

- the geographic area and locations that can be managed in practice;
- the costs of joining and the ongoing fees that will be charged to members;
- management situations compatible with group membership (for example, a group might only deal with forests run directly by their owners or, conversely, may only include forests managed completely by the resource manager).

10.2.4 INTERPRETING THE STANDARD

Forest management standards often need interpretation and explanation before it is clear what the forest manager needs to do, in practice, to meet the standard. The amount of interpretation needed will depend upon which certification standard is being used. Standards that are produced nationally or regionally may be more clearly adapted to the local situation, making reference to national legislation, policies or guidelines. International standards that have not been adapted to the local situation may require a considerable amount of interpretation and guidance before it is clear to the forest manager how they should be applied.

This is often particularly the case when trying to apply forest management standards to small forests – often the main membership of group schemes. Most forest management standards apply more clearly to large forest organizations. They have often been written with large forests in mind and include requirements covering issues such as environmental and social impact assessment; consultative processes; the setting aside of protected areas; and the promotion of local industries. Such requirements are relevant to large enterprises, but may be excessive when applied to small forests unless appropriately interpreted.

An important part of the group manager's role is to provide this interpretation of the standard for small forest members. This can be a challenging task, particularly where there are no locally defined standard nor previously certified groups. Group managers are likely to find themselves in one of three situations, as outlined in Box 10.2.

Box 10.2 Interpreting standards for group members

Where certification is against a national standard, this may be detailed enough to be used directly by forest managers. However, this is not always the case: in many countries, even national standards have to be written to deal with a range of forest types and management approaches. With Forest Stewardship Council (FSC) certification, the forest management standards available for a particular country may be at one of three stages of development: complete, in development or not yet begun. This will have an influence on the group manager's task of interpreting the requirements for group members:

- *A local standard exists.* It may be possible to simply give this standard to members without further guidance. In most cases, however, some discussion or extra information will be needed, which might take the form of guidance notes, an annotated copy of the standard or practical discussion of the standard during field visits.
- *A draft local standard exists.* The group manager can provide the draft standard or an interpretation of the draft to group members, making it clear to them that it may change and that they will have 12 months in which to comply with the finalized standard once it is approved.
- *There is no local standard.* The group manager will need to base an interpretation on the international standard. Considerable adaptation of the standard may be required in order to apply it within the context of the group members' forests.

In all situations, the group manager needs to tailor his or her interpretation or guidance to the scale of the group members' forests, to their forest type and management practices, and to their abilities to understand technical documents. If most members' forests are managed by professional foresters, they may be happy to interpret the standard for themselves. In this case it is important to ensure that the standard is being interpreted similarly in different forests. Where the members are predominantly non-professional, or have no forestry training, greater assistance with interpretation may be needed.

10.3 Membership requirements

The group manager is responsible for ensuring that all group members comply with the certification requirements. In order to do so, it is important to clearly define the requirements and procedures for joining the group, leaving the group and the conditions under which a member can be expelled from the group.

10.3.1 JOINING THE GROUP

The group manager needs to ensure that the forest management of any new member complies with the standard before being allowed to join the group and that it has documented procedures describing this process.

When a forest owner applies to join a certified group, the owner should be provided with adequate information about the group membership requirements. It may be helpful for the group manager to develop an application pack that summarizes the membership requirements. In some cases, it will be more appropriate to provide information verbally, through meetings or workshops. Box 10.3 outlines the type of information that needs to be provided to prospective members.

If the forest owner decides to join the group, the next step is for the group manager to find out whether the owner meets the group requirements, as well as the requirements of the standard.

Box 10.3 Information for group applicants

Forest owners or managers applying to join a certified group need to be provided with the following information:

- general information about the group scheme – for example, who it is run by, affiliations and the certification scheme under which it is certified;
- group membership requirements, including joining, leaving, expulsion and monitoring requirements;
- the standard or an interpretation of the requirements;
- a summary of the certification process;
- the certification body and certification scheme's rights to enter and monitor the forest;
- any requirements for making information available to the public;
- the complaints procedure;
- the costs of group membership, including joining, ongoing and monitoring fees, as well as the likely costs of dealing with any problems that may be identified.

Firstly, the group manager needs to check that the prospective member meets the basic requirements of the group, such as maximum or minimum size, forest type, location or ownership. It may be helpful to check that the applicant has legal (or adequate customary) rights to manage the forest. Once basic details have been confirmed, forest management practices need to be checked through the pre-entry inspection.

A pre-entry inspection (or initial assessment) allows the group manager to check that forest management is adequate. It can be carried out by the group manager or by another inspector authorized by the group manager. In order to make sure that pre-entry inspections are carried out consistently and thoroughly, it is often useful to develop a checklist that covers all of the relevant issues. The checklist can fulfil two functions:

1 It is a reminder of the areas that the inspector needs to check during the visit, including all of the requirements of the certification standard.
2 It is a record of the results of the inspection, including areas that need to be followed up after the inspection.

If forest management does not meet the requirements of the standard, it is important that the inspector explains where the problems are and that the forest owner or manager understands how to address the problem.

It is essential that inspectors and the group manager maintain written records of the problems that they encounter, the process for resolving them and the time frame in which to do so, and any re-inspection visits or documentation checks that they make. This process is checked by the certification body during an assessment and forms the basis of the group system for ensuring that only forests which meet the requirements of the standard are allowed to join the group.

Once the inspections have been carried out and the group manager is satisfied that the applicant meets the membership requirements, it is important to formally document the new membership status. The new member and the group manager usually sign a formal agreement that sets out the group requirements, the member's commitment to the requirements and the manager's confirmation that they have undergone a pre-entry inspection and meet the initial requirements. This may, for example, be accomplished by:

- the group manager signing and dating the pre-entry inspection checklist, confirming that the applicant meets the requirements; and
- the group manager and the new member signing a membership agreement confirming acceptance as a member; or
- the group manager sending a letter to the new member, confirming acceptance.

Records of the formal acceptance of new members into a group must be kept on file by the group manager. It may also be necessary to inform the certification body when new members join the group (for example, the FSC stipulates that the group manager must inform the certification body within one month of new members being accepted into the group).

10.3.2 LEAVING THE GROUP

Sustainable forest management is a long-term activity and certification generally requires that the forest is managed in accordance with the standard over the long term. Certification requirements, such as maintaining good relations with local communities and stakeholders, conservation, appropriate silviculture techniques and monitoring, are applicable throughout the growing cycle, not only during harvesting. As a result, certification schemes are concerned with ensuring that forests do not enter certification at harvesting time in order to facilitate the marketing of timber, and then withdraw from the scheme once harvesting is complete. This is particularly of relevance to group schemes, where small forests predominate, which may have a short harvesting period, interspersed by long intervals with no harvesting. The group manager therefore needs to ensure that group members are committed to belonging to the group and managing their forests in accordance with the long-term standard. Groups who have unusually high membership turnover rates are likely to be carefully scrutinized by certification bodies.

However, under some circumstances it may be completely legitimate for members to need to leave the group. These reasons might include:

- *Sale of the forest*: where a forest is sold, the new manager may not wish to continue group membership, or may not comply with the membership requirements.
- *Poor sales of certified products*: certification does not necessarily provide market access for products, particularly from small forests. Where sales of certified products are not as high as expected, a member may not be able to justify the cost of group membership.
- *Natural disaster*: forests that are affected by natural disasters, such as fire, hurricanes or floods may not be able to continue with normal forest management for many years. Especially in small forests, contingency planning for such disasters may not be possible. Continued membership of the group may not be feasible or desirable in this situation.

Under FSC certification, the group manager must inform the certification body within one month of a member leaving, describing the reasons for departure.

10.3.3 EXPULSION FROM THE GROUP

It is the group manager's responsibility to ensure that all group members meet the requirements of the standard. If one member does not meet the requirements, and does not take adequate action to rectify the situation, the entire group's certification will be threatened. Therefore, the group manger needs to have a system for expelling members who do not meet group requirements. This system should be developed and documented when setting up the group so that all members understand and accept that this system can be applied to them when they sign the membership agreement.

It is useful to have a procedure in place that sets out the process to be followed when it is necessary to expel a member. This can be given to new members when they join, ensuring that there is no dispute later about the process and that the group manager consistently follows the same procedure. The elements of an expulsion procedure are outlined in Box 10.4.

10.4 Consultation and complaints

Different certification schemes make quite different requirements about consultation, so the responsibility of the group manager for carrying out consultation will depend upon the scheme requirements. However, it is often the case in forest certification schemes that consultation with local communities and other stakeholders needs to be carried out as part of ongoing forest management and as part of the certification process.

10.4.1 CONSULTATION AS PART OF FOREST MANAGEMENT

Many scheme standards (for example, the Forest Stewardship Council and the Canadian Standards Association) make specific requirements for forest managers to carry out consultation as part of forest management (see Chapter 3). Under group certification schemes, the responsibility for consultation can be split between the group manager and the members. Exactly how responsibility is divided up will depend upon the circumstances of the group; but it is likely that group members will focus on consultation locally, with immediate neighbours and people directly affected by their forest

Box 10.4 **Elements of an expulsion procedure**

The expulsion procedure should cover the following elements:

- *The circumstances under which a member can be expelled*: these might include non-compliance with membership requirements and failure to take action to comply within a specified timeframe; failure to pay group fees; or gross and irreparable contravention of the certification standard.
- *How the member will be informed of problems before expulsion*: this needs to allow the member the opportunity to address the problem before expulsion, but should follow a formal process and time frame. The group manager might inform the member by letter, specifying the problem, setting out a time limit for addressing the problem and describing how the actions will be checked.
- *How the member will be informed of expulsion*: if the issues are not addressed, the member needs to be formally expelled. The expulsion must be documented, usually by letter. The letter should explain that the member is being expelled and specify that the member can no longer claim that their forests (or products) are certified. The right to appeal and the process and timeframe for doing so should be explained.
- *The appeals process*: the procedure should explain how long the member has in which to lodge an appeal and how long the process will take. The appeals process should be described; this might include formation of an appeals panel or committee, which should be impartial and have had no involvement in the decision to expel the member. Some groups make use of professional forestry associations as arbiters in such situations.

management, while the group manager might be involved in wider consultation, perhaps with local and national governmental, as well as NGO and regional stakeholders.

10.4.2 CONSULTATION FOR CERTIFICATION

Under some certification schemes, the certification body is required to carry out consultation as part of their assessment (see 'Stakeholder consultation' in Chapter 9). This aims to ensure that the certification body is aware of any issues, positive or negative, between stakeholders and the forest manager, and to check that consultation is being carried out as part of ongoing management, as required by the standard.

In order to carry out consultation for a group certification, it is important that stakeholders associated with all group members are aware that the certification process is happening and that they have an opportunity to comment to the certification body if necessary. The certification body contacts some stakeholders directly; however, the group manager and group members are responsible for informing local stakeholders that the certification is taking place. The means of doing this will depend upon local circumstances: it might be appropriate to send letters, place an advertisement in a local newspaper or other media, put up signs on local community notice boards or make an announcement at a public meeting.

10.4.3 DEALING WITH COMPLAINTS

Most certification schemes include requirements about dealing with complaints; for example, the FSC standard includes a requirement for an appropriate mechanism for resolving grievances and complaints. Complaints in a group scheme are likely to be of two types:

1 complaints from stakeholders about group members and the way in which their forests are being managed;
2 complaints from members or stakeholders about the group manager and the way in which the group is being run.

The group manager needs to have a procedure that defines how each of these types of complaints will be dealt with. This should cover who will deal with a complaint, within what timeframe and what they will do to follow up complaints. This often includes setting up an appeals committee, similar to that established for membership appeals, in order to consider complaints that cannot be dealt with by the group manager and members.

10.4.3.1 Complaints about a group member

These should initially be investigated by the group management, who will report the findings to both the member and the complainant. If the complainant is not satisfied, they should be provided with information on how to make an appeal to the appeals committee.

10.4.3.2 Complaints about the group manager

A mechanism for this type of complaint must exist so that the complaint can be investigated independently of the group manager. This could involve an independent third party or the appeals committee.

Any complaints received, and the actions taken to resolve them, need to be clearly registered and documented.

10.5 Monitoring group members

One of the most important roles of the group manager is to ensure that group members comply with the certification standard. For group schemes where members manage their own forests, this is through the development and implementation of a monitoring programme. For resource managers, the situation is slightly different, since it is the group manager who implements forest management. Therefore, resource managers tend to have a less formal approach to monitoring, focusing more on ensuring that their own activities ensure consistent compliance with the standard in all of their certified forests.

The main elements of setting up and running a monitoring programme are:

- organizing the monitoring programme;
- developing a monitoring checklist;
- recording, registering and following up non-compliances.

Each of these is discussed briefly below.

10.5.1 ORGANIZING THE MONITORING PROGRAMME

A monitoring programme may be very simple or may comprise a range of people, sampling levels and communication networks. The complexity of the monitoring programme will depend upon the size and complexity of the group. Two main questions need to be considered for all monitoring programmes:

1 *Who will carry out the monitoring?* This might be conducted by a single individual (for example, the group manager) or by a monitoring team authorized by the group manager. It may be the same person who carried out the pre-entry inspection. The inspector(s) need to be independent of the group member being checked and should have appropriate experience and training.

Box 10.5 **The monitoring checklist**

The exact contents of the monitoring checklist will vary from one group scheme to another, but most monitoring checklists should cover:

- *Administrative information*: this includes the name of the forest owner/manager; location of the forest; person carrying out the inspection; date of inspection.
- *Information from previous visits*: this entails issues identified at previous visits that need to be followed up, as well as information from stakeholders that needs to be checked.
- *Changes reported by the member*: new operations, planting, harvesting or road building should be checked in the field; changes to ownership or to the scope of the member's forest must be reported.
- *Group requirements*: this involves continued compliance with group requirements, including the certification standard.

2 *When will monitoring be done?* Some groups monitor each member each year. However, if the group is made up of very small forest owners where there are long periods when nothing happens in an individual ownership, this may be inappropriate and the frequency of monitoring may be decided based upon whether harvesting, thinning, road building, planting or other forest operations have been carried out. Frequency may also be defined by the certification scheme. Where the monitoring programme does not include every member every year, it may be necessary to use sampling. Issues related to sampling are discussed in Chapter 4, Section 4.1.2.2.

10.5.2 DEVELOPING A MONITORING CHECKLIST

During monitoring visits, the inspector aims to check that the member meets all of the group requirements. This will include ensuring that forest management meets the requirements of the standard, that any previously identified problems have been addressed, that any comments raised by stakeholders are checked and that any other group requirements are complied with. As with the pre-entry inspection, it is useful to develop a monitoring checklist to cover all of these issues.

The checklist can act as a planning tool prior to the inspection, a guide during the inspection and as a record of the results after the inspection. It is therefore worth ensuring that the checklist is carefully thought out, with an easy-to-use format and adequate space for comments. The contents of the checklist will depend upon the forest certification standard being used, any interpretation of the standard provided by the group manager and any specific group requirements (see Box 10.5).

10.5.3 RECORDING, REGISTERING AND FOLLOWING UP PROBLEMS

The monitoring checklist can provide a record of observations about each of the requirements that the member must meet, but where non-compliances with these requirements are identified, there needs to be a specific mechanism to ensure that these are reported, brought to the attention of the member and followed up. Each group scheme needs to have a mechanism for doing this, which includes:

- *Recording and communicating the non-compliance*: this comprises a mechanism to ensure that the problem is clearly recorded and that it is communicated to the member.

- *Registering and reviewing non-compliances*: as well as being communicated to the member, the group scheme requires a mechanism to ensure that there is a central register of all outstanding non-compliances and a mechanism for making sure that these are followed up and addressed.
- *Confirming resolution of non-compliances*: finally, there needs to be a mechanism for confirming that a non-compliance has been resolved and for removing it from the register. This may require a repeat monitoring visit to the member, so it is worth considering the cost implications of this and who should pay.

When an identified non-compliance is not adequately addressed by a member, the group manager needs to take action urgently, culminating in expulsion, as discussed in Section 10.3.3.

10.6 Group system documents and records

In order to manage the group, to keep on top of membership, monitoring and follow-up activities and to deal with the certification body, it is essential for the group manager to maintain good group system documents and records. Group system documents are the standard documents that help the group to function, such as application forms, information packs, checklists and procedures. Records are the information that is collected about group members and their activities, such as completed membership forms, completed monitoring checklists and details of non-compliances.

10.7 Training and information

Two types of training and information provision will be important in setting up and running a group: training for group management staff and information for group members. Both are important and can make an important difference to the success of the group.

10.7.1 TRAINING FOR GROUP MANAGEMENT STAFF

Group management may be carried out by an individual group manager, or may be conducted by a range of staff members, who may have other duties and responsibilities unrelated to the group. Whoever is responsible for running the group, it is essential that they are adequately trained and experienced to do so. In addition to the forest management skills that are essential to understanding how group members' forests are run, group management staff need to understand two key areas:

1. *How the group scheme works*: all group management staff need to understand the group's objectives, procedures and requirements.
2. *How to carry out inspections*: inspections are a fundamental part of the control of the group, ensuring that the requirements of the standard are being implemented on the ground. Auditing skills are fundamental to these inspections; such skills can be learned through participating in audits with an experienced auditor and through audit training courses.

10.7.2 INFORMATION AND TRAINING FOR GROUP MEMBERS

The group manager is responsible for ensuring that certification requirements are met in members' forests. The manager will therefore need to ensure that all group members are aware of and are able to meet the requirements. In order to achieve this, the group manager may need to provide some basic information to members and may wish to offer wider training and advice on forest management in order to help members meet the requirements.

10.8 Chain of custody and claims

Forest owners or managers in a certified group scheme are entitled to sell their forest products as certified. The group manager must ensure that members who sell their forest products as certified are:

- implementing adequate chain-of-custody controls (chain of custody and group chain of custody are described in Chapters 11 and 12);
- meeting all of the requirements of the certification scheme for the control of claims and use of the logo.

Group managers need to develop their own controls for chain of custody within their group, and to ensure that they are regularly monitored as part of group monitoring activities.

10.9 Regional certification

The Programme for the Endorsement of Forest Certification Schemes (PEFCS) enables endorsed national certification schemes to develop regional certification schemes. Not all national schemes endorsed by PEFCS have done so: some have developed group certification schemes similar to those described above. However, the regional model has been applied in both Finland (FFCS) and Germany.

Regional certification aims to reduce the barriers to participation in forest certification for small forest owners by submitting an entire geographical region to certification. Certification is applied for by a forest organization, which must represent the forest owners or managers of more than 50 per cent of the forest area in the region. Individual forest owners may make a commitment to achieve certification, or the commitment may be made by the forest owners' association on behalf of the owners and managers whom they represent in the region.

The assessment of the regional certification scheme by a certification body is carried out at both the regional level and, through sampling, of individual forest owners/managers. The forest organization which applied for certification is responsible for ensuring that individual forest owners/managers comply with the certification requirements.

There are clear benefits of regional certification in terms of getting large amounts of raw material certified. However, the approach remains controversial for two reasons. Firstly, some models have used the approach that all forest owners in a certified region are automatically certified unless they actively opt out, rather than the usual approach of requiring forest owners to actively choose to be certified. This raises the question of what level of commitment is required to meet certification requirements from forest owners who do not necessarily know that they are certified.

Secondly, the mechanisms for ensuring that individual forest owners and managers in a region implement an adequate standard of forest management on the ground are not always clear. This is particularly so for those owners and forests not under the control of the organization which sought regional certification.

Note

1 Detailed guidance on setting up and running a group certification scheme can be found in the ProForest publication *Group Certification for Forests: A Practical Guide*, available free from www.ProForest.net/Publications.

11 Chain-of-Custody Certification: Implementing the System

In the previous chapters, we examined certification of forests. In this chapter we turn to the process of linking certified forests to the products that they produce. This is an extremely important component of forest certification schemes because it provides the basis for any claim that is made about a product. The most widely used term for this process is *chain of custody*. The theory of chain of custody and claims is discussed in detail in Chapter 6. This chapter focuses on implementing a chain of custody in practice.

As discussed in Chapter 6, linking a claim to a certified forest is straightforward; but demonstrating the link between a product and a certified forest is more difficult. The wood products sector often has complex production chains so that material may go through multiple processing, change ownership several times and cross international borders between leaving the forest and becoming a final product. In order to make any claim about the product, it is necessary to have a mechanism in place to trace certified material all the way through this process.

11.1 Who needs to implement chain of custody?

Chain of custody requires the control of certified material through the entire supply chain, from the forest to the final product. Each organization in the chain must implement a chain-of-custody system and will require a chain-of-custody certificate (see Figure 6.2). This means that every organization which processes the material or takes legal ownership has to implement chain of custody.

For example, a chain-of-custody system must be implemented by a harvesting contractor who buys standing timber in a forest, by all processors throughout the chain, by agents who buy and sell the material and by distributors who buy in large quantities and then break bulk and sell on in smaller lots.

If the wood or fibre is handled by an organization without an active chain-of-custody certificate, the certified status of the material is lost and it cannot be regained because there is no independent guarantee that the integrity of the certified material has not been compromised by mixing with material that is uncertified. Therefore, any organization which wants to be able to sell certified material or products needs a chain-of-custody certificate.

11.2 What type of chain of custody?

As discussed in Chapter 6, there are a number of approaches to chain of custody depending upon the type of claim that will be made for the final product. The approach used has some significant implications for the way a chain-of-custody system is designed and implemented.

Therefore, the first stage in developing a chain-of-custody system is to decide which approach to use. The two main alternatives are production of 100 per cent certified products using exclusively raw material from certified forests, or the production of percentage-based products using a mixture of certified and uncertified raw material. If the percentage-based approach is used, then there are a number of different options.

Almost all certification schemes allow more than one approach, but most do not allow all approaches. Therefore, it is important to check which of the approaches outlined below is allowed by the certification scheme being used before making a decision about the one to adopt.

11.2.1 100 PER CENT CERTIFIED

In some supply chains, it is possible to use only material from certified forests in the final product. This requires the complete segregation of certified material at every stage from the forest to the final product, and the exclusive use of this certified material for all certified and labelled products.

11.2.2 PERCENTAGE-BASED CLAIMS

In practice, because it is often extremely difficult to use exclusively certified material (see Chapter 6, Section 6.3 for further discussion), most schemes allow the use of percentage-based production and claims. There are four approaches to the way in which percentage production can be controlled, each of which has implications for the way the chain-of-custody system is implemented:

1 *Percentage in individual products*: one approach to percentage labelling is to control the proportions of certified and uncertified material in each individual product. This requires the same degree of control over the chain of custody as for 100 per cent-certified products since it is necessary to maintain the identity of certified and uncertified material throughout the process until all material is combined in the final product. This final stage must be sufficiently controlled to ensure that the precise amount of certified material (by weight or by volume) in an individual product is known.
2 *Percentage in a product line or process*: in many production processes it is not possible to control the percentages in an individual product, and control has to be undertaken at the level of a product line or process. In this case, it is necessary to control the proportions of certified and uncertified material entering a product line.

> For processes that operate using batch production, this is normally relatively straightforward and input into each batch can be controlled. For processes operating continuous production, it is more complex. In this case it is necessary to base the control of raw material input on a 'nominal batch'. A nominal batch is a defined period of time, such as a day, a week or a year. The average amount of certified raw material entering a process over the nominal batch period must be sufficiently controlled to ensure that it always exceeds the minimum allowed (see Chapter 6, Section 6.3.1.2 for further discussion of nominal batches).

3 *Percentage in–percentage out*: an alternative approach, also referred to as input–output or volume accounting, is to certify a proportion of output equivalent to the proportion of certified raw material used in the production process. Thus, if 10 per cent of the raw material entering the process is certified, 10 per cent of the products can be considered and sold as certified. For this approach to chain of custody, the amount of certified material entering the production process must be known.
4 *Processor certification*: processor certification focuses on the certification of each organization in the production chain, rather than on the control of certified material. The requirements will depend upon the rules of the certification scheme.

For all approaches to percentage-based chain of custody, the origin of the *uncertified* material in the production process is also very important. This is discussed further in Section 11.4.2.

11.2.3 SELECTING THE BEST APPROACH

Where a certification scheme allows more than one approach to chain of custody, an organization contemplating certification will need to decide which approach to use. The choice will depend on three factors:

1. *Which approaches are recognized by the certification scheme being used?* It is important to check carefully which approaches are allowed. It is also worth thinking about whether it is likely that chain-of-custody certification under a second scheme may be sought in future. If so, it may be sensible to select an approach recognized by both schemes.
2. *What type of product claim does the market require or prefer?* While most approaches are equally accepted by large parts of the market, there may be particular organizations or market segments which have preferences. For example, if purchasers favour 100 per cent products over those that are per cent labelled, this is important to consider.
3. *Which approach best suits the situation of the organization implementing the chain of custody?* Percentage-based approaches will need to be used if:
   - it is not possible to obtain some parts of a product from certified sources;
   - it is not possible to purchase exclusively certified material and it is not practical to segregate certified and uncertified raw material from entering a process.

   However, whenever contemplating a percentage-based approach, it is important to bear in mind the requirements for control of the *uncertified* material in the process (see Section 11.4.2). In some cases, it is easier to produce 100 per cent-certified products than to try to implement a system that controls the source of uncertified material.

11.3 Developing a chain-of-custody system

In practice, chain of custody has to deal with two things:

1. control of certified material *within* each certified organization in the supply chain;
2. control of certified material *between* each certified organization in the supply chain.

Control within the organization requires managing the material through internal processing. The way in which controls are designed and implemented will depend upon the approach to chain of custody that is adopted.

Control over certified material moving between different organizations in the chain is usually managed through the control of sales and dispatch from the supplying organization, together with control of purchasing and goods inwards at the receiving organization.

In addition, there is a need to have in place systems to control any claims or labelling of the product.

In practice, this means that any organization implementing a chain of custody will need to have a system that ensures:

- adequate control of purchasing and goods inward to ensure that only certified material is purchased and accepted as certified and that requirements for uncertified material for percentage-based production are met;
- proper control of the internal processing of the material;
- adequate control of sales and dispatch of final products to ensure that only certified product is sold and dispatched as certified;
- control of all claims and labelling to ensure that they meet the requirements of the certification scheme.

FIGURE 11.1 The four elements of a chain-of-custody system

These four elements are shown in Figure 11.1 and each is discussed in detail in the following sections. 'Chain-of-custody case studies' provides practical examples of how chain of custody can be implemented for three real processes: a sawmill, a timber stockist and an MDF manufacturer.

11.4 Product sourcing: Purchasing and goods in

The first element of a chain-of-custody system is to control the purchase and acceptance of raw materials:

- Purchase of certified raw materials needs to be adequate in order to provide the link to the previous stage in the chain.
- Purchase of uncertified raw materials for percentage-based production must ensure that they meet the certification scheme requirements for uncertified material in certified products.

Each of these is discussed below.

11.4.1 SOURCING CERTIFIED MATERIAL

11.4.1.1 Purchasing

Identifying certified suppliers
Certified raw material can only be supplied by a chain-of-custody-certified supplier. Therefore, *before* an order is placed it is important to check that the supplier is certified. This can be done by obtaining a copy of the supplier's chain-of-custody certificate and checking to make sure that:

- It is still valid.
- Its scope covers the material to be purchased. Many organizations produce a combination of certified and uncertified products, so it is important to ensure that the chain-of-custody certificate covers the type of product being purchased as certified.

Alternatively, if the certification body or certification scheme keeps a register of certificate holders on a website, it is possible to use this to check if a supplier is certified. This is also a good way of checking that the copy of the certificate is accurate and up to date.

> There needs to be a procedure or work instruction which ensures that the person or department responsible for purchasing raw materials confirms that a supplier holds a valid chain-of-custody certificate before placing an order for certified raw material. If an ISO 9000 system is in place, this can be included as part of the procedure for approving suppliers.

11.4.1.2 Specifying certified product

Even if a supplier holds a valid chain-of-custody certificate, they may trade in both certified and uncertified materials. Therefore, it is essential to specify in the purchase order or contract that certified material is required.

> Purchasing procedures should include the need to specify certified material. If purchase orders are produced electronically, then the system should automatically query whether certified material is required and add the request to the purchase order. If there is a manual system with pre-printed documents to fill in, the possibility of ordering certified material can be added – for example, as a tick box.

11.4.1.3 Receiving goods

Received material should be checked that it conforms to the specification requested, including its certified status. There are two ways in which material should be identified as certified, both of which should be confirmed:

1 The invoice and any other relevant documentation, such as a delivery note or transport documents, should specify that it is certified and provide the chain-of-custody certificate number of the supplier.
2 Where it is practical, the material should be labelled or physically identified as being certified.

> Procedures for accepting deliveries of raw materials should include a requirement to check both the documentation and the identification of certified material, and to reject or quarantine if either is inadequate, exactly as for any other failure to meet specifications. If an ISO 9000 system is in place, this can be incorporated within existing goods-inwards procedures. Procedures for processing invoices should ensure that confirmation of certified status is included before authorization to pay is given.

11.4.2 SOURCING UNCERTIFIED MATERIAL FOR PERCENTAGE-BASED PRODUCTION

As discussed in Chapter 6, when percentage-based production and labelling is used, this means that the products produced contain material that is derived from forests which were definitely *not* certified. Since certified products are generally purchased as a means of implementing policies to exclude purchase of any wood or paper products originating from illegally or badly managed forests, it is important to ensure that, even if the material was not certified, it was not from one of these unacceptable sources (see 'Dealing with the uncertified component' in Chapter 6). Therefore, most certification schemes have requirements for the control of uncertified material used in the percentage-based production of certified products.

The requirements of each scheme vary, so it is important to check exactly what is required and, based on these requirements and the guidance below, to develop an appropriate system.

11.4.2.1 Confirm what sources are not permitted

Each scheme has a slightly different definition of which sources are not allowed. Almost all include wood from illegal sources, but many have additional exclusions. Box 11.1 provides a summary of the main excluded sources.

Box 11.1 Sources of wood or fibre that are not permitted in percentage-based production by one or more forest certification schemes

Uncertified material used in the percentage-based production of certified products should not be from:

- illegal sources;
- protected areas or forests that have been proposed for national parks but have not yet been formally gazetted;
- forests that have a particularly high conservation value (for example, biodiversity hotspots, high conservation-value forests, forests of outstanding conservation value and endangered forests), unless they are certified or demonstrably well managed;
- forests where there are serious tenure disputes, particularly where these involve the failure to respect the customary rights of indigenous or local people;
- forests that are being inappropriately converted to other uses.

11.4.2.2 Develop a system to check that sources are acceptable

In order to meet the requirements of certification schemes to exclude wood and fibre from prohibited sources, it is necessary to develop and implement a system that checks and controls sourcing. There is no single way of doing this; but the process outlined below is practical and is used by a range of different organizations.

Establish the source of uncertified raw material

Since all of the prohibited sources relate to the management of the original forest, it is the *forest of origin* that must be established. This process is relatively straightforward for processors at the beginning of the supply chain, but can be more complicated for organizations further down the chain since it involves establishing where their suppliers obtained the raw material. In some cases, this is reasonably easy regarding the country of origin, but extremely difficult regarding sourcing to an individual forest. The importance of this depends upon the level of risk that the material represents.

Undertake a risk assessment

For each source, undertake an assessment of the risk that it is from an unacceptable source. Where information is available regarding sourcing to the level of a forest, this can be done for the forest in question. Where information is only readily available for a country or region, then the risk relating to that country or region needs to be assessed. Where it is high, then it will be necessary to establish the forest of origin and to check whether it is managed appropriately. Where it is low, it is probably acceptable to manage sourcing at a country level.

Discontinue high risk or prohibited sources

Once the assessment is complete, all high-risk or prohibited sources of raw material must be discontinued, at least for certified product lines. If it is not possible to purchase uncertified raw material from acceptable sources, then it will *not be possible* to produce percentage-based certified products. The exclusion of prohibited raw material is as essential to the overall chain of custody as the reliable sourcing of certified material.

Once the initial assessment has been undertaken and some experience is gained of the process, a procedure will be needed to ensure that every purchase of uncertified raw material for percentage-based production of raw material is adequately controlled. This may be done through:

Approved suppliers: it is often most practical to limit the suppliers of uncertified material to those who have been screened and will definitely provide acceptable material.

Approved countries or regions of origin: where purchasing is widespread and changes regularly, it may be necessary to implement control by allowing purchases only from countries or regions assessed as low risk, or by requiring a case-by-case appraisal for countries which are high risk.

11.4.2.3 Commitment

Most certification schemes ask for a commitment or statement that is available to customers and, in some cases, a public statement that all uncertified material used for certified products is from an acceptable source.

It is absolutely critical that this statement is only produced and used once the organization has implemented controls to ensure that the commitment is being achieved in practice.

11.4.2.4 Specifying and receiving uncertified material

Once the system for controlling uncertified sources has been developed, specifying, purchasing and receiving uncertified materials can all be undertaken in a similar way to the process for certified material set out in Section 11.4.1.

11.5 Production process: Implementing internal controls

Control of product sourcing should take care of the chain of custody between the previous organization in the supply chain and the organization implementing its own system. The next stage is to develop a system to control chain of custody internally. Internal chain-of-custody controls are usually based on the identification and management of critical control points.

11.5.1 IDENTIFYING CRITICAL CONTROL POINTS

Critical control points (CCPs) are all the points in the process where it might be possible to mix certified and uncertified material. An analysis of the process needs to be undertaken to identify each of these points.

Examples of critical control points include a log yard used to store both certified and uncertified logs, a production line used to produce both certified and uncertified furniture and a warehouse used to store both certified and uncertified products.

The critical control points will depend not only upon the process, but upon the type of chain of custody being implemented. If 100 per cent-certified products are being produced, then certified material will need to be traced through the entire production process. If percentage labelling is used based on either a batch or a percentage in–percentage out approach, then certified material will only need to be traced to the point where it enters the production line.

11.5.2 MANAGING CRITICAL CONTROL POINTS

For each critical control point, it is necessary to find the best way of ensuring that there is no mixing of certified and uncertified material. The most appropriate way to do this will depend upon both the process and the organization. However, it will always be based on a combination of segregation, identification and documentation.

11.5.2.1 Product segregation

One of the most effective ways of preventing mixing of certified and non-certified materials is by ensuring that certified material is always kept physically separate from uncertified material. Opportunities for segregation should be considered at each critical control point, including:

- *Storage*: certified raw materials and products can be stored in separate areas from non-certified material and products.
- *Production*: production runs for certified products can be undertaken on separate production lines (physical separation) or, if this is not possible, carried out at different times from production runs for non-certified products on the same production line (separation in time).

Practical examples of segregation include separate areas in the log yard for certified logs, separate bays in the warehouse for certified raw materials or products, particular days designated for certified production, and different hoppers for certified and uncertified part-assembled products.

11.5.2.2 Product identification

Another straightforward way of preventing mixing of certified and uncertified products is through physical marking of certified products:

- *Raw materials*: certified raw materials such as logs, sawn timber, plywood or paper are clearly marked as certified.
- *Work in progress and finished products*: identification can include different job cards or work-in-progress cards where these are used; physical marking or labelling of part-assembled items; different coloured pallets or hoppers for storage; and unique identification numbers, from which it is possible to trace the material to a certified source.

Practical examples of identification methods include paint marks on logs, bar-coding of logs, a paper label on stacked sawn timber being kiln dried, or an obviously different colour for job cards for certified products.

11.5.2.3 Documentation

Good documentation is an essential part of good chain of custody, whether this is in a paper form or computer based – for example:

- procedures that set out the controls for each identified critical control point;
- records of all types, including accurate production records from which it is possible to identify the source and quantity of materials input, and volume or number of goods manufactured; and stock records of raw materials and finished product, incorporating (where appropriate) annual stock-taking results.

11.5.3 DEVELOPING THE SYSTEM

Once all of the critical control points have been identified, and the best way to implement control for each one has been identified, it is relatively simple to develop a full chain-of-custody system for the process.

11.5.3.1 Systems for 100 per cent-certified production

The system implemented must ensure that, for every critical control point, the combination of segregation, identification and documentation is sufficient to prevent mixing of certified and uncertified material.

One approach is to develop a tracing system that provides sufficient traceability to make it possible to link a finished product to the exact raw material used to make it. However, this is not essential if it is possible to demonstrate that adequate controls are in place to ensure that only certified material is used in the manufacturing process.

For example, a sawmill produces 100 per cent-certified sawn, kiln-dried timber from a dedicated sawing line. Either of the options below provides an acceptable chain of custody:

Option 1: complete traceability is maintained by recording the log identification numbers of all logs sawn and stacked on an individual pallet. The numbers are printed on a label that is attached to the pallet throughout drying and shipping.

Option 2: records are kept of the identification numbers of all logs entering the certified line, and all pallets of timber produced are labelled as certified; however, no link is made between individual logs and particular pallets.

Where it is not possible to trace back from final product to raw material, it is particularly important to have good records of raw materials used, conversion ratios and products produced (see Section 11.5.4 below).

11.5.3.2 Systems for percentage-based production

Where a percentage-based approach is used, the control of certified material is required until the point where it is mixed with uncertified material. An important additional requirement is the calculation of the percentage in the product. Most certification schemes have very specific rules for this. Whichever approach is used, the system must include accurate record-keeping, showing exactly what raw material was used.

Individual product-based percentage

Control of certified and uncertified material will be needed throughout the production process. The percentage in each product must be calculated based on either the weight or volume of certified raw material used for the individual product.

Product line or process-based percentage

Control of certified and uncertified material is required to the point where the raw material enters the production process. If the certification scheme or the product label or claim specifies a minimum proportion of certified material that must be in the product line, then the controls must be adequate to ensure that this is delivered. The percentage of certified material for batches is based on the proportions of raw materials used for the batch. The calculation for continuous processes has to be done based on either a nominal batch or a rolling average approach (see Box 11.2).

Where there is a rapid turnover of raw material, the control of proportions of certified and uncertified material will probably have to be achieved in close cooperation with the purchasing

Box 11.2 Chain-of-custody controls for percentage-based production in continuous processes

Where controls of the percentage in a product line are being implemented for a continuous process, a time period has to be defined during which the percentage of certified material in the product line can be calculated. This can be achieved through either a nominal batch or a rolling average.

Nominal batches

The calculation of the proportion of certified and uncertified material has to be made at the end of each nominal batch period, whether this comprises a day, a week or a year.

Provided that the proportion of certified material used meets the requirements of the certification scheme, then all of the products made during that period can be considered certified. If any label or claim states a percentage, it must not be greater than the actual proportion of certified material used.

In practice, it is sensible to manage the proportions of raw material entering the process to ensure that the minimum required percentage is achieved. If products are labelled or sold before a nominal batch period ends, then control of the material entering the process is mandatory.

Rolling averages for continuous processes

In a rolling average approach, the average proportions of certified and uncertified raw material used are calculated regularly at a defined frequency that is shorter than the nominal batch period. For example, for a one-month nominal batch period, a rolling average might be calculated weekly, as shown below.

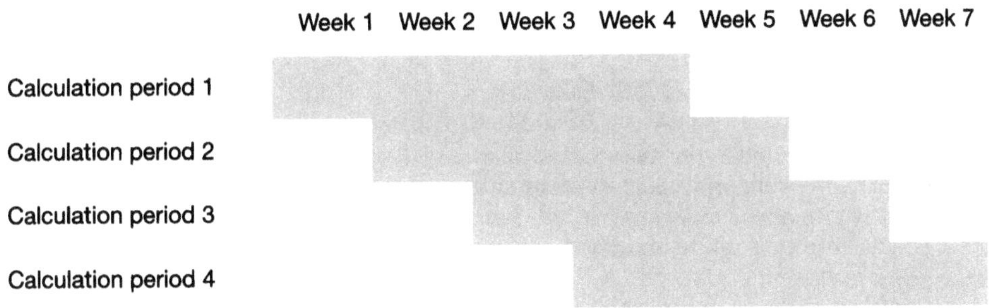

For a rolling average approach, it is necessary to control the input of raw material in order to ensure that the rolling average of certified raw material never drops below the minimum allowed.

department or buyers who will need to ensure that the quantity of certified material being ordered and delivered is always sufficient to meet the minimum permitted. Where there is more scope for storing raw materials and selecting what will be used, then control can be carried out within the process.

It is important to have a mechanism for reacting quickly if the average is at risk of dropping below the minimum threshold. If the average does drop below the minimum threshold, *none* of the product produced during that nominal batch period can be labelled or sold as certified.

Percentage in–percentage out

This approach requires the control of the quantities of certified and uncertified material entering the manufacturing process, providing the basis for calculating the proportion of product that can be allocated as certified.

11.5.4 CONVERSION RATIOS AND RECONCILIATION OF QUANTITIES

In addition to managing critical control points, the other key element of internal chain of custody is the reconciliation of certified material in and certified material out based on conversion ratios. The conversion ratio is the quantity of final product that can be produced from a defined amount of raw material. Thus, the conversion ratio of a process indicates the efficiency of conversion or, looking at it the other way round, the losses associated with the processing of wood pulp.

For example, a typical round wood sawmill loses 40 to 60 per cent of volume between round and sawn wood. In a particular mill, for every 100 cubic metres of round wood that are processed, 50 cubic metres of sawn timber are produced. The conversion ratio in this case is 50 per cent.

For any process, if the volume or quantity of certified raw material entering the system is recorded and the conversion ratio is known, it is possible to calculate the theoretical quantity of certified finished product that should have been produced. This can then be compared to the actual quantity of finished products produced to ensure that there is no major discrepancy.

For example, records show that a sawmill with a conversion ratio of 50 per cent has processed 750 cubic metres of certified round wood and produced 378 cubic metres of sawn timber, which is very similar to the theoretical production of 375 cubic metres, providing assurance that there are no obvious problems with the chain of custody.

11.6 Final product and sales

The control of the final product and sales is critical in providing both the end point of the internal chain of custody and the link to the chain-of-custody system of the next organization in the chain. It is also the point where labels and many of the claims about the product need to be managed. Therefore, systems must be in place in order to ensure:

- *Traceability*: there is sufficient information about the final product to ensure that it can be linked back to the certified production.
- *Identification*: the product must be clearly identified as certified with the correct packaging or labelling. This should ideally indicate the quantity and specification of the material, as well as its certified status, not only to prevent mixing at the production site, but also to inform the next customer. This is necessary whether the product is an intermediate one that will be processed further or a final product.
- *Documentation*: it is important that when certified products are sold, the accompanying documentation specifies the certified status. This can be done by ensuring that sales invoices and other sales documentation include:
    - the organization's chain-of-custody certificate number in the format specified by the certification body;

      for example, the Forest Stewardship Council requires certificate numbers in the format of: [certification body code]-CoC-[certificate number], such as ABC-CoC-1234;

    - confirmation that the specific product being supplied is certified.

Many organizations simply write or print this information on their sales documents. Other organizations have printed special sales documents that carry their certificate number and are used only for sales of certified products.

11.7 Managing claims and labels

Once an organization is chain-of-custody certified and is producing certified products, it is eligible to make claims and to use the certification scheme logo. Certification scheme logos are copyright material and usually involve an internationally registered trademark. There are strict legal terms and conditions associated with their use.

A company holding a certificate is able to use the logo in accordance with the permitted terms and conditions as long as all uses are authorized. Some schemes require each use to be checked by the certifier in advance, while others require use to be checked at surveillance visits.

Logos and related claims can be used in two ways:

1 *On product*: the logo is used to designate that the wood in a labelled product is derived from a certified source. The logo may be used on the product itself or on the labelling or packaging.

> Most schemes have more than one form of the logo or claim depending upon whether the product contains 100 per cent-certified material or was produced using some form of percentage-based approach. It is extremely important to make sure that the right form is used for each product.

2 *Off product*: the logo is used to promote or advertise forestry or forest products that meet the requirements of the certification scheme. The logo can be used in brochures, leaflets, reports, organizational training materials, advertisements and commercials.

A system must be in place to ensure that all labels and claims, both on and off product, meet the requirements set out by the certification system. It is usually a good idea to nominate a particular person to be in control of all claims and labels within the organization and to develop a procedure to control use of the logo and related claims.

It is especially important to ensure that there is adequate control over areas that are handled externally, such as development of a company website or design of an annual report where the use of the logo or a claim may be included.

11.8 Implementing the chain-of-custody system

11.8.1 TRAINING

Training is almost always an essential element of implementing a chain of custody. All personnel must be adequately trained in understanding and implementing the requirements if the system is to work effectively. This is important even in very small organizations:

- A management representative should be identified, and should have the responsibility and authority to implement and maintain the chain of custody.
- All staff should understand their specific responsibilities in managing chain of custody and should have adequate training in order to fulfil their assigned tasks.
- Records of training and experience should be maintained and should be appropriate to the scale of operations, identifying previous training and likely additional training needs.

11.8.2 INTEGRATION WITH EXISTING SYSTEMS

As discussed above, whenever possible the elements of a chain-of-custody system should be integrated with any existing controls that a company may already have via systems such as ISO 9000, standard operating procedures or computerized systems. This is important for two reasons:

1 It reduces the amount of work needed to develop and implement a chain-of-custody system.
2 It is more likely that the system will be reliably implemented if it becomes part of normal working practice, rather than an additional requirement, resulting in a more reliable chain of custody.

11.9 Chain-of-custody case studies

Probably the easiest way of understanding chain-of-custody systems is to look at some real examples. The case studies in this section include a typical chain-of-custody system for three different processes:

- a sawmill;
- a timber stockist;
- an MDF manufacturer.

All are based on Forest Stewardship Council (FSC) chain-of-custody requirements, but would meet the requirements of most other certification systems.

11.9.1 SAWMILL CHAIN-OF-CUSTODY CASE STUDY

A sawmill processes some 17,000 cubic metres of softwood roundwood into a variety of sawn products, principally sawn carcassing, but also planed and finished materials. All products sold are kiln dried. At present, only 60 per cent of its raw material is certified. The remaining 40 per cent is uncertified and the sources are not monitored. The mill operates complete segregation and only sells 100 per cent-certified products.

Purchasing

If certified timber is purchased:

- A copy of the supplier's certificate must be held prior to any order being made. This should either be a forest management/chain-of-custody certificate from the landholding if that is the vendor or a chain-of-custody certificate if purchases are from a dealer or haulier.
- Purchase orders specify that certified timber is required.
- When invoices are received, they are only approved for payment if they specify that the timber supplied was certified and the chain-of-custody certificate number is quoted.

Goods in

At the weighbridge:

- All roundwood must be weighed and checked in order to see that it meets mill purchase specification (for example, size, quality and legality/certification). Loads not meeting the specification are rejected and sent away.
- For certified loads, the transport documentation is checked for confirmation that the load is certified. It must include the purchase order number. If the documentation is inadequate, the load is rejected.
- For non-certified loads, the felling licence and transport documentation is checked to make sure that the origin is within the region.

Loads are assigned to either certified or non-certified holding bays in the storage yard, as appropriate. Weighbridge tickets record load specification, including certified or non-certified status.

Sorting

- Roundwood is sorted on a batch certified/non-certified basis. The sorting line is cleaned between certified and non-certified runs. No material must remain on the line prior to a new batch being put through.

Sorting

- Sorted material is put into bays by size and certified status. All certified bays are marked 'CERTIFIED', with the size in green letters, and non-certified bays note size only in red letters.

Milling

- All milling is conducted on a batch certified or non-certified basis. Saw line is cleaned between batches; no material must be left on the line after cleaning.
- For certified runs, internal order must specify certified nature of material and the roundwood bay from which it is taken. Certified internal orders are printed on green paper, uncertified on yellow.
- Sawn material is banded in packs for kilning when finished. Green bands denote 'certified', red bands 'uncertified'. All finished material must leave the mill banded. Reject material for re-sawing must be banded or sprayed green or red to indicate its certified or non-certified status.
- Volumes of material produced on each batch are recorded and reconciled against roundwood inputs to give a conversion ratio.
- All packs must be labelled with the batch number (green crayon denoted 'certified'/red crayon denotes 'uncertified'), which should be the same number as the internal sawing order.

Kilning

- Sawn timber will only be accepted for kilning if it is securely banded. Packs will not be broken. Since packs are clearly identified as certified and non-certified, segregation within the kiln is not required.
- All packs leaving the kiln are labelled with volume, number of pieces, size, batch number and certified status, including the sawmill's chain-of-custody number.

Planing

- Material for the planer is processed on a certified/non-certified batch basis as per internal order (green denotes 'certified'/red denotes 'uncertified'). Only packs banded green can be taken from the kiln/sawn storage areas for certified runs.
- All packs must be labelled with the batch number (green crayon denotes 'certified'/red crayon denotes 'uncertified'), which should be the same number as the internal planing order.
- Job sheets with each batch should record the pack numbers of sawn material used and the pack numbers of finished material, and should reconcile volumes of raw material in and volumes of finished material to give conversion ratios.
- All finished planed material must be packed, wrapped and banded (green denotes 'certified'/red denotes 'uncertified'). All packs are labelled with volume, number of pieces, size, batch number and certified status, including the sawmill's chain-of-custody number.

Finished storage

- Sawn and planed material should be put into the certified area or the non-certified segregated area.

Sales

- Sales orders should raise a picking order. Items indicated as certified should only be picked from the certified area and should be clearly identified by green banding/labels. If the appropriate material is unavailable, the picking order should be returned to sales and re-issued as appropriate. Delivery/invoice notes are based on the completed picking document and record the certified/non-certified status of each pack, together with the sawmill's chain-of-custody certificate number.

Records

- Copies of suppliers' certificates must be kept intact.
- Volumes of certified material bought and sold, together with conversion ratios, must be recorded and reconciled on a six-monthly basis.
- Training records of staff in chain-of-custody procedures must be kept intact.
- All records pertaining to certification must be kept for a minimum of five years.

Staff training

- For each process, one member of staff is nominated to be responsible for that area's chain of custody. The mill manager has overall responsibility for chain of custody.
- All staff with any connection with chain of custody must receive training in chain of custody, not only in their particular area, but regarding overall procedures.

11.9.2 TIMBER STOCKIST CHAIN-OF-CUSTODY CASE STUDY

A timber stockist purchases a variety of sawn timber and panels and sells them on, typically to small- and medium-sized builders, often breaking packs to meet a particular order. Otherwise the stockist does not alter the nature of the material at all. If non-standard sizes are required for a bulk order, they will be ordered from a supplier as appropriate.

Purchasing wood and panel products

1. Check whether the supplier has a valid chain-of-custody certificate. Make sure there is a copy on file.
2. A certified product code must be created before a purchase order is sent to a certified supplier, and the product description must clearly state the requirement for a certified product. The minimum percentage that applies to the product must also be clearly stated under the description.
3. All Forest Stewardship Council (FSC) products must have 'CERT' as a suffix to the code in order to distinguish them from non-certified products.
4. Make sure that the supplier has sufficient quantities of certified product in order to complete the order and that the packs are clearly labelled 'CERTIFIED'.
5. The purchase order should request that the supplier provide the following:
   - the order number on all correspondence;
   - the description, quantity and volume of all products on the delivery note, as well as identification that the products are certified.
6. Goods-in staff must ensure that the correct goods are delivered, that the condition is acceptable and that goods are certified (with a delivery note/labelling). All certified boards and timber must have their edges sprayed green for identification prior to storage. Green marking should be clear in order to identify material at all times during storage.

Processing orders

1. If the order from the customer specifies certified timber, ensure that the dedicated product code is used and that the job action notes that are sent to the warehouse for picking clearly state that only certified products may be used. These must be identified with green paint and should only be picked as certified if marked as such.
2. If the order does *not* specify certified material, a certified product may still be sent and (as above) all paper work must clearly state 'certified'.

3 When both certified and non-certified products are supplied on the same order, it must clearly show on all paper work which are certified and which are non-certified.

Delivery notes, stock control and invoicing

1 All delivery notes will clearly show which items are certified and the minimum percentage that applies for each separate product.
2 Stock control will distinguish certified items from non-certified.
3 All invoices will clearly identify all certified items and will clearly show the chain-of-custody number and the minimum percentage that applies for each separate item.
4 A record of all certified products will be maintained through the stock-control system. The volume and quantity of all certified transactions will be recorded, allowing the total certified product sold over a six-month period to be calculated.
5 The chain-of-custody certificate number will be included on all delivery notes and invoices for certified material as a matter of course. Notes and invoices will also be accompanied by the company's policy statement:
This company is able to source and supply certified products. The letters FSC and the percentage given in the product description indicate the minimum percentage of wood used in making the product that comes from well-managed forests independently certified according to the rules of the Forest Stewardship Council.
6 Records of certified material purchased, sold and in stock will be kept on a six-monthly basis.

Labels and stencils

1 Ensure that labels from existing packs are always kept with the remaining material when packs split. Ensure that remaining material is still identified by green paint.

11.9.3 MDF MANUFACTURER CHAIN-OF-CUSTODY CASE STUDY

A company makes MDF board and has a percentage-based certificate that states that at least 30 per cent of the virgin raw fibre in the board is certified. Chain-of-custody procedures are based on maintaining a rolling mean over a nominal batch length of 30 days. The rolling mean is based on weighbridge figures of incoming material. The company's procedures contain a commitment to avoiding controversial sources of non-certified wood, which it meets by only purchasing from forests in the region where it has confirmed that management is acceptable.

Purchasing

- Certified raw material: check that each supplier of certified material has a valid forest management/chain of custody or chain-of-custody certificate. Make sure that there is a copy on file and that all purchase orders specify the requirement for certified material.
- Uncertified raw material: purchases must be made either directly from forests in the region or from approved suppliers who have agreed in writing only to supply from regional forests. The policy on avoiding controversial sources should be signed by the site manager and copies should be made publicly available.

Goods in

- At the weighbridge, all roundwood should be weighed and checked to see whether it meets mill purchase specifications for size, quality and certified/local basis. The timber delivery note must specify forest of origin, the supplier's chain-of-custody certificate number or, if not certified, the forest of origin and the felling licence number.
- Loads that do not meet specification should be rejected and sent away. Weighbridge tickets should record load specification, including certified or non-certified status.
- Weighbridge tickets must be returned daily to the timber-buying department.

Production

- All raw-material input streams, including non-wood elements, should be recorded on a daily basis, and volumes should be sent daily to the timber-buying department.
- All sheet materials produced should be individually stencilled during production with a certification scheme logo and a per cent claim. Pack wrapping should repeat the claim.

Documentation

- Volumes of certified virgin raw fibre entering the factory, sufficient to meet current certified requirements, must be calculated against conversion ratios from actual MDF production.
- Volumes of raw material delivered must be calculated daily to ascertain per cent certified.
- If at any point during a 30-day batch volumes of certified material drop below the required threshold (30 per cent virgin raw fibre), increased certified volumes must be brought in to allow the overall batch to remain above 30 per cent.
- If it is not possible to maintain the required certified per cent in a batch, all labelling of the product will cease.
- Reconciliations for each batch will be kept and six-monthly resolutions of all inputs–outputs will be retained.
- A list of all loads of virgin raw fibre will be kept, together with their origin, certificate number or felling licence and volumes received.

Sales

- All delivery notes/invoices must carry the company's chain-of-custody certificate number and will clearly show which items are certified and the minimum per cent that applies for each separate product.

11.10 Setting up a chain-of-custody group scheme

Although chain-of-custody audits are more straightforward and therefore much cheaper than forest management audits, they are, nevertheless, a significant cost for very small organizations. This may result in a significant barrier to certification, which has implications for both the equity of the certification system and the overall flow of certified materials, especially when the industry may rely at certain stages on small companies (for example, during harvesting).

In order to address this problem, some certification schemes have developed group chain-of-custody schemes that are specifically aimed at organizations such as small sawmills, craftspeople, artisan woodworkers and small harvesting companies and hauliers. A group chain-of-custody is a system in which members operate under one certificate that is organized and administered by a group manager (this can be an individual manager, a co-operative or an association). These schemes are generally only available to groups of small enterprises and may have eligibility requirements.

For example, the FSC currently defines the eligibility for individual members to join a group scheme as:

- having no more than 15 employees (including full-time, part-time and seasonal staff); or
- having no more than 25 staff with an annual turnover of US$1 million, with turnover being defined as total annual revenue from goods and services.

This definition is temporary and is a reflection of the difficulty of applying rules worldwide across nations, between which there are large variations in labour costs and technologies. The definition certainly encompasses the small organizations which it was set up to help, but could also include, for example, timber agents and other potentially large companies who employ few staff.

The group manager manages the scheme to ensure that its members comply with the requirements of the certification scheme and deals directly with the certification body. It is the responsibility of the group manager to ensure that chain of custody is properly implemented by each member of the group in a transparent and affordable way and in accordance with the chain-of-custody guidelines and rules of the group. The group manager must also carry out an annual monitoring of members in accordance with the scheme rules. This is very similar to the requirements for group certification for forests discussed in Chapter 10.

While the requirements for maintaining a chain of custody are equally rigorous whether it is implemented by a group member or an organization holding an individual certificate, once a group scheme is established, only a sample of the members is included in the annual surveillance by the certification body. This allows significant savings over individual chain-of-custody certification.

A further advantage of the group scheme is that the group entity will act as a consultant to members, helping them to achieve certification.

12 Chain-of-Custody Certification: Getting Certified and Making Claims

12.1 Preparation

There are two elements to the chain of custody process. The first is a company's preparation of a robust chain-of-custody system, which was discussed in Chapter 11, and the second is the audit process itself, which is addressed in this chapter.

The first step for any organization seeking to develop and implement a chain of custody is to review existing arrangements and identify weaknesses or gaps that could allow the security of the chain of custody to be breached. Organizations which already have detailed procedures and strict stock-control mechanisms may only need to carry out minimal modification of their existing process. Where organizations are less organized or have a more informal method of working, new structures will have to be developed and implemented, as discussed in Chapter 11.

It is very useful to identify a management representative within the organization who will have defined responsibility and authority for developing and implementing the chain of custody. It is essential that this individual is provided with adequate authority to set up, develop and ensure implementation of the necessary chain-of-custody framework.

There are two particularly important facts to remember when developing a chain-of-custody system:

1 Use existing systems wherever possible, particularly if there is already an ISO 9000 management system or a stock-control system in place. However, if existing systems are *not* adequate, then they must either be adapted or new systems must be developed and implemented.
2 All staff members must be properly trained so that they understand what they have to do and how important it is to maintain the chain of custody.

Once the chain-of-custody system has been implemented, it is often useful to carry out an internal audit to check that it is working. If needed, it may also be possible for a certification body to carry out a pre-audit visit to help identify any remaining gaps; however, unlike forest certification, this is not routine for chain of custody and usually only happens for particularly large or complex organizations.

12.2 Choosing a certification body

Chain-of-custody certification must be carried out by an accredited certification body. Each certification scheme should be able to provide information on the certification bodies which they have accredited. As with forest management certification, it is worth contacting different certification bodies and finding out what they charge, how quickly they could carry out the assessment, how long the gap between assessment and certification is likely to be, and what support they can provide with information on logo use and claims.

There is an additional point that is worth considering for those organizations that are unsure of which certification scheme they need to be certified against. Several certification bodies now have accreditation from more than one certification scheme and are able to carry out a single assessment to provide certification for two or more schemes. If the certification body is accredited to carry out

ISO 9000 certification, it is also possible to carry out a combined ISO 9000 and chain-of-custody audit.

12.3 The assessment

Chain-of-custody certification is a much more straightforward process than forest management certification. Generally, it involves only one or, at the most, two assessors and takes only one day, except for large or multi-site organizations.

The audit itself will involve an inspection of both documents and the physical process. The job of the auditors is to collect objective evidence in order to show that:

- There is proper control of purchasing certified material; if a percentage-based approach is in operation, there is also proper control of purchasing uncertified material.
- The system in place for internal chain-of-custody controls meets the requirements of the certification scheme and is working in practice.
- In terms of a percentage-based scheme, all of the requirements for minimum thresholds and mixing have been met.
- There is adequate understanding and implementation of all requirements for labelling and claims.
- The system for sales and dispatch of products meets the requirements of the certification scheme and is properly implemented.

In order to do this, the assessors will need to look at the following areas.

12.3.1 SYSTEMS

Assessors will examine the documented plan for chain of custody, including relevant procedures, purchasing systems, stock-control systems and sales systems.

12.3.2 IMPLEMENTATION

Assessors will examine the production area and relevant departments in order to check segregation, product identification, implementation of procedures, training of operators and general implementation of the system.

12.3.3 RECORDS

Assessors will check a range of records, such as:

- purchase records, including purchase orders, contracts, invoices and lists of approved suppliers;
- goods-inwards notes and records of receipt of goods;
- stock records of raw materials and finished products, including, where appropriate, annual stock-taking results;
- production records;
- sales orders received and invoices issued.

The auditors may also want to check that there is a procedure, appropriate to the scale of the operation, for identifying, collecting, filing, storing, maintaining and disposing of all relevant records, which specifies how long records will be kept.

Where the system is based on tracing through the entire process, the auditors may wish to select some final products at random and check that it is possible to trace back to the materials used.

12.3.4 RECONCILIATION OF QUANTITIES

An important part of the control of chain of custody is monitoring quantities of certified material bought and certified products sold in order to ensure that the two figures match. Therefore, records must be kept of:

- quantities of raw materials purchased from each supplier, as well as the total for each type of raw material;
- conversion ratios calculated for each process;
- sales details of certified products, including quantities sold to each customer, total quantities sold and a description of the products.

Summaries of certified raw material used and product produced should be produced regularly, at a minimum of every six months.

12.3.5 UNCERTIFIED MATERIAL

Where products contain a percentage of uncertified material, the auditors will need to check that there is a system in place to control the origin of the uncertified material that meets the requirements of the certification scheme.

They will also need to confirm the actual quantities and origins of material purchased. A summary should be produced detailing the sources and volumes of uncertified material used.

12.3.6 LABELLING

The auditor will need to check all use of labels and any claims being made in order to ensure that they are accurate and conform to the strict criteria laid down by certification schemes. It is often useful to prepare in advance a list of all uses, including on products, in marketing material and brochures, on websites and in advertising.

12.4 Findings and CARs

During the course of the assessment the auditor or audit team will look for *objective evidence* that the requirements for chain of custody are being fully implemented. This evidence will come from the documents they review, their observations and the people with whom they talk.

If they find requirements that are not being implemented, this is classified as a *non-compliance*. Whenever a non-compliance is identified, a corresponding *corrective action request* (CAR) is raised. A CAR sets out the details of the non-compliance and requires that action is taken to resolve the problem. There are two types of CARs (see also Chapter 4, Section 4.1.2.5):

1 Major CARs (also called *pre-conditions*) are raised when there is complete failure to comply with a requirement of the standard or a systematic failure to implement plans and procedures. If a major CAR is raised, it must be adequately addressed *before* certification can proceed (this is why it is sometimes called a pre-condition because it has to be addressed pre-certification).

> Typically, where there is a strong possibility that certified material might be contaminated, a major non-conformance will be raised that precludes the issue of a certificate.

2 Minor CARs (also called *conditions*) are raised when there is partial compliance with a requirement or a non-systematic failure to implement plans or procedures. If a minor CAR is raised,

certification can continue but only on the condition that the non-compliances are addressed within an agreed time frame (hence the alternative name of condition). It is important to remember that if they are not addressed within the time frame agreed, minor CARs will automatically be raised to major CARs.

If a major CAR is raised, this will need to be addressed and the auditor will need to confirm that the actions taken are adequate prior to certification. This may involve a follow-up visit, which usually involves additional costs.

12.5 Certification decision and surveillance

If no major non-compliances are identified, then a certification decision and the issue of the certificate should be fairly rapid. The certificate should set out the scope of the certification (that is, the products or production lines that are included) and is usually valid for five years, conditional on regular surveillance visits.

Surveillance visits are very similar to the initial audit, though they may be slightly shorter. They typically take place once every 6 to 12 months, depending upon the requirements of the certification scheme and the complexity of the organization holding the certificate.

12.6 Making claims

For most certificate holders, the reason for undergoing chain of custody is to allow them to make claims about the products they sell. These claims can take a range of forms, including:

- labels on products, including the logo of the certification scheme;
- publicity material to accompany products at the point of sale, such as posters or brochures;
- information in catalogues or brochures indicating the certified status of products available;
- statements in annual reports or other company documents;
- information or advertising on the company website;
- publicity at trade shows or other similar events.

For a certification scheme, ensuring both the accuracy of all claims and the appropriate use of the scheme logo are fundamental to maintaining the credibility of the scheme. If claims are inaccurate or misleading, this will discredit the scheme as a whole, undermining the value of any claims made by other certificate holders.

As a result, most certification schemes have strict rules and detailed guidance about what claims can and cannot be made, and about the use of the scheme logo. Certification bodies are generally required to ensure that these rules are adhered to as part of the ongoing certification and surveillance process.

Part Three

Existing Forest Certification Schemes

13 Introduction to Existing Schemes

As discussed in Parts One and Two, over the ten years since forest certification first emerged, a number of certification schemes have been developed. Appendix 1 provides detailed information on seven of the main schemes, providing a unique insight into each scheme as seen by its own supporters and developers. The seven schemes reviewed are:

1 CertforChile – national certification scheme in Chile;
2 Canadian Standards Association (CSA) – Canada's national standard for sustainable forest management;
3 Forest Stewardship Council (FSC) Scheme – an international programme;
4 Lembaga Ecolabel Indonesia (LEI) – Indonesian sustainable production forest management certification scheme;
5 Malaysian Timber Certification Council (MTCC) Scheme;
6 Programme for the Endorsement of Forest Certification (PEFC) Schemes – international programmes;
7 North American Sustainable Forestry Initiative (SFI).

Other forest certification schemes exist in addition to these seven – for example, the Brazilian Cerflor scheme and the Ghanaian and Australian national schemes. However, the seven schemes listed above provide a comprehensive overview of the range of schemes in existence.

This chapter provides a brief comparative summary of the seven schemes, examining:

- the type and scope of the scheme;
- the date the scheme was set up;
- the structure and governance of the scheme;
- the standard, certification approach, accreditation arrangements and control of claims;
- requirements for consultation or public information;
- arrangements for small forest owners;
- the current status of the scheme;
- perceptions of the scheme.

13.1 Type of scheme and scope

As discussed in Chapter 2, there are a number of different ways in which certification schemes can be organized. The seven schemes reviewed can be divided into three types:

1 international schemes developed by a dedicated organization;
2 national or regional schemes developed by a dedicated organization;
3 national schemes developed by the existing national standards organization.

As can be seen from the summary below, all of the schemes were developed by dedicated organizations with the exception of the Canadian scheme, which was developed by the existing Canadian Standards Association (CSA), though many of the schemes make use of existing national institutions for accreditation.

Two of the schemes – the FSC and the PEFC – are international in scope. However, they differ in approach. The PEFC acts as a mechanism for the recognition and endorsement of national schemes that meet PEFC requirements. The FSC is a global scheme with a set of international principles and criteria, as well as global requirements. One result of this difference is that the FSC is, in principle, applicable anywhere in the world, whereas the PEFC, though international, is only applicable to countries with an endorsed national scheme.

All of the other schemes are national in scope with the exception of the SFI scheme, which has been used in both the US and Canada, making it a regional scheme.

International scheme developed by dedicated organization	National scheme developed by dedicated organization	National scheme developed by national standards organization
Forest Stewardship Council (FSC) Programme for the Endorsement of Forest Certification (PEFC)	CertforChile Lembaga Ekolabel Indonesia (LEI) Malaysian Timber Certification Council (MTCC) Sustainable Forestry Initiative (SFI)	Canadian Standards Association (CSA)

Several of the national schemes are also specific to particular forest types – for example, the CertforChile scheme currently applies only to plantation forests, while LEI was initially developed for natural forests and has only recently developed standards for plantations and community managed forests.

13.2 Date set up

The oldest of the schemes, the FSC, was set up in 1993 with the first certificate awarded by an FSC-accredited certification body in 1995. The most recent of the schemes to be set up was CertforChile in 2002.

1993	1994	1995	1996	1997	1998	1999	2000	2001	2002
FSC CSA	LEI	SFI				PEFC		MTCC	CertforChile

13.3 Ownership and governance

CertforChile CertforChile is an independent, non-profit membership organization that is legally recognized in Chile. Application for membership is open to anyone. It is overseen by a board elected by the membership and day-to-day running is undertaken by a secretariat.

CSA CSA is an independent, non-profit, national standards-setting Canadian organization. It was initially approached by a variety of stakeholders (federal and provincial governments, industry and associations) to develop the standard according to the internationally recognized standards development process. Control over the standard is through a sustainable forest management (SFM) technical committee made up of four chambers representing the academic/professional/practitioner sector; general interest/the environment; government; and business interests, with a staff member from CSA coordinating the standard.

FSC FSC is an independent, non-profit membership organization registered in Mexico and based in Germany. Application for membership is open to anyone except governments. The membership is the ultimate decision-making body through a three-yearly general assembly. It is overseen by a board elected by the membership, made up equally of environmental, social and economic members. Day-to-day running is undertaken by a secretariat. National initiatives and contact people have been endorsed in several countries to represent the FSC locally.

LEI Lembaga Ekolabel Indonesia (Indonesian Eco-labelling Foundation) is an independent foundation that was legally established in Indonesia. It is overseen by a board of eight trustees who represent a range of interests. The board of trustees appoints a three-member board of directors from among its number to oversee day-to-day running of LEI. The board of directors currently appoints the executive director, though this may change when LEI becomes a constituent-based organization, which is planned for 2005. The executive director, supported by a secretariat, is responsible for running LEI.

MTCC MTCC is a company registered in Malaysia, set up specifically to develop a national standard for Malaysia. It is governed by a board of nine trustees comprising a chairman and eight other members, including two representatives each from academic and research organizations, the timber industry, NGOs and government. Day-to-day running is undertaken by a chief executive officer and a secretariat.

PEFC The PEFC council is a non-profit membership organization based in Luxemburg. Two types of members are recognized – ordinary members, who are national governing bodies running forest certification schemes in individual countries, and extraordinary members, who are international organizations which support the objectives of the PEFC council.

The highest decision-making body is the general assembly, where voting rights are distributed among national governing bodies according to the size of the forest sector in the country; decisions are made by simple majority. The general assembly appoints a board of directors to oversee the organization. Day-to-day running is undertaken by a secretary-general and a secretariat.

SFI The SFI programme was initially developed and operated by the American Forest and Paper Association (AF&PA). In 2001, governance of the standard and certification procedures was shifted to the Sustainable Forestry Board (SFB), which became an independent non-profit organization in 2002. The SFB has 15 members, comprising conservation and environment organizations (one third), professional and academic experts (one third) and the forest industry (one third). Day-to-day running of SFI is undertaken by a director supported by a secretariat.

13.4 Standard

As discussed in Chapter 3, it is important to consider both the development process and the content of forest management standards:

- The development process is crucial because it is always necessary to make decisions about how to deal with incomplete information or conflicting requirements. Both the composition of the group making the decisions and the way in which the decisions are made will largely determine the eventual balance of the standard.

- The content of the standard is equally important as it sets out the requirements that will actually be implemented in a certified forest.

The schemes have a variety of ways in which standards are developed, ranging from multi-stakeholder consensus-based processes to technical working groups, combined with consultation. The content is also variable, to some extent reflecting the differences in forest type and socio-economic content, but also the documents upon which the standards were based.

	Development process	Content
CertforChile	• The standard was drafted on behalf of CertforChile by a technical committee overseen by a superior council, which represents a range of interests. Two public consultation meetings and field testing were included, with final decisions made by the superior council.	• The standard applies to plantations, with a standard for natural forest under development. • The CertforChile standard draws on Montreal Process criteria and indicators (C&I) (see Box 14.1) and FSC principles and criteria (P&C).
CSA	• The CSA standard was developed and subsequently revised by a 32-member technical committee made up of four chambers representing the academic/professional/practitioner sector; general interest/the environment; government; and business interests. • Public consultation was also included since it is part of the requirements for national standards development.	• The standard includes three key requirements: systems, performance and public participation. • Systems requirements are consistent with ISO 14001. • Performance requirements are based on the Canadian Council of Forest Ministers criteria, which are derived from the Montreal Process. • Participation is required to define the precise performance values for a particular forest company.
FSC	• National standards are developed through multi-stakeholder consensus-based working groups. Three chambers – economic, social and environmental – each have equal weight. • National standards are field-tested prior to finalizing and submission to the FSC for endorsement.	• All FSC national standards have to be based on the FSC's global P&C. The P&C are predominantly performance requirements, but include some systems elements, such as management planning and monitoring.
LEI	• Standards are developed by a working group representing a range of stakeholder interests and informed by a number of consultation workshops.	• Three forest standards have been developed – natural forest, plantations and community-based management. • The standards draw on the International Tropical Timber Organization (ITTO) criteria, the FSC P&C, ISO 14001 and the Tropenbos Hierarchical Framework.
MTCC	• Standards are developed by a technical working group who consults with a wide range of stakeholders.	• The standard covers the main types of natural forest found in Malaysia (dry, peat swamp and mangrove). The current standard was developed

MTCC (cont.)		based on the ITTO criteria. A revised standard, due for completion by January 2005, is being developed based on the FSC P&C.
PEFC	• A forum is created and all stakeholders must be invited to participate; but the process can proceed in the absence of groups who choose not to attend. All input must be documented. Consensus is an objective, but not a requirement, of development. The final draft is available for comment for 60 days for finalizing.	• Standards must be based on the Pan-European Operational Level Guidelines, be consistent with national law and incorporate the core International Labour Organization (ILO) conventions.
SFI	• The original SFI Principles and Implementation Guidelines were developed for the American Forest and Paper Association (AF&PA) by a technical committee of professional foresters and scientists in a process that included consultation with a range of stakeholders. The standard was subsequently revised in 2001 in order to bring it more in line with ISO, and again in 2002 as a result of input from the Sustainable Forestry Board (SFB).	• The standard is not based on any C&I set, but draws upon the outputs from the 1992 United Nations Conference on Environment and Development (UNCED) and covers technical, environmental and social issues.

13.5 Certification approach

The certification process is at the heart of any certification scheme. The International Organization for Standardization (ISO) provides a considerable amount of guidance about the way in which it should proceed (see Chapter 4); Chapter 9 outlines the way in which the process is undertaken generally.

Below is a summary of the approaches taken by different schemes. The main differences are the degree to which consultation is included in the process; the requirement for independent peer review of the certification report; and the way in which the final certification decision is made. Almost all of the schemes separate the certification decision from the audit team in line with ISO requirements. Some allow the certification body to make the decision (this is the common ISO approach, as discussed in 'The decision-making process' in Chapter 4), while others continue to make the final decision themselves or, as in the case of LEI, appoint an independent panel to make the decision.

CertforChile	• Audits are undertaken by teams from auditing companies accredited by the National Standards Institute of Chile (the Instituto Nacional de Normas, or INN). Information is collected from documents, field visits and consultation. • A report is produced presenting scores for each principle and criterion, which are peer reviewed by independent specialists. • Under the transition system, the final certification decision is made by the CertforChile superior council on a consensus basis. In the future, the certification decision will be made by the accredited certification body.

CSA	• Certification is undertaken by accredited certification bodies (registration bodies). All auditors must be certified auditors with the appropriate education, background, work experience and credentials. Objective audit evidence is collected from documents, field visits, discussion, observation and consultation. The audit procedure follows defined International Organization for Standardization (ISO) requirements. • The final certification decision is made by the certification body independent of the audit team.
FSC	• Certification is undertaken by accredited certification bodies. The audit must be preceded by a four-week consultation period. Information is collected from documents, field visits, discussions and consultation. A report is produced by the audit team that is peer reviewed by independent specialists. • The final certification decision is made by the certification body, independent of the audit team.
LEI	• Certification is undertaken by accredited certification bodies. Audits must be carried out by auditors registered by LEI after completing an LEI training course. Consultation, facilitated by provincial communication forums, is a requirement for each audit, together with document review and field visits. • The final certification decision is made by an independent panel.
MTCC	• MTCC processes applications and arranges for audits to be carried out by registered assessors (assessment organizations). Based on their report, peer review comments and comments of the assessed organization, MTCC prepares an evaluation report. This is submitted to the Certification Council, which makes the final certification decision.
PEFC	• Certification is undertaken by an accredited certification body. Auditors undertake document review and field visits; but there is no general requirement for consultation or peer review of the reports produced. • The final certification decision is made by the certification body independent of the audit team.
SFI	• Certification is undertaken by an audit team led by a registered and adequately experienced lead assessor. Information is collected from documents, field visits, interviews and, at the discretion of the audit team, from consultation with external organizations. • The audit team makes the final certification decision.

13.6 Accreditation arrangements

Accreditation was discussed in detail in Chapter 5. In most countries, an existing national accreditation body has traditionally provided accreditation, and this is the approach adopted by several of the schemes. An alternative that has recently become more common is to have an accreditation body set up specifically for a particular standard. This approach has also been used.

An interesting note is that most of the schemes are still in the process of developing or improving accreditation, or include a mechanism to allow it to be developed over time.

CertforChile	It is planned that this will be undertaken by the National Standards Institute of Chile (INN) starting in 2005. In the interim, there is no formal system; but auditors are required to have international experience of forest auditing against performance standards.
CSA	Accreditation is undertaken by the national accreditation body, the Standards Council of Canada (SCC).
FSC	Accreditation is undertaken by a unit of the FSC. Currently, this is part of the main FSC organization, though it may become a separate entity in future in order to provide a clear separation between accreditation and standard-setting functions.
LEI	Accreditation is undertaken by LEI, which both accredits certification bodies and trains and registers auditors.
MTCC	Currently, a certification committee established by the board reviews applications for certification based on reports submitted by independent assessors, and also registers independent assessors and peer reviewers. It is planned that this will be replaced by accreditation provided by the Malaysian national accreditation body (the Department of Standards Malaysia) by the end of 2004.
PEFC	Accreditation is provided by national accreditation bodies in each country. Since this can take some time to develop, certification bodies are allowed to operate for up to four years on the basis of a generic systems accreditation before they are required to have a specific forestry-related accreditation.
SFI	There is currently no accreditation for certification bodies. However, lead SFI programme auditors must be registered environmental management system lead auditors or the equivalent with either the US Registrar Accreditation Board (RAB) or the Canadian Environmental Auditing Association (CEAA), and must meet applicable training, education and experience requirements established by the American National Standards Institute (ANSI).

13.7 Requirements for consultation and transparency

As discussed in Chapter 4 (see in particular Section 4.2.3.2), consultation and the provision of public information on the certification decision is important in providing transparency. This, in turn, can contribute to the credibility of a certification scheme. There are significant differences between the seven schemes in their requirements for both consultation and the provision of public information.

Some schemes require wide-ranging consultation, both by the forest organization as a requirement of the standard, and by the certification body as part of the certification process, while other schemes require little or no consultation by either.

Similarly, some schemes require summary public certification reports that provide relatively detailed information on objective evidence of compliance, as well as any non-compliances observed, while others do not require any public information.

All of the schemes have mechanisms for dealing with complaints about both certified organizations and certification bodies or audit teams.

	Information and consultation by forest organization	Consultation by certification body	Public information on certification decision
CertforChile	• Summary of the management plan and monitoring results must be available.	• Consultation with a wide range of stakeholders is required.	• Audit reports are published on CertforChile's website.
CSA	• Public participation is mandatory for refining the performance requirements at the local level. • Policy statements, performance requirements, sustainable forest management (SFM) plan and annual report must be available.	• Consultation with a wide range of stakeholders is required, particularly locally but increasingly nationally.	• Audit reports must be publicly available, summarizing the performance against the requirements of the standard, as well as details of non-compliances.
FSC	• Ongoing consultation with interested parties is required. • Summary of management plan and monitoring results must be available.	• Consultation process must be initiated four weeks prior to the main assessment, and should include all interested parties.	• Summary of audit reports must be publicly available, and should include both results for each principle and details of non-compliances.
LEI	• Forest managers are required to consult with local groups and communities.	• Public announcement is required at least 30 days before assessment, followed by public meetings with all interested parties.	• Summary of results must be publicly available and further information is offered on request.
MTCC	• Reference is made to the degree of participation by indigenous people and local communities.	• Some consultation with local communities is required as part of the assessment.	• Some information on the results should be provided on the MTCC website.
PEFC	• There are no specific requirements, although consultation is included by some national schemes.	• Consultation is at the discretion of the certification body.	• There are no requirements other than ISO requirements for the name and location of certified organizations.
SFI	• Consultation is at the discretion of the forest organization.	• Consultation is at the discretion of the certification body.	• There are no requirements other than ISO requirements for the name and location of certified organizations.

13.8 Arrangements for small-scale forest owners

A significant proportion of the world's forests are owned and managed by individuals or communities. Such owners often only have a few hectares; therefore, they are disproportionately affected by anything that involves significant costs and bureaucracy. Yet, the involvement of small-scale forest owners and communities is very important because forests are extremely important for their livelihoods and because, collectively, they have a major impact on forests.

Therefore, it is important that any scheme which operates in a country or region where there are small-scale forest owners makes sure that there are no inequitable barriers in the way that the scheme is operated and run. Group certification, one of the most common mechanisms for addressing this issue, is discussed in detail in Chapter 10.

CertforChile	Group certification is allowed. Each individual member of the group must meet the requirements of the standard.
CSA	Although the current standard can be used by small-scale forest owners, the CSA sustainable forest management (SFM) technical committee has recognized the need to improve the practical application of the standard in this area and aims to have new provisions by the end of 2004.
FSC	Group certification is encouraged through group or resource manager schemes. Each individual member of the group must meet the requirements of the standard.
LEI	The community-based forest management system has been developed specifically for small-scale forest management, administered either by individuals or community groups.
MTCC	There are no specific arrangements since all natural forest in Malaysia is state owned and management is not undertaken on a small scale.
PEFC	The PEFC allows group and regional certification, both of which aim to ensure that certification is available to even the smallest forest owners in a cost-effective way.
SFI	During 2000, the SFI programme entered into mutual recognition with the American Tree Farm System, which provides certification for small forest owners and is examining similar agreements with other certification schemes that are designed for small forest owners.

13.9 Current status

The status of each scheme can be measured in many ways, including the total hectares certified, the number of forest management certificates issued, the number of chain-of-custody certificates issued, or the market share or global coverage of certified products. Figure 13.1 shows the relative number of hectares certified by each of the seven schemes in December 2003. It is probably the most common measure used; but it does not give any information about the quantity of wood being produced and sold as certified, the number of forest owners and managers involved, or the ecological or social value of the forests certified, all of which are relevant aspects of scheme performance as well.

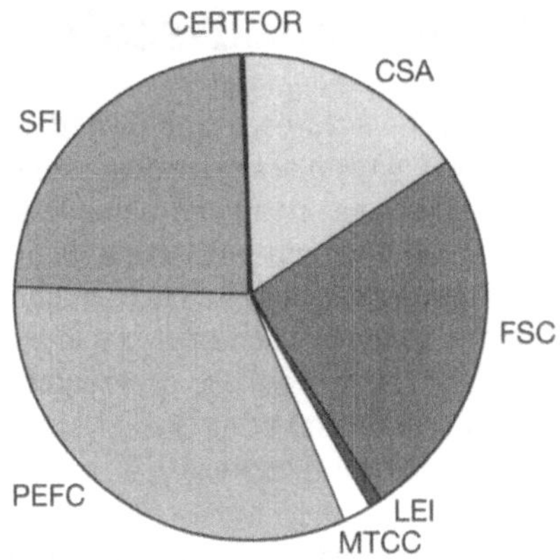

FIGURE 13.1 **Relative number of hectares of forest certified per scheme in December 2003**

13.10 Control of claims

All of the schemes have a logo and operate some form of tracing. In most cases, this is a standard chain-of-custody approach. All of the schemes allow mixing of certified and uncertified material; but there are differences in what type of mixing is permitted and the way in which the uncertified material is controlled.

	Chain of custody and logo	Percentage claims
CertforChile	• Chain of custody is included in the scheme. • The logo can be used on or off product.	• There are three types of accounting for certified material: 100 per cent separation, input–output and percentage-based claims. • Uncertified material must be controlled to exclude illegal and other unacceptable sources, including endangered species, areas with unresolved tenure disputes, reserves and national parks.
CSA	• Chain of custody was added to the scheme in 2001 and was called CSA PLUS 1163. • The logo can be used on or off product in accordance with CSA international logo-use specifications.	• There are two types of accounting for certified material: 100 per cent separation or percentage-based claims. A minimum of 70 per cent CSA certified content is required in order to use the CSA sustainable forest management (SFM) logo. • Uncertified material must be controlled in order to exclude illegal and controversial sources.
FSC	• Chain of custody is included in the scheme. • The logo can be used on or off product according to FSC rules for logo use.	• There are two types of accounting for certified material: 100 per cent separation and percentage-based claims. Input–output is likely to be

		added in 2004. • Uncertified material must be controlled in order to exclude unacceptable sources, including illegal, inadequately managed, high conservation value forests (HCVF), forests with social conflicts and genetically modified material.
LEI	• A chain of custody standard was developed in 2002. • The logo can be used on or off product according to LEI rules for logo use.	• LEI allows up to 30 per cent of the material used to come from uncertified sources. • Uncertified material must be from legal sources; but there are no procedures specified for checking this.
MTCC	• The scheme includes a chain-of-custody standard. • The logo can be used on or off product according to MTTC rules for logo use.	• There are two types of accounting for certified material: 100 per cent separation and percentage-based claims. • There are no requirements for control of uncertified material in percentage-based production.
PEFC	• The scheme includes a chain-of-custody standard. • The logo can be used on products with more than 70 per cent content or off product.	• There are three types of accounting for certified material: 100 per cent separation, percentage-based claims and input–output. • Uncertified material should exclude material from illegal sources or protected areas.
SFI	• SFI does not operate conventional chain of custody, but uses the certification of processors under the SFI programme.	• Certified processors must utilize at least one third SFI-certified material (from either SFI certified forests or processors) and must check that the rest of the material comes from forest sources that are compatible with SFI goals. Timber from illegal sources and biodiversity hotspots should be excluded.

13.11 Politics and perceptions

Forest certification has been a very contentious issue, with strong feelings on all sides about the strengths and weaknesses of different schemes. The summary below provides a brief overview of the main supporters and critics of each scheme.

Appendix 2 also provides a list of some of the main public documents that analyse or criticize different schemes. It is important to note that schemes often question or disagree with many of the findings of the reports, and some have published responses on their websites.

CertforChile	• The scheme is supported by government, foresters and larger forestry companies, and also has support from academics and some NGOs. • Critics, including some NGOs, see it as a source of competition with the FSC. • Since the scheme is relatively new, there has been little assessment to date of what it delivers on the ground.
CSA	• The scheme is strongly supported by the government and industry in Canada, and has received positive feedback from local interest groups, environmental groups and First Nations, who have been part of the process of implementation and certification for a local forest organization. • Critics include some national and international environmental NGOs who argue that the use of general requirements with local interpretation results in too much variability between different certified forest organizations.
FSC	• The FSC is strongly supported by most international environmental and social NGOs. It also has strong support among some sectors of the industry, particularly those for whom brand protection is important, such as retailers and investors. Some governments are also supportive of the FSC, both for forest certification and for procurement. • The main critics of the FSC scheme have been forest owners and managers and their associations, who are concerned that it is too strongly dominated by NGO interests. Some governments, particularly in producer countries, are also critical of the FSC for being too demanding. Conversely, some NGOs have criticized the scheme for not being demanding enough.
LEI	• As the national scheme, the LEI is supported by many key players in Indonesia, including representatives of government, industry and NGOs, and has been supported by many international NGOs and retailers, particularly because of its close relationship with the FSC. • Its main critics have been local and international NGOs opposed to any form of certification in Indonesia until current uncertainty about land rights has been resolved.
MTCC	• The scheme is supported by the Malaysian government and industry. It also has some support from local groups representing indigenous people. • The main critics have been environmental NGOs and indigenous peoples' groups, who claim that they have inadequate influence over the development and interpretation of the certification standard.
PEFC	• The PEFC scheme is strongly supported by small forest owners' associations in Europe, as well as many national governments and parts of the forest industry. With the extension of its scope to become a global scheme, it is also backed by many of the supporters of national schemes who seek endorsement. • The main critics have been environmental and social NGOs whose key concerns are that environmental and social interests have inadequate influence, and that there is insufficient consultation and provision of public information in the certification process to provide adequate levels of transparency.
SFI	• The SFI is strongly supported by the US industry and many in the Canadian industry. It has been building support among national environmental NGOs in the US, including membership of the Nature Conservancy on the sustainable forestry board. • Critics include national and international environmental NGOs who claim that the requirements of the standard are not stringent enough (though there is a recognition that they are improving), and that the lack of consultation or public information in the certification process means that the scheme is not transparent.

13.12 Information about the schemes

Detailed contact information for each scheme is provided in Appendix 1; Table 13.11 lists the individual website addresses.

TABLE 13.1 Websites of the seven schemes

CertforChile	www.certfor.org
CSA	www.certifiedwood.csa.ca
FSC	www.fscoax.org, www.fsc-info.org
LEI	www.lei.or.id
MTCC	www.mtcc.com.my
PEFC	www.pefc.org (This website is particularly useful as it also provides links to all participating national schemes.)
SFI	www.aboutsfi.org, www.aboutsfb.org, www.sfiprogram.info

Part Four

Forest Certification in Context: Policy, Progress and Remaining Issues

14 Policy and the Institutional Context of Certification

The original purpose of forest certification was twofold (cf Baharuddin and Simula, 1994):

1 to improve the social, environmental and economic quality of forest management, thereby providing a tool to contribute to the achievement of sustainable[1] or responsible forest management;
2 to allow the market to reliably differentiate and purchase products coming from responsibly managed forests and to provide the managers of these forests with improved market access for their products.

However, as discussed in Chapter 1, over the last decade several additional uses have emerged, including:

- investment risk reduction and a mechanism for helping to raise funds for good forest management;
- a reduced need for law enforcement and higher collection rate of forest taxes and royalties;
- improved efficiency in forestry, logistics and marketing of forest products;
- verification that donor funds are resulting in projected improvements in management.

Elliott (1999) proposed a broader objective for forest certification: to promote and facilitate policy-oriented learning amongst actors in the forest policy domain so that acceptable standards of forest management, covering economic, social and environment issues in a balanced manner, can be defined and used. This encapsulates and emphasizes the policy process view of certification.

14.1 Forest certification as a policy instrument

14.1.1 CERTIFICATION AS A TOOL OF FOREST AND ENVIRONMENT POLICIES

Policy tools and instruments are often referred to as *hard* or *soft*. Hard policy instruments are those which force implementation, such as regulations or fines. Soft policy instruments are those which encourage implementation, such as incentives, voluntary approaches and market mechanisms.

In the forestry context, certification (and related labelling) falls into the group of soft policy instruments and is probably one of the most powerful tools (see Figure 14.1). It has usually been market driven and voluntary, involving a broad range of stakeholder groups beyond market actors. This broad level of participation both contributes to the legitimacy of the instrument and enables it to influence forest and environmental policy (Bass, 1999). However, it must be emphasized that certification cannot substitute for the hard policy instruments of regulation and enforcement.

Certification alone cannot result in responsible forest management; but it can play an important complementary role within a wider policy mix, together with regulation, incentives and other instruments, provided that a number of pre-conditions are met. As a market-based instrument, certification has some potential to contribute to internalizing the full costs of responsible forest management. Certification also influences the policy development process as it can change traditional relationship structures between stakeholder groups involved in forestry (Elliott, 1999).

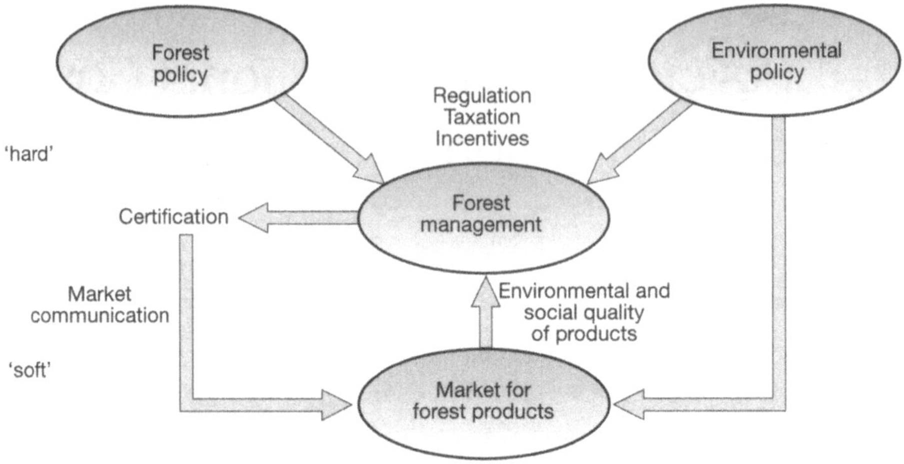

FIGURE 14.1 Policy context of market-oriented certification

In spite of these contributions, certification has limitations. It is a targeted tool, not a panacea for solving all of the problems of poor forest management worldwide. In addition, it is not an instrument to halt deforestation or a substitute for regulation and enforcement.

As already discussed, certification is among the most controversial current issues in the forestry sector. Two main factors have contributed to this controversy:

1 In order to be credible, the development of certification criteria (that is, the requirements specifying how forests should be managed) has to take place through a participatory process involving a range of stakeholder groups. This shifts power to environmental and social groups who have not had similar access to the decision-making process before and away from forest owners, industry and governments who have traditionally dominated forest policy.
2 As in the case of all policy instruments related to the trade of forest products, the impacts of certification are not evenly distributed. As a result, some actors – for example, tropical countries and smallholder forest owners – have been particularly concerned about the equity impacts of certification.

In addition, the first international scheme, the Forest Stewardship Council (FSC), was perceived to be a monopoly dominated by non-governmental organizations (NGOs), a perception which made the FSC unacceptable to forest owners and industry in a number of countries. However, with the development of other schemes, there is no longer a question of monopoly.

A foundation of policies and laws that fosters sustainable development and livelihoods, including good business practice, is fundamental to good forestry. If this is in place, the challenge is to build technical capacity to implement sustainable forestry management that reflects society's expectations. There are several ways of defining this type of management – a certification standard is one of them. The certification process is then the means to verify that the required standard has been achieved in practice (see Figure 14.2).

In many situations, certification can contribute to the development of both good governance and the capacity to undertake responsible forestry. A review of international policy instruments concluded that certification has been one of the major policy innovations to improve private-sector forestry (Landell-Mills and Ford, 1999).

Nevertheless, new instruments of this kind face a range of difficulties related to the policy context, and they are likely to be more effective where there are conducive policies, established tenure of forest land, secure rights, supportive institutional arrangements, consistent policy signals, an

FIGURE 14.2 The three elements essential to the progress of forest certification

adequate information base, stakeholder support, and strong implementation capacity. These are often lacking, particularly in developing countries. Three questions, then, arise (Bass and Simula, 1999):

1 *How much of the governance and forestry elements need to be in place before certification is viable?* Most countries with larger areas of certified forests already have reasonable policies, forest institutions and capacities, which are necessary for certification. In other countries, certification may have limited impact without fundamental policy changes in several areas – for example, in Indonesia (Elliott, 1999).
2 *Can certification 'fast-track' the development of the policy environment where it is not yet conducive?* In many countries, certification appears to be contributing to the improvement of national or local forestry standards, to wider policy debate, to the legitimizing of marginalized stakeholders, and – in some cases – to a greater coherence of policies and instruments. This impact is greater where certification is developing national standards (rather than isolated certificates against global standards). Such national processes can be viewed as processes of mutual learning, which should ultimately influence public policy (see, for example, Elliott, 1999).[2] However, as discussed earlier, certification is not a substitute for hard policy instruments.
3 *How can certification fit with the regulatory regime?* The distinction between certification and regulatory regimes that are based on performance standards may be narrow.[3] Both offer producers a choice between meeting standards or facing possible penalties. In certification, penalties are generally financial. In the case of regulation, they are administrative or judicial.

The answer to these questions is likely to vary depending upon the context in different regions and countries. This is discussed further in Chapter 15, which examines in detail the enabling conditions required for certification to succeed.

Although certification is nominally a market-based instrument, the environmental objectives that it encapsulates are not determined within a closed market system, but by public debate, including by groups who may not participate formally in the market. This is what makes certification particularly credible from a policy perspective, especially in countries where civil society approaches are considered legitimate (DFID, 1999). The fact that certification and regulation are potentially synergistic, rather than antagonistic, perhaps supports the case for their closer integration and provides a clearer role for governments in developing certification.

Certification has proved to be valuable as an instrument to guide the *transition* to responsible forest management, especially through broad national processes, rather than the mere accumulation

TABLE 14.1 **Selected policy issues identified with relevance to certification**

Issue	Link with certification
Responsible or sustainable forest management	Certification contributes to sustainable management, making the concept operational and providing incentives for its adoption.
Balance between economic, social and environmental concerns	Certification standards can express this balance in concrete terms for forest management.
Illegal harvesting	Only timber coming from forests that are managed and utilized according to legal requirements can be certified. Legal compliance is a baseline requirement in all certification standards.
Conservation of biodiversity	Certification standards can address this issue in the management of production forests that play a key role in most national biodiversity conservation strategies.
Poor enforcement, corruption and weak institutional setting	Certification cannot replace supervision or enforcement, but it can complement it. Certification increases transparency of the supply chain and it reinforces forest organizations' management and control capacity.
Limited involvement of local people, and low level of awareness and participation	Certification offers access to all stakeholder groups to participate in standard-setting, which results in criteria for how forests should be managed. It also improves transparency in assessing compliance with the standards set.
Restitution and other forms of privatization of forestry operations	Certification can put small private forest owners at a disadvantage, but it can also provide benefits for them. It can strengthen private forest owners' organizations and help make individual owners aware of how their forests should be managed.
Timber markets	The impact of certification on supply is not known and depends upon local conditions; the impact on demand (export and domestic) is expected to be positive.
Distribution of benefits	Through improved transparency and participation of local people, it can be expected that certification has positive (although probably indirect) distributional impacts.

Source: Indufor-Eco (2000)

of individual certificates. Certification looks likely to continue to broaden the base of policy and planning decisions, and may find other roles. For example, there is a discernible global trend towards more intensification of forest management for fibre production, leaving extensive areas available for environmental services. Certification can help to guide this in a way that improves the sustainability, productivity and equitability of the result.

However, once the transition to responsible management has been made, it seems likely that the role of certification will be more routine, fully integrated with other policies and regulations.

14.1.2 CERTIFICATION AND POLICY ISSUES

Because of its focus on the three elements of sustainability (economic, environmental and social), certification has links with a whole range of issues, which are addressed through national policies. Some of the most important are summarized in Table 14.1.

14.1.3 CERTIFICATION IN THE POLICY MIX

Forest certification is only one of the policy instruments in the trade and environment interface related to responsible forest management. It can only have a complementary role, and where it is driven by market demand its impact will be constrained by the fact that only a relatively small share of forest production is traded internationally due to the bulky nature and low unit value of many forest products.

In countries where certification is required or promoted for domestically traded forest products, the situation is different. In developing countries, there are often limitations to the introduction of such instruments in the short and medium term due to other consumer priorities and the nature of local markets (for example, high share of informal trade). Optimum policies to achieve environmental and trade objectives are likely to be composed of several instruments where certification and labelling are accompanied by adequate macro-economic policies, as well as sector-specific regulatory measures, incentives and taxation, within the provisions of instruments such as multilateral environmental agreements.

With regard to the application of certification, country situations vary considerably. As discussed in Chapter 1, in some cases the driving force will be the expected market benefits: certification can be an efficient tool to communicate the quality of forest management in a credible and transparent way to buyers, consumers, investors, local communities and other stakeholders. In other situations, the driving forces may be internal (for example, improvement of current forest management standards or their enforcement).

There is often lack of clarity on goals and objectives – that is, whether certification is mainly seen as a tool for improving forest management or as a means of market communication. This depends upon the current level of forest management quality (how significant are the improvements required?) and the dependence of the country's producers upon environmentally sensitive markets where certification and related labelling are recognized by buyers of forest products. Nevertheless, these goals are not mutually exclusive and, indeed, should reinforce each other.

In conclusion, the role of certification in the overall policy mix to achieve sustainable management of forests needs to be assessed individually within the context of a specific country. This is discussed further in Chapter 15.

14.2 Criteria and indicators for SFM and certification as complementary policy tools

One of the main policy developments that has been occurring in parallel with forest certification is the development of international sets of criteria and indicators (C&I) for sustainable forest management (SFM). These have been formulated largely through intergovernmental processes in response to commitments, such as those made at the United Nations Conference on Environment and Development (UNCED), to monitor and report on the state of national forests. Therefore, they include criteria such as the total area and the condition of the forest, as summarized in Box 14.1.

14.2.1 INTERGOVERNMENTAL C&I PROCESSES

A recent global stocktaking on the development and use of C&I for SFM has revealed that approximately 150 countries are involved in nine regional processes (see Table 14.2), and the focus on implementation is shifting from national level to local levels, including individual forest management units (CICI, 2003).

Box 14.1 **The main elements identified by international C&I processes that contribute to sustainable forest management**

The principle elements that contribute to sustainable forest management are:

1 extent of forest resources;
2 biological diversity;
3 forest health and vitality;
4 productive functions of forest resources;
5 protective functions of forest resources;
6 socio-economic functions; and
7 legal, policy and institutional framework.

Source: CICI (2003)

14.2.2 DIFFERENCES AND SIMILARITIES

Certification and C&I are two different tools aimed at promoting a similar goal: the sustainable management of the world's forests. In spite of sharing many similarities, there are also considerable differences between these two concepts. Similarities concern the broad common goals and general approaches. Both tools are voluntary approaches towards promoting SFM. They incorporate key elements of sustainability as defined internationally and they are based on the use of data collection for establishing evidence.

Significant distinctions between C&I applications and certification systems concern scale, purpose, outcome, target groups and participating actors. The scale of C&I frameworks ranges from the national to the forest management unit (FMU) level. Many government-led C&I processes focus on national-level frameworks. Forest certification, on the other hand, is primarily concerned with the FMU level.

The purpose is also different. While the C&I sets represent a descriptive approach, forest certification is, essentially, based on prescriptive standards. Certification standards are often conceptually based on, or referenced to, the descriptive C&I sets, but go beyond them in terms of prescription and detail. Consequently, the outcomes are different. C&I contain no targets or performance expectations, while certification is an assessment against performance standards.

Due to the different scale and purpose, significant differences are found with regard to the target groups of the two tools and the typical groups who participate in their development. While the elaboration of C&I sets is generally led by governmental and semi-governmental bodies, forest certification standards and systems are often set up by processes that involve a much wider range of stakeholders, with limited government involvement.

National-level C&I are an important reference basis for forest certification; but they have primarily been developed for reporting on forest conditions at the national level. They are not intended for assessing the performance of forest management at the level of the FMU (Rametsteiner and Simula, 2001).

14.2.3 FMU-LEVEL C&I AND CERTIFICATION STANDARDS

Developing a set of C&I at the FMU level requires substantial adaptation of international or national C&I. Some indicators written for the national scale may not be applicable at an FMU scale – for

TABLE 14.2 International initiatives and processes on criteria and indicators

Initiative/process	Number of countries involved	Region (vegetation zone/geographic area)
Ministerial Conference for the Protection of Forests in Europe (MCPFE or Pan-European Process)	41[a]	European boreal and temperate forests
Montreal Process	12[b]	Temperate forests in the Americas, Asia, the Pacific
International Tropical Timber Organization (ITTO)	31[c]	Tropical natural forests
Tarapoto Proposal	8[d]	Amazon Basin
African Timber Organization (ATO)	14[e]	Tropical forests of Africa
African Dry-zone Process	30[f]	Sub-saharan Africa
Near East Process	30[g]	Near East
Dry Forest Asia Initiative	9[h]	South Asia and Mongolia, China, Myanmar, Thailand
Lepaterique Process	7[i]	Central America
Total number of countries involved	**149**	

Notes:

a Russia is also under the Montreal Process and Turkey is under the Near East Process.

b China is also under the Dry Forest Asia Initiative.

c Producing member countries; the total number of members is 57.

d All countries are also ITTO-producing member countries.

e In the ATO process, nine countries are ITTO-producing member countries and five countries belong to the Africa Dry Zone: Angola (Africa Dry Zone); Cameroon (ITTO); Central African Republic (ITTO); Congo (ITTO); Côte-d'Ivoire (ITTO); Democratic Republic of Congo (ITTO and Africa Dry Zone); Equatorial Guinea; Gabon (ITTO); Ghana (ITTO); Liberia (ITTO); Nigeria, São Tomé et Principe and Tanzania (Africa Dry Zone); and Togo (ITTO).

f Four countries belong to the Near East Process, as well.

g Four countries are also African Dry-zone Process members (Djibouti, Mauritania, Somalia and the Sudan), one is a MCPFE member (Turkey) and one is an ITTO-consuming member country (Egypt).

h Five countries are also ITTO members: three producing-member countries (India, Myanmar and Thailand) and two consuming-member countries (China and Nepal); China is also a Montreal Process member.

i Three countries are also ITTO-producing member countries (Guatemala, Honduras and Panama).

Source: Simula (2003)

example, criteria that are related to global carbon sink contribution, land reservations for protected areas or the sectoral socio-economic contributions of forestry (such as to gross domestic product). Most of the national-level indicators, however, can also be applied at the FMU level, provided that they are appropriately adapted.

Several of the government-led processes, such as the International Tropical Timber Organization (ITTO), the Tarapoto or the African Timber Organization (ATO) proposals, have elaborated C&I sets for the FMU level in tandem with those aimed at the national level. The Pan-European Forest Process (the Ministerial Conference on the Protection of Forests in Europe) produced operational-level guidelines based on its C&I as a second phase of development.

Other work on global C&I has been undertaken by the Centre for International Forestry Research (CIFOR), which has focused research on C&I at the FMU level, assisting countries in field-testing and providing a toolkit for developing locally applicable C&I (Prabhu et al, 2002). The United Nations Food and Agriculture Organization (FAO), the ITTO, the World Conservation Union (IUCN), the International Union of Forest Research Organizations (IUFRO), the United Nations Environment Programme (UNEP) and the United Nations Development Programme (UNDP), as well as NGOs, communities and the private sector, have also supported work on the development and field-testing of national- and sub-national-level C&I.

Over 30 countries worldwide are working on, or have completed, national standards and schemes for forest certification. Among the leading forest certification initiatives, the FSC offers the only existing global set of standards that was explicitly elaborated to function as a reference for certification of forest management: the FSC principles and criteria (P&C). Therefore, unlike other schemes, FSC national standards are not developed from any of the existing C&I sets; rather, they are based on the FSC P&C.

Other schemes have used existing C&I sets. The Programme for the Endorsement of Forest Certification (PEFC) initiative has used the criteria and indicators, together with the operational-level guidelines, of the Pan-European Forest Process as the basis for national standards. The Malaysian Timber Certification Council (MTCC) used the ITTO C&I as a framework for its certification standards, but revised them later in order to correspond to the structure of the FSC P&C. The sub-national-level set of principles, criteria, indicators and verifiers of the ATO is planned to be used as a common framework for a regional certification standard in Africa. The Canadian Standards Association (CSA) standard draws on the Montreal Process and its national interpretation in Canada.

There are also a number of certification schemes, such as the US Sustainable Forestry Initiative (SFI), that were developed independently of any existing C&I sets.

It is apparent that work conducted on national and sub-national criteria provides a solid basis for establishing a common global understanding of what constitutes SFM. A global set of common core elements for criteria that could guide future efforts in this field have already been identified and are listed in Box 14.1 (CICI, 2003). However, while all of these criteria are usually covered, to some extent, by forest certification standards, *indicators* are less systematic and vary depending upon local ecological and socio-cultural conditions, reflecting their diversity.

Given the relatively early stages of standards development, it is foreseen that the various interpretations of criteria and indicators will converge as C&I continue to improve, and that C&I frameworks will be increasingly integrated with forest certification standards (CICI, 2003). Progress in this area will be important for regionally adapted, yet globally recognizable, certification standards.

14.2.4 SETTING THRESHOLDS

While the coverage of issues addressed is likely to continue to converge, the performance levels required may need to remain separate. Because of the heterogeneity of local forest conditions, it may not be desirable to specify common threshold levels – for example, across forest types and socio-economic contexts.

Such specifications are better developed through a participatory process at a local or national level, taking into account the relevant internationally agreed SFM framework, as well as the work conducted by neighbouring regions in order to make the resulting standard internationally acceptable. This approach has the added benefit of taking into account the full range of forest values for society (Rametsteiner and Simula, 2001).

In conclusion, certification and C&I are voluntary instruments and rely upon the same data. The C&I sets can provide a framework for certification standards (goals and approaches) since they tend to be descriptive, while prescriptive certification standards define performance requirements at the FMU level. These tools are used for different purposes and their target groups and development processes also tend to differ. Due to the heterogeneity of local situations, threshold values and other performance requirements are best developed through a participatory process that draws upon relevant C&I sets and takes into account existing standards in the region.

14.3 Forest certification and verification of legal compliance

Many countries have made considerable progress in developing forest laws and regulations and in improving planning for the use of forest land and resources. However, they have often been less successful in controlling and monitoring implementation, and in addressing the fundamental problems of land tenure and poor governance.

Traditional government 'command and control' through direct public involvement is still the rule in many developing countries, despite the inherent weakness of using civil services for the control of forest operations. Particularly in developing countries, they all too often lack proper expertise, resources and appropriate management systems, and they frequently have a low level of cost efficiency. The absence of a clear separation between the approval of concessions and management plans and law enforcement, on the one side, and verification of compliance, on the other creates conflicts of interest and the potential for rampant corruption (de la Rochefordière and Mitchell, 2001).

Certification or independent verification of legal compliance is potentially a useful tool for enforcement in conditions where corruption is common and illegal operations are rampant. It could, in fact, be used to monitor and control a range of aspects, such as forest land occupation and conversion, management practices, logging and wood transportation and trade.

Verification can either take the form of a mandatory approach developed specifically for the country and task in question, or be achieved through the use of existing voluntary schemes. Some countries have considered incorporating voluntary forest management standards within their legal requirements.

The increasing use of independent certification or verification could eventually lead to the contracting-out of enforcement. Such an arrangement would have major policy implications that have not yet been analysed in any detail for the forestry sector, though such an approach has been used successfully in other industries for many years.

14.4 Role of governments

Governments can be very effective in driving forward the implementation of responsible forest management and certification. A distinction, therefore, needs to be made between those governments who show political will and commitment and are making real efforts towards achieving sustainable management, and those who are not. Those who are working to achieve sustainability should be assisted in every possible way. The role of governments is often crucial, not only because of their role as legislators and enforcers, but also because public forest ownership is dominant in many countries, particularly in the tropics (White and Martin, 2002). Therefore, there is a need to make a distinction between the roles of governments as regulators, as promoters of public policy and, in many countries, as forest owners (IPF, 1997).

The government's role in certification is closely related to the overall interest in promoting the sustainable management of forests for their multiple economic, social and environmental benefits and utilities. Given that certification is primarily a voluntary instrument that is implemented by the private sector, there is no need for governments to endorse particular certification schemes or standards; but they may need to define the criteria for acceptable schemes.

On the other hand, governments have a variety of powerful tools and approaches to promote sustainable forest management. Examples include (Simula et al, 2003):

- international agreements (bilateral and multilateral);
- ensuring the long-term secure tenure of forests;

- management of lands under public or state ownership (protection forests, production forests, parks and natural reserves) where governments are responsible for establishing forest management policies and practices, and for ensuring their implementation (including long-term land leasing or concession terms and conditions affecting public forest lands);
- establishment of laws and regulations that affect the forest sector (including forest and environmental legislation; land-use planning laws and zoning regulations; forest practice rules and regulations; and worker health and safety laws and regulations) and ensure effective enforcement of the rules and regulations;
- provision of economic incentives and subsidies (for example, taxes and tax policies, and financial assistance for improved forest management practices);
- research (for instance, wood utilization, forest management, and ecological and social aspects);
- forest inventory and assessment (including periodic inventories of forest area, growth, mortality and other conditions, as well as assessments of the economic, social and environmental dimensions of the forest sector);
- information and education (including technical assistance to forest landowners and forest communities, as well as general public education).

Policies that are detrimental to good forest management or create unnecessary disincentives should be revised or eliminated.

While governments have not been central in the development of the FSC, they have been much more active in promoting the development of national certification schemes in countries such as Canada, Finland, Norway, Malaysia and Ghana. Another example is Mexico, where the recent forest law obliges the government to promote certification. In Russia, the forest law has also made a provision for certification.

Where certification is market driven (as it has been in most cases, to date), in principle it should be able to operate without government intervention or support, and this has proved to be the case in many circumstances. Yet, in the process of developing and promoting certification, government involvement is desirable for a number of reasons (Bass and Simula, 1999):

- *Getting the development of certification started*: because certification is a 'soft policy' instrument, NGOs and private-sector bodies can be good catalysts for its development. However, where these bodies are weak or ill organized, government can help to get certification started.
- *Ensuring an enabling framework for certification, including specific incentives*: the widespread adoption of certification will be difficult without a legislative framework that encourages the development of enabling conditions (see Chapter 15). Therefore, a number of government-related activities have to be integrated with the development of certification. Removing perverse policies and ensuring policy coherence are obvious starting points. Any regulation that is incompatible with voluntary, positive, third-party certification should be revised. Furthermore, government interventions may be needed to help with the trade of timber from sustainably managed forests.
- *Addressing and mitigating problems of certification related to equity*: in many cases, some government action might be required – for example, monitoring and addressing equity impacts, ensuring balanced stakeholder representation and providing support to marginalized groups. Support to small-scale forest owners may be needed to achieve certification in order to avoid putting them at a disadvantage compared with large-scale resourceful operators. This is discussed further in Chapter 19.
- *Setting the ground rules for efficiency and credibility in the market*: this reduces the risk of a proliferation of parallel schemes, which could create confusion amongst consumers, and ensures that international trade rules are properly considered.
- *Ensuring consistency with national policy aims*: the development of the forestry policy and institutional framework should not be undermined by an inappropriate approach towards certification. Purely market-driven decision-making in forest resource management will not ensure sustainability:

temporal scales and the value priorities of private forest owners and managers need to be reconciled with national objectives. Some national schemes have also been able to take forward particular national policies.

- *Developing nationally applicable criteria and indicators for sustainable or responsible forest management and support to national standard development processes*: national C&I provide a tool to identify the components of responsible forest management in the specific country conditions, and therefore serve as a useful reference for the development of national certification standards.
- *Providing targeted support and incentives to certified forest management units*: this could help to overcome hurdles that forest owners and managers are faced with when making the initial investment to meet the standard requirements and is discussed further in Chapter 16.
- *Supporting the market promotion of certified products*: many governments are providing support for promoting wood and wood products in domestic and export markets. Since these products are often presented as environmentally friendly materials for building and construction, certification can be used as a tool to provide assurance to buyers that wood raw material has been harvested in a way that has not put the sustainability of the forest resource at risk.
- *Applying public procurement criteria that make provision for certification*: some governments (for example, the UK, France and Belgium) have, or are in the process of specifying in their purchasing policies, a preference for certified wood-based products.

As the Intergovernmental Panel on Forests (IPF, 1997) has concluded, governments have a particular role to play in encouraging transparency, the full participation of interested parties, non-discrimination and open access to voluntary certification schemes.

Governments are large forest owners in many countries, or the ultimate custodians of forest lands owned by the people or local communities. Government agencies often implement aspects of forest management on lands under their direct responsibility. Where the private sector undertakes forest management on such lands, the government grants utilization rights and defines appropriate conditions. In general, supervision and control of forest management on all types of lands are the responsibility of the government's forestry administration.

Governments will have to decide on the feasibility of certification for their own forests on a case-by-case basis. Participatory processes to develop certification criteria could help to resolve conflicts on the use of state land. Third-party certification/verification could establish credibility for organizations that manage these lands. It could also reduce verification costs, in comparison to audits by government agencies. This involves issues of amalgamating existing government standards and audit protocols with certification standards and procedures.

Certification can also be considered in relation to privatization processes, applied as a condition to ensure that environmental and social performance is not being compromised by the designated manager. An example of this is the case of the privatized South African Forestry Company (SAFCOL) in South Africa.

Notes

1 There is some disagreement about the use of the term 'sustainable forest management' because of the uncertainty about what constitutes sustainability. Some groups prefer to use 'responsible forest management', while others point out that 'sustainable forest management' and the abbreviation SFM are already widely used and understood as a shorthand for forest management that balances social, economic and environmental imperatives. We have used both 'sustainable forest management' and 'responsible forest management' interchangeably to reflect this debate.

2 Elliott (1999) describes how the national certification processes in Canada, Indonesia and Sweden have involved fast-track 'private' policy development, which can be contrasted with the slower public policy

process. Questions may then arise as to whether certification is by-passing democratic processes or utilizing them better. Perceptions of this will depend upon the country in question.

3 A standards-based approach does not have to be linked to certification. For example, standards-based *legislation* can also be efficient. Many countries have mandatory standards of performance for forest management.

15 The Enabling Conditions Framework

As discussed in Chapter 14, the many different conditions that contribute to an enabling environment for forest certification can be broadly divided between three elements (shown schematically in Figure 14.2):

1 the governance and regulatory framework;
2 the understanding of, and technical capacity to implement, responsible forest management;
3 the demand for, and capacity to undertake, certification.

It is widely accepted that for certification to be implemented successfully, it is important to have the right policy and regulatory framework in place, together with the capacity to undertake sound forest management. However, as discussed in Chapter 14, it is also now recognized that certification can contribute to the achievement of both responsible forest management and a better policy and regulatory environment.

However, it is not always clear how these three sets of requirements will interact in a particular context, or how they impact upon each other. Yet, understanding and analysing these interactions is important for a number of reasons:

- *Adapting purchasing policies*: an ever-increasing number of companies and governments are developing purchasing policies that specify certified forest products. However, there are some countries or regions where there are significant barriers that make progress slow. It is important for purchasers to have a mechanism for identifying these barriers so that they can ensure that they are realistic in their expectations of progress.

> For example, if there is no mechanism for harvesters to have any long-term management control over the state forest that they utilize, this is likely to form a barrier to certification that can only be overcome with the support of the state government. Therefore, it would be unrealistic to demand that all wood from these forests must be certified.

- *Supporting the development of certification*: many donors and non-governmental organizations (NGOs) see certification as a useful mechanism for improving forest governance and management, and undertake projects to promote certification. In this case, it is important to accurately identify the main barriers to certification so that project activities can be focused on these key areas.

> For example, if an analysis shows that the main barrier to certification is the lack of capacity to develop a national standard, there is little point in a project that focuses on training auditors.

- *Identifying where certification could be a useful tool to support the policy and governance framework*: in some cases, the development of certification has acted as a catalyst for improvements in forest policy or regulation at a national level.

> For example, the implementation of best management practices in certified forests has provided a model that government can use to demonstrate that such practice is possible.

The development of multi-stakeholder standard-setting processes in many countries has also resulted in greatly increased dialogue between different interested parties and much greater involvement of a range of stakeholders in developing and implementing forest policy.

As a result, many donor projects now see certification as a possible tool for improving forest policy and governance. Such projects require a mechanism for confirming that certification is the most appropriate tool to use in any particular situation.

- *Understanding the successes and failures of certification*: there have been a number of different reviews and studies of certification that examine its successes and failures (see Chapter 16). Each provides a series of useful insights; but the lack of a common framework for identifying and analysing the key issues in each situation makes it difficult to compare the results.

Therefore, it is very practical to have a framework that systematically sets out, for each of the three elements, the detailed requirements or *conditions* which are useful or necessary in creating an *enabling* environment for forest certification (see Figure 15.1). This can form a basis for analysis which ensures that each condition is systematically considered.

The Enabling Conditions Framework[1] (ECF) is one such tool, originally developed for the World Bank–World Wide Fund for Nature (WWF) Alliance for Forest Conservation and Sustainable Use (World Bank–WWF, 2003) to inform the development of their capacity-building strategy.

15.1 Introduction to the tool

The Enabling Conditions Framework (see Figure 15.1) is simply a mechanism for guiding a systematic analysis of each of the three elements that are required if certification is to succeed. For each element, the important conditions that are essential or important for progress have been identified. This allows the user to consider each condition, thereby building up a coherent picture of a particular situation. The identified conditions for each of the three elements are discussed below.

15.2 Governance and regulatory framework

It is widely recognized that without an encouraging governance and regulatory framework, responsible forest management is difficult to achieve. As a result, donors, development banks and governments have all focused on supporting and improving governance structures and legislation in an effort to improve forest management. The Pyramid, an analytical tool developed for use by the World Bank, identifies key elements that need to be in place (Mayers, 2003). Seven conditions are important in ensuring an enabling governance and regulatory framework:

1. clear tenure and resource rights;
2. protection of the forest resource;
3. market and investment conditions and absence of disincentives for responsible forest management;
4. recognition of the roles of different stakeholders and institutions;
5. coherent forest policy and legislation;
6. an effective set of instruments for implementation; and
7. adequate supervision and control.

Each of these is discussed below.

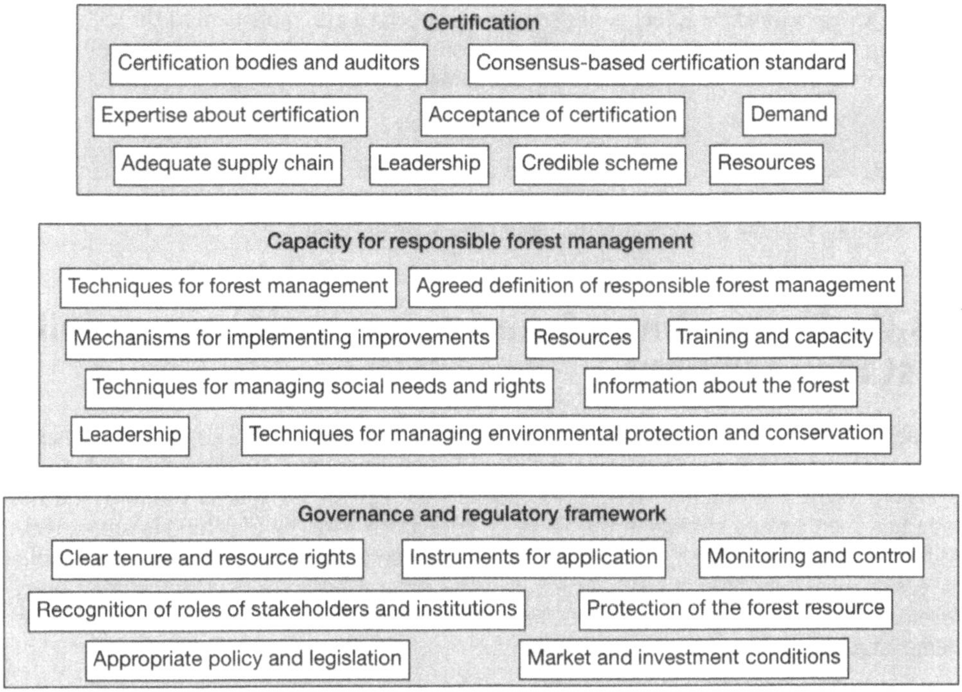

FIGURE 15.1 Schematic representation of the three elements of the Enabling Conditions Framework, together with their conditions

Clear tenure and resource rights An essential foundation for good forest management is to have clarity and security about tenure and resource rights to the forest. All rights, including those of the forest owner and the forest manager, as well as any customary and other rights, need to be clearly recognized and maintained. It is also essential to have a mechanism or a process in place for resolving tenure conflicts that is accepted by most stakeholders.

Resource protection Encroachment, illegal logging and the trade in stolen timber are major threats to forests and responsible forest management in many countries. Therefore, it is important that there are adequate controls to prevent widespread illegal logging and clearance of forests.

Market and investment conditions and lack of disincentives It is important that the prevailing conditions allow for the type of long-term planning and investment needed for good forest management. Absence of perverse incentives that promote inappropriate alternative land-use or poor forest management is also important.

Recognition of the roles of different stakeholders and institutions The forestry sector often involves a range of different groups, including government, industry, local communities, indigenous people and NGOs. It is important that the roles and rights of each group are negotiated and recognized in order to ensure that there is adequate representation of different interest groups in policy development and forest management, and that the rights and responsibilities of each group are understood.

Coherent forest policy and legislation It is very important that there is an adequate policy and legislative framework in place that requires and supports sustainable forest management. In addition, there must be a mechanism for adjusting laws and policies in changing conditions.

Instruments for implementation Effective legislation and policies are implemented through a range of instruments, including regulatory control and enforcement, tax incentives and support services. Strong institutions need to be in place to encourage and enforce implementation of laws and policies.

Adequate supervision and control Laws and policies are more likely to be routinely implemented if there is a process in place for monitoring, supervising and controlling compliance, and if there are penalties for organizations that fail to comply, with culprits being prosecuted.

15.3 Understanding of, and capacity for, responsible forest management

Governance and regulation are essential in creating an environment that is conducive to responsible forest management. Equally important is the technical capacity to understand and implement responsible forestry in practice. This is not always easy. Forests are hugely variable, and the last decade has seen a rapid change in the range of goods and services that forests are expected to provide. Nevertheless, there is a huge amount of experience and information available, collected over tens or even hundreds of years, which provides a good basis for developing good practices. Nine elements have been identified as important in providing an enabling environment for implementing responsible forest management:

1 adequate information about the forest and its response to management;
2 definition of responsible or sustainable forest management;
3 management and operational techniques;
4 delivering environmental protection and conservation;
5 integration of social needs and rights;
6 training and capacity;
7 mechanisms for implementation;
8 resources; and
9 leadership.

Each of these is discussed below.

Information about the forest All forest management should be based on the best possible understanding of the forest and how it responds to management interventions. The necessary information is not limited to trees, but includes the values and functions of the forest.

Agreed definition of responsible or sustainable forest management In order to implement responsible management, it is necessary to know what it is. Over the last decade it has been widely agreed that this encompasses an appropriate combination of economic, social and environmental criteria; but there is often less agreement about exactly how these three elements should be combined and balanced in practice.

Management and operational techniques It is necessary to know how to manage the forest in order to produce the desired range of goods and services. This requires developing appropriate operational techniques for activities such as silviculture, harvesting, road building and other management interventions.

Delivering environmental protection and conservation Implementing responsible management requires practical techniques to plan, implement and monitor measures to maintain and enhance biodiversity and minimize environmental impacts in managed forests. This requires an understanding of both what conservation and environmental protection is needed, and how the necessary measures are implemented.

Integration of social needs and rights, including consultation Implementing responsible management requires practical techniques for identifying and meeting social needs and rights. This requires a range of information, including workers' needs and rights, consultation methods, participatory approaches and community development.

Training and capacity Responsible management can only be achieved if the people in charge of implementation – forest managers, field staff, consultants, trainers and academics, governments, communities and all other relevant stakeholders – are adequately trained and experienced. Training should be considered as a continuous process.

Mechanisms for implementation Experience has shown that in most countries forest managers are much more likely to implement the requirements of good forest management if they are provided with adequate support in doing it. This typically involves extension services, fiscal incentives and other support. One important way of providing this support that has recently been developed is the concept of a phased approach to implementing the full requirements of forest management standards (see Chapter 17).

Resources The cost of implementing responsible management in terms of both human and financial resources can be very high, and this constitutes one of the main barriers to implementation in many places. It is often a particularly serious issue for small and community enterprises; but may be a problem for forest operations of all sizes and types. This issue can be divided into two parts: covering the costs of improving management to the desired level and covering the costs of maintaining improvements once they have been implemented. The former is an investment cost, while the latter is an operational cost.

Leadership Experience from development projects in general, and from places where responsible forest management and certification have made progress, suggests that there is often much greater success when there are a few key individuals or organizations who take on a leadership role and push the responsible forestry agenda forward. This might include people from government, an NGO or an industry representative.

15.4 Standard development and certification

Certification verifies the implementation of responsible forest management, so it clearly depends upon the two previous elements. In addition, there are nine additional conditions that are useful or necessary for certification to be successful:

1 a credible certification scheme;
2 a multi-stakeholder consensus-based standard;
3 availability of certification bodies and trained auditors;
4 expertise in certification;
5 acceptance of certification;
6 demand for certification;
7 sufficient resources;

8 strong committed leadership; and
9 a supply chain with the ability to deliver market requirements.

Each of these is discussed below.

A credible certification scheme Certification has been much more successful where a credible certification scheme operates in the country or region. Credibility of the certification process is required to address both the needs of those seeking certification and the requirements of those creating the demand.

A Multi-stakeholder consensus-based standard Successful certification needs to be based on a standard for responsible forest management that is widely supported and well understood. This is best achieved through a national multi-stakeholder standard-setting process based on consensus.

Availability of certification bodies and trained auditors Certification cannot proceed without certification bodies and trained auditors. For larger companies, it may be possible to bring this in from elsewhere; but for medium- and small-sized enterprises, the absence of local certification expertise is likely to prove a barrier to progress.

Expertise in certification It is not always essential, at first, to have a certification body fully operational in a country; but it is important to have trained people who can both provide support to the forest managers in understanding and preparing for certification, and provide the expertise for certification audit teams who operate in the country.

Acceptance of certification A major barrier to certification is a lack of understanding or acceptance of the concept from one or more of the key groups who are involved in implementing certification or affected by forest management. Awareness-raising on the benefits of certification can overcome this barrier.

Demand for certification It is unlikely that certification will progress unless there is a demand for independent verification or demonstration that responsible forestry has been achieved. There are a number of different sources of demand for certification (see Chapter 1 and Box 15.1).

Sufficient resources For many forest operations, the cost of complying with the standard, rather than the cost of the certification itself, is the bigger financial barrier to certification. However, for other organizations the cost of certification is a significant barrier. This is particularly likely to be the case for small-scale and community operations where management may be good but poorly documented and where income is low. Some certification schemes and certification bodies are seeking to address this through group schemes, simplified auditing requirements or certification support programmes (see Chapter 19).

Strong committed leadership Certification is often more successful where there is a particular individual or group who actively promotes and supports the process. It is also helpful to have a critical assemblage of well-trained, committed supporters of certification in companies, NGOs, support agencies and government agencies.

Ability of the supply chain to deliver market requirements Even if there is a demand from the market for certified wood products, many forest enterprises cannot get access to these markets because the processors and manufacturers in the supply chain cannot meet the demands for quality, price, volume or reliability. Certification cannot guarantee successful access to markets if the supplier is not otherwise competitive.

Box 15.1 Various sources of demand for forest certification

Demand for certified wood and paper products

One of the strongest drivers of forest certification to date has been market demand from buyers and retailers, predominantly in Europe and North America, for wood and paper products from certified forests. This has had a tendency to stimulate certification in the countries or regions which supply these markets, although this has not always been the case since in many countries the barriers resulting from the absence of some of the conditions discussed above were too great.

Where market demand is lacking, forests whose primary product is timber are unlikely to pursue certification. Therefore, one of the ways in which certification can be driven is by stimulating demand for certified products among the main customers for a particular forest, whether they are domestic or international.

The following examples illustrate this point:

- There were significant increases in certification after, first, B&Q in the UK and then Home Depot in the US announced that they would give preference to certified wood, thereby creating a clear demand.
- 80 per cent of the wood from the Brazilian Amazon is used in Brazil. There has already been some success in creating national demand for certified wood through the work of the Brazilian Forest and Trade Network.
- The Tropical Forest Trust (TFT) uses fees from garden furniture retailers in Europe to provide technical assistance and support to producers to move them towards certification, providing a double incentive of both assistance and demand for the certified product.

Demand for other goods and services

Although early progress in certification was primarily driven by traditional market demand, recently there has been an increase in the use of certification to demonstrate the delivery of other goods and services such as carbon, water protection or conservation of biodiversity (see Chapter 21). This type of demand may be of interest in stimulating certification in forests where the conventional market-based certification will not have an impact – for example, forests being financed for management as protected areas or carbon sinks. The same idea might be developed for forests being financed to produce predominantly subsistence goods or under community management.

Investors and donors

Another area in which there is demand for certification is through the mechanism of demonstrating to investors or donors that the forests in which they are investing are being well managed. This type of demand may come from the private sector (banks or investment funds), donor organizations (multilaterals, bilaterals or foundations) or government, where they are investing in or subsidizing domestic forestry.

Examples of this include the following:

- The World Bank-funded PRODEFOR project in Mexico financially supports forest certification for community forests, and the government has incorporated certification in the forest law.
- The UK government is introducing a requirement for some woodlands in England to obtain certification in order to receive grant aid for management.

Governments

There is already some discussion about the role of certification in providing governments with an independent mechanism for monitoring whether forest law and other requirements are being met – for example, as part of concession agreements or community management agreements. Certification can be particularly useful when government lacks the capacity to undertake effective monitoring.

15.5 Understanding the interactions between the governance, forestry and certification elements

One of the most useful things that the Enabling Conditions Framework allows is a systematic review of the interactions between the three elements, identifying both:

- the potential for certification to support the development of governance and responsible forest management; and
- the impact on certification if a condition is not met.

This is summarized below.

15.5.1 GOVERNANCE AND REGULATORY FRAMEWORK

Condition	Implications of absence or potential for certification to support
Tenure and resource rights: are legal and traditional tenure and resource rights clearly recognized and maintained, or is a process in place that is accepted by most stakeholders for resolving tenure conflicts?	Certification is difficult to achieve where tenure rights are not clearly defined.
Resource protection: is illegal logging and forest clearance adequately controlled?	Good forest management and, therefore, certification is difficult to achieve where illegal activities are widespread or uncontrolled.
Roles of different stakeholders and institutions: are the roles of all stakeholders and institutions in forestry negotiated and recognized?	Certification contributes to the better recognition of roles and responsibilities through the standard-setting process and requirements for participation and consultation.
Coherent forest policy and legislation: is there a coherent framework in place that requires and supports responsible forest management and an absence of laws and policies that prevent or conflict with its implementation?	Certification is difficult to achieve where policies and laws are confused or inconsistent with responsible forest management. However, in some cases certification may trigger the recognition of problems and provide an opportunity for change.
Coherent set of instruments: are there mechanisms in place to encourage and enforce implementation of laws and policies?	Certification promotes legal compliance but is unlikely to be sufficient on its own.
Monitoring: is there a mechanism for the supervision and control of legal and other requirements for responsible forest management?	Certification can be a useful tool to assist or support monitoring of legal compliance.

15.5.2 UNDERSTANDING OF, AND CAPACITY FOR, RESPONSIBLE FOREST MANAGEMENT

Condition	Implications of absence
Understanding of the forest resource, including forest dynamics, standing volume, growth and yield, and response to management.	Where this information is not available, only large forest enterprises are likely to be able to develop it independently; even for them it will take time.
Knowledge and agreement about what responsible or sustainable forest management entails.	Where this has not been achieved, a multi-stakeholder national standard-setting process can be a good way of developing a widely supported definition of responsible or sustainable forest management for a country.
Techniques for forest management, including management planning, harvesting, silviculture and road building.	Certification is generally much slower where techniques for good management need to be developed or imported from other regions, particularly for small- and medium-sized enterprises.
Environmental protection, conservation planning and identification, protection and monitoring of endangered species and forests of high conservation value.	Certification promotes the development of approaches and expertise in conservation planning. However, where these are not available, there may be considerable costs associated with their development.
Integration of social needs, including access to resources, workers' needs and rights, and community development.	Certification has driven the development of approaches for integrating social needs in forest management. Since this tends to be more important for large organizations, small forest enterprises are less strongly affected by the absence of expertise.
Capacity of forest owners, managers and field staff to understand and implement the requirements of responsible forest management, including adequate training and support.	Where capacity already exists, certification can proceed much more quickly, though for larger organizations it can drive the development of training programmes and internal capacity-building.
Sufficient resources (people and money) to invest in improving forest management to the level required for responsible management.	Lack of resources is a serious barrier for many small- and medium-scale enterprises, as well as some large organizations in developing countries.
Strong, committed leadership: sufficient numbers of well-trained, committed supporters of responsible management in government, NGOs, companies and support agencies, or a strong lead organization.	Certification is likely to be slower in the absence of local leadership.

15.5.3 DEMAND AND CAPACITY FOR CERTIFICATION

Condition	Implications of absence
Credible certification scheme applicable to the region.	Certification is much easier to proceed with when a scheme with wide support is available locally and is accepted by relevant markets.
Standard is based on a multi-stakeholder consensus-based standard-setting process.	Where there is no national standard, certification under a national scheme is not possible. It can still proceed under the international Forest Stewardship Council (FSC) scheme, based on the generic principles and criteria (P&C); but where there is no consensus about what constitutes responsible forest management, there is often considerable disagreement about appropriate local interpretation.
Availability of certification bodies and trained auditors.	The lack of local certification bodies is a barrier for small organizations. Larger organizations are better able to procure services from abroad; but this still adds complexity and cost to the process.
Availability of local expertise in understanding and implementing standards.	The lack of local expertise to explain, train and support certification has slowed down progress in many places. It is particularly important if there are no local certification bodies.
Understanding and acceptance of certification by forest owners and managers, indigenous groups and other key stakeholders.	Where forest owners or managers, or other key groups such as indigenous people, government or industry, do not support certification, this can cause a barrier to progress.
Demand for certification from markets, investors, donors, governments or other actors able to influence forest managers.	Certification has generally been a market-driven tool, whether it is those purchasing forest products or those investing in the sector which create the demand. Where there is no demand, certification is very limited since there is no incentive to be certified.
Ability of the supply chain to supply the type of goods or services demanded by the market.	A key prerequisite for certified suppliers is the ability to deliver, at competitive prices, the type and quality of product or service required by the segment of the market demanding certification. Where this is absent, certification on its own is unlikely to be sufficient and therefore the demand for, and benefits of, certification will disappear.
Sufficient resources (people and money) to cover the costs of certification.	This is most likely to be a constraint for small forest owners. For medium- and large-sized organizations, certification bodies consistently report that very few are put off only by the cost of the certification process.
Strong, committed leadership: sufficient numbers of well-trained, committed supporters of certification or a strong lead organization.	Certification can progress without leadership, but moves much quicker where it exists.

Note

1 The Enabling Conditions Framework was developed in collaboration with Richard Donovan and Rebecca Butterfield.

16 The Impacts, Costs and Benefits of Forest Certification

16.1 What should certification deliver?

Before beginning a review of the impacts of certification, it is useful to consider what range of impacts can reasonably be expected. This is important because it is only possible to assess the extent of the success or failure of certification if there is a clear understanding of what it should (and should not) be expected to deliver.

A whole range of potential benefits has been suggested by various authors (see, for example, Bass and Simula, 1999; Vogt et al, 2000; Bass et al, 2001; Elliott, 2001) and is listed in Box 16.1. However, some assessments of forest certification have been critical of its failure to achieve impacts that are outside the scope of potential benefits which it can reasonably be expected to deliver.

For example, a number of commentators have criticized certification because it has not halted global deforestation. If this was one of the goals or expectations of certification, then this would certainly be a failure. However, as discussed in Chapter 1, certification was developed as a market tool for the forest products industry and was never envisaged as a solution for the wider problems of forest conversion to other uses. Therefore, it is inaccurate to treat this as a failure or weakness.

The original intentions behind the development of certification and, therefore, the impacts that it can be expected to deliver, are:

- the implementation of responsible forest management on the ground, particularly in forests producing commercial forest products;
- the provision of a mechanism to allow the market to purchase and promote forest products from well-managed sources.

In addition, as discussed in Chapters 14 and 15, it is now recognized that certification has had an influence on governance and the regulatory framework in some situations.

There has also been significant concern about the potential impacts of certification on small and community-managed forests since they are perceived to be the most vulnerable to an inequitable instrument.

As a result, this chapter considers the impacts of certification on:

- the implementation of improved forest management;
- the forest products market;
- forest governance; and
- community forests.

In addition, the economic costs and benefits of certification are examined. These are extremely important since financial and economic implications are likely to have a strong influence on the long-term sustainability of forest certification as a viable tool.

Box 16.1 Potential benefits from forest certification

The list below summarizes the potential benefits derived from certification, proposed by a number of different commentators. It should be noted that different stakeholder groups will attach different degrees of importance to each benefit, and that developed and developing countries are in varying situations with regard to profit from the potential benefits.

Potential forest management and economic benefits

Management and economic benefits include:

- improved performance standards;
- enhanced control of resources;
- improved management systems, including internal mechanisms of planning, monitoring, evaluation and reporting;
- reduced regulatory control;
- permanent economic viability and opening of new markets;
- improved market access and occasional higher prices; and
- improved enterprise image and business ethics.

Potential social benefits

Social benefits include:

- addressing the public's environmental and social concerns in forest management;
- balancing the objectives of forest owners, other stakeholders and society;
- empowering the poor and less favoured;
- poverty alleviation; and
- community participation.

Potential environmental benefits

Environmental benefits include:

- environmental conservation;
- maintenance and enhancement of biodiversity;
- maintenance and enhancement of high-conservation value forests; and
- improved workers' rights and living conditions.

16.2 Constraints

There are a number of constraints to any review of the impacts of certification:

- Firstly, it is a relatively new tool with the first certificate issued less than a decade ago and many of the more recent schemes operational for only two or three years. Therefore, the information available on its impacts is still limited and incomplete, particularly for tropical forests where the progress of certification has been slowest.
- Secondly, much of the analysis undertaken to date has focused on the Forest Stewardship Council (FSC) scheme, with less information on the impacts of other schemes; therefore, it is difficult to judge the extent to which conclusions will apply to all certification schemes.
- Thirdly, forest standards are complex, the structure and composition of forests are extremely variable and, as discussed in Chapter 1, the policy context in which forests are being managed is dynamic. As a result, it is not always easy to identify with certainty the precise cause of a particular change in forest management.
- A further issue is the difficulty of obtaining systematic information about the actions taken by forest managers as a result of certification. One approach has been to review the requirements for improvements made during certification audits (corrective action requests, or CARs). While this provides some extremely useful information and is the basis for much of the discussion below, it does not identify any of the improvements undertaken by forest organizations prior to seeking certification.
- Finally, as certification progresses, the impacts change; therefore, any assessment will only be temporary and will require regular updating.

Despite all of this, there is now a reasonable amount of information available on the effects of certification in a range of forests, and it is extremely useful to consider what some of the main impacts, both positive and negative, have been in order to ensure that future use of certification is effective.

16.3 The impacts of certification

16.3.1 IMPACTS ON FOREST MANAGEMENT

One of the major criticisms of forest certification after a decade of development is that the majority of certified forests are temperate and boreal, with relatively slow progress in tropical forests. One reason for this is that many of the first companies to become certified were those that had to make few changes in their management practices in order to meet the requirements of certification standards (Rametsteiner, 1999).

An additional point is that certification was developed as a market-driven tool, and the huge majority of wood and fibre traded internationally is from temperate and boreal forests, so it is logical that the initial impacts should occur predominantly in these forests.

Therefore, much of the information available about the impacts of certification relates to European and North American forests, which were already reasonably well managed prior to the advent of certification. However, there has also been some certification in tropical regions, and although this may be a relatively small proportion of the total tropical forest area, it nevertheless provides information on both the actual and the potential impacts of certification.

The impacts of certification on forest management can be divided into two types:

1 impacts on the people managing forests; and
2 impacts on the forests themselves and those working or living in and around them.

Each of these is discussed below.

16.3.1.1 Impacts on forest managers

One of the most important impacts of forest certification has been through the development and implementation of forest management standards that incorporate and give equal weight to economic, technical, social and environmental requirements. The broad scope of these standards, addressing issues such as labour relations, occupational safety and health, resource use rights, employment and community participation, in addition to traditional 'forestry' issues, such as inventory, silviculture and harvesting, has changed the way in which many forest managers view their roles and responsibilities.

This impact has extended well beyond the boundaries of certified forests. Most forest managers in larger organizations, wherever they operate and however good or bad their management is, now have some awareness of these issues.

Certification has also introduced third-party auditing to forest management. This has initiated improvements in internal auditing and monitoring in forest organizations and also provides an impartial external view to forest owners on the management status of their forests. This is particularly important for those owners who are not themselves managing their forests, whether they are governments leasing out concessions or small forest owners contracting out management to a resource manager (Baharuddin and Simula, 2001).

16.3.1.2 Impacts on forests

Management techniques

Detailed case studies made of FSC certifications on a global scale have demonstrated a wide variety of improvements in certified forests, sometimes minor, but sometimes involving radical departures from the previous management style in a region (Muthoo, 2001). For example, some certified tropical forests in parts of the Amazon Basin and South-East Asia are conspicuous examples of management that complies with national and international standards in striking contrast to many neighbouring operations.

However, based on a review of corrective action requests summarized in public summary reports of FSC certifications, many of the improvements undertaken as a result of certification relate to the management processes of organizations, especially in planning and monitoring (Rametsteiner, 1999; Thornber, 1999). In almost one out of two certification assessments, CARs were raised that concerned management plans. An interesting question is whether this is because many organizations improved their practice on the ground in preparation for certification, but did not document all of the changes made.

Conservation of biodiversity and environmental services

In Europe, one of the main impacts of certification has been to encourage management that returns forests closer to their natural vegetation state (Rametsteiner, 1999). This is the result of increasing the diversity of native trees and mixed stands, improving the protection of rare and threatened species and their habitats, and reducing the use of chemicals in forest management. The expectation is that, as a result, there will be an increase in biodiversity in certified forests (Lindhe, 2004). This needs to be verified by rigorous research and monitoring in certified forests in order to assess what the actual impacts on biodiversity are.

Examining the impacts of certification more widely, improved conservation of biodiversity appears to be a consistent benefit (Thornber, 1999; Rametsteiner, 1999). Corrective action requests raised in FSC certifications included requirements for increased protection of representative ecosystems and rare, threatened or endangered species, and for more rigorous assessments of environmental impacts This finding is supported by Gullison (2003), who shows that FSC certification can lead to the establishment of significant areas of protected set-aside reserves within certified forest estates.

The exact extent to which these measures improve ecosystem functions, increase biodiversity or lead to better survival of endangered species is something that conservation biologists continue to debate, and in the absence of reliable techniques for assessing true ecological sustainability, any definitive answer is unlikely (Ghazoul, 2001). However, it is clear that certification is having a significant impact upon the way in which forest managers think about and implement conservation measures within their forest areas.

Gullison (2003) goes on to point out that certification has done little to reduce the incidence of deforestation or the destruction of high conservation value forests (HCVF) in tropical areas. This is predominantly because the combination of high discount rates, political and economic uncertainty and insecure land tenure in many areas favours land clearance against long-term management for sustainable timber production. This confirms the point made earlier that certification is not an instrument to combat deforestation. However, it does appear to be a useful tool for promoting conservation of biodiversity and other environmental values within production forests.

Social impacts

Social impacts have been more difficult to quantify. Anecdotal evidence from certification bodies indicates that there has been a range of benefits, including improvements in health and safety, greater respect for workers' rights, and increased capacity for consultation and collaboration with local communities. It is less clear how consistent these impacts are between countries and between schemes.

Additionally, in many regions, certification has highlighted the problems of land rights (Fern, 2004), but it has not necessarily contributed to solving them. As with the problems of deforestation, it appears that land rights comprise an issue that needs to be addressed at the level of national governance, rather than through certification.

In conclusion, it is clear that certification has had positive impacts upon the management of production forests, particularly in boreal and temperate regions. It is likely that the impacts on tropical forests will become more significant over the next few years since certification seems likely to accelerate in the tropics as a result of:

- the development of national standards that are locally applicable and supported;
- the recent emergence and growing acceptance of phased approaches to certification (see Chapter 17), which allow forest managers to access market benefits during the implementation of improvements; and
- practical support for forest managers through initiatives such as the World Wide Fund for Nature (WWF) Producer Groups (GFTN, 2004) and the Tropical Forest Trust programme (TFT, 2004).

16.3.2 IMPACTS ON THE TIMBER MARKET

Most of the early certifications were awarded to companies that were already implementing good forest management. Some commentators have treated this as a problem or failure. In fact, one of the initial reasons for setting up forest certification was to provide a mechanism to allow buyers to differentiate between products from well-managed or unsustainably managed forests. Therefore, in allowing certain producers to differentiate themselves from their poorly managed competitors, the rapid certification of existing well-managed forests was a certification success.

Globally, certified forest products still account for a very limited subset of the international timber trade. Thus, the impacts of certification on the trade could still be described as minor overall, though in some sectors of the trade they are relatively major.

There are a number of reasons for this relatively slow progress from a trade perspective, the most important being that the environmentally sensitive markets which are demanding certified products (certain European Union countries and, to a growing extent, the US) account for a relatively small proportion of the timber trade.

In Europe, it is retail companies – notably in the DIY sector, where there is high visibility to the consumer – that have been the strongest supporters of certification. These companies have had a major influence in driving forward certification and the marketing of certified timber. On the other hand, in sectors such as the construction timber market, which is not particularly visible to the consumer, supply chain pressure has not worked as effectively to promote certification amongst suppliers. However, this is changing in some European countries, particularly as a result of government procurement policies.

Some national and many sub-national government bodies are adopting procurement policies that promote the purchasing of 'green' products wherever possible. Public bodies can have a significant effect on the timber trade. In the case of the UK, the public sector accounts for around 15 per cent of the construction timber market so that the policy shift towards requiring verifiably legal and sustainable timber (discussed in Chapter 14) has a significant influence on the market. However, this driver is a more complex one than private-sector purchasing because governments are wary of supporting individual certification schemes and have to be careful about the way in which they specify the buying of certified timber in order to avoid contravening World Trade Organization (WTO) regulations.

The impact on the trade of tropical timber has been particularly limited. Of the three main tropical timber-producing regions, South-East Asia, South America and Africa, Africa supplies the highest percentage of its output to European markets. Sensitive European buyers therefore have greater potential to 'leverage' change in the African forest industry than in South America or Asia. Demand for certified timber from these areas still outstrips supply, so market pressure should be effective. Despite this, in early 2004 there were still no certified natural forests in the major exporting countries in Africa; instead, Africa's export to Asia has expanded rapidly. However, the situation may change as a number of forest companies in the region begin to make public commitments to progress and even to certification, as discussed in the previous section.

On the other hand, a number of producers in other tropical regions, notably South America, have used the demand for certified products as a mechanism to access high-value European and North American markets with considerable success. As a result, certification has made reasonable progress in the natural forests of South America.

One of the initial expectations from many of those involved in forest certification was that the marketplace would induce change in forest management through the incentive of a price premium. Bass et al (2001) highlight that price premiums paid to producers of certified wood have been weak and, in some cases, non-existent.

Market benefits, in sensitive markets, have often been reaped by the retailers who have promoted certified products in order to protect their corporate reputation and market share. Any premiums that have materialized have been driven more by a shortage of certified products at the retail end than a conscious willingness on the part of the purchasers to pay a price for sustainability (Rametsteiner, 2002). In general, producers have not benefited to the degree expected. In an FSC survey undertaken as part of the development of its percentage claims rules, respondents were asked whether they had ever received a price premium for their certified products. None of the responding forest managers reported a premium, whereas almost half of processors and two-thirds of retailers at least occasionally receive a premium (FSC, 2002).

Consequently, as changes begin to filter through the timber trade, it has been the fear of exclusion from markets that has driven many producers to pursue certification, rather than a price premium for their products (for example, for South Africa, see Frost et al, 2003; Eba'a Atyi and Simula, 2002).

16.3.3 IMPACTS UPON FOREST GOVERNANCE

Perhaps the greatest, and least expected, impact of forest certification to date has been in the arena of forest policy and governance.

National forest agencies initially resisted the concept of market-based regulation through certification, due to its inevitable implication that state regulations are either inadequate or ineffectively enforced. Furthermore, state forestry bodies, just like private companies, often oppose public scrutiny of their operations and, therefore, may not wish to pursue the certification of state forest lands.

However, the process of developing national standards, and the involvement of government bodies in these processes, has had beneficial effects on the overall understanding of sustainable forest management and its regulation. This has led some national bodies to harmonize their own management standards with those of the certification scheme, and to perceive the schemes as less of a threat to their own integrity.

Where this has occurred, there is potential for governments to differentiate supervision and control intensity between certified and non-certified forests (Vogt et al, 2000; Molnar, 2003). This issue still needs careful consideration, and only in rare cases would it be justified to replace government control by third-party certification (See 'Confidence in accreditation' in Chapter 5).

The development of national standards has also provided a forum for involving a far wider range of stakeholders than has traditionally been provided with access to forest policy development. This has served two very important functions:

- Firstly, it has changed the balance of power, giving more influence to environmental and social interests and reducing the dominance of government and economic interests.
- Secondly, it has provided a mechanism for learning and engagement, where factions who may have disagreed for many years about forest management can spend time understanding each others' views and, as a result, are finding ways of compromising and moving forward.

There is growing anecdotal evidence which suggests that this process of engaging a wide range of stakeholders may be fundamental in combating some of the wider problems faced by the forest sector, such as corruption, deforestation and illegal logging.

However, it is also important to note that the uptake of certification has been slow in areas where corrupt, unsustainable and illegal practices are common. There is no doubt that, as discussed in Chapter 15, sound existing governance is an important enabling condition for certification (Rametsteiner, 1999); nevertheless, certification can also contribute to the development of sound governance.

16.3.4 IMPACTS UPON COMMUNITIES

Molnar (2003) and Bass et al (2001) have reviewed the impacts of certification upon forest management in community forests. While it is generally expected that community forestry enterprises, which are not as well linked to high-value forest product export markets, have less to gain from the process of certification, it appears that there can be some benefits.

About 50 community forests have been certified by the FSC (Molnar, 2003). Among the impacts reported for certified communities from countries such as Mexico and Guatemala were the use of certification as a basis to secure land rights for communities; achievement of a higher level of recognition and involvement in political dialogue; and the ability to attract increased amounts of donor support.

According to Bass et al (2001), communities who have undergone certification have felt the following impacts:

- more scientifically rigorous forest management techniques, though at the possible expense of local or indigenous management practices;
- more businesslike methods and tighter management of financial resources, together with the enhanced professional status of the community enterprise;
- increased community involvement in management and more equitable sharing of benefits;
- increased focus on export markets, which has, in some cases, enabled the exposure of lesser known timber species on the international stage.

However, certification can only be a complementary tool in assisting community forests and community enterprises which process forest products. Firstly, they have to produce a product of the right type and quality at a competitive price, and only then can they make effective use of certification as a marketing tool for the product. This has frequently not been the case; therefore, some community forests have discontinued certification when external support, such as donor funding, has ended because they are unable to benefit from their certified status.

16.4 Economic costs and benefits of forest certification

The benefits do not come free: implementing the standard and undergoing certification add costs. In addition, some of the requirements of the standard can lead to foregone benefits for forest owners.

To what extent potential benefits can be achieved, in practice, and how costs can be minimized will vary from one local situation to another depending upon how certification is promoted and implemented. It is important to consider carefully where the expected benefits will exceed the costs as these are the situations in which certification is most likely to be appropriate.

16.4.1 BENEFITS OF CERTIFICATION

The main economic benefits of certification are price premiums and market access. As discussed earlier in Section 16.3, price premiums tend to have been captured by the processing chain and are, in any case, likely to be short term as the supply of certified products increases. Nevertheless, this has been an important benefit for some of the pioneers of certification, who were at least able to recoup their costs. For example, price premiums of between 5 and 65 per cent were reported by some producers of tropical sawn timber and plywood where supply is very limited.

Market access has been a more obvious benefit for some players. A good example is the South African paper sector, which sought certification early and successfully captured a share of the market for certified paper in Europe (particularly the UK, The Netherlands and Germany). Several South American companies have had similar experiences with the production of certified plywood, doors and garden furniture where the ability to supply certified products provided access to a high-value market that offered an economic return on the investment in certification.

However, for many producers and suppliers of temperate and boreal timber, certification is becoming a baseline requirement and buyers strongly resist any pressure to pay any extra for certified products, even though certification adds value to the product in the sense that it provides information on the environmental quality of the product.

16.4.2 COSTS OF FOREST CERTIFICATION

Costs of certification can be divided into direct costs and indirect costs.

The main direct costs are the costs of forest management and chain-of-custody certification. Direct costs of forest management certification are often relatively higher for tropical forests than temperate forests, partly because many certifiers are located in temperate countries and partly because tropical forests are complex, both ecologically and socially. Costs are also relatively higher for small organizations than for large ones.

Indirect costs are those related to compliance with the standard, which involves upgrading forest management and/or the wood processing systems in order to meet the requirements of the certification standard. Such efforts can be relatively minimal in cases where forest management was already good enough before certification, which is the case in many temperate situations. In contrast, the indirect costs of forest certification become very high if the company does not practice good forest stewardship, as is the case in many forest concessions in tropical countries.

Costs of chain of custody depend upon the management system of the enterprise, particularly control measures and records. Many timber processing companies produce both certified and non-certified products, which implies additional costs related to the separation of the two types of raw materials and products. Some internationally operating companies are certified under two international systems (for example, the Forest Stewardship Council or the Programme for the Endorsement of Forest Certification), which also has an impact upon the costs. However, the costs of chain-of-custody certification are generally only a fraction of the costs of forest management certification.

16.4.3 SHARING THE COSTS AND BENEFITS OF FOREST CERTIFICATION

In general, certification costs tend to be much greater for primary producers than for processors, while the benefits of certification, which relate primarily to market access, tend to be reaped by actors further down the supply chain. Therefore, at present, the main financial winners from forest certification appear to processors and retailers rather than forest owners or managers. This may be one of the barriers preventing the more rapid and extensive uptake of certification and suggests that, in the absence of other incentives for forest managers, lack of direct financial benefits may continue to act as a disincentive.

17 Phased Approaches to Forest Certification

17.1 The need for a phased approach

As the market for certified timber grows, particularly in the higher-value European and North American markets, lack of certification has begun to act as a barrier to entering these markets. Often, this involves tropical timber and timber from other developing countries where certification has progressed very slowly. Therefore, it is important to understand what is causing the slow progress towards certification, and to find solutions that address both the need to increase the area of certified forest and the need to minimize interim market barriers for uncertified timber from forests where there are active efforts to improve forest management.

It is now widely accepted that many impediments to certification in developing countries cannot be resolved overnight. Many of the enabling conditions discussed in Chapters 15 and 16 are absent or inadequate, making the process of achieving certification a long and challenging one. Countless barriers relate to the governance and regulatory framework, and these are difficult for individual forest managers or concessionaires to address. However, there are also a number of constraints at the level of the forest management unit (Simula et al, 2003):

- In many countries, the implementation of forest management systems is still incipient and their key elements (for example, resource assessments, forest management plans and monitoring systems) are not yet in place or are inadequate.
- Considerable resources are required to implement the requirements of a certification standard; but developing countries face many institutional, social, human resource and financial constraints, which means that such resources are often scarce.
- The process of implementing the standard can be very lengthy, often taking several years. Many factors slow down the development, which easily stalls for external reasons. A mechanism for periodically assessing the progress made could help forest managers to meet deadlines or commitments involved in the certification process.
- Another serious hurdle is uncertainty about the benefits of certification (see Chapter 15). If there are no intermediate incentives available for forest managers who undertake the long and costly process of standard implementation until full compliance is achieved and a certificate is obtained, the continued investment may seem difficult to justify.
- Forest managers are often overwhelmed by a large number of activities to be undertaken in order to meet the standard's requirements (see, for example, Gretzinger, 2003).

A phased approach to certification can help overcome these problems. By dividing full compliance with the standard into a series of phases, it is possible to focus the limited resources available onto one or two tasks at a time, instead of trying to begin all of the necessary activities at once. In addition, external support can be efficiently focused on priority activities.

A formal mechanism for implementing the standard through a series of phases, particularly if it is linked to some form of verification, can also make it is much easier to assess progress towards responsible forest management. This, in turn, can provide a basis for providing incentives to forest managers who are making real progress even before full certification is achieved. For example:

- As there is shortage of certified timber in some markets, buyers can be encouraged to purchase from forests undergoing phased implementation of the certification standard, thereby providing initial preferential access to these markets. In order to recognize forests that should be included in this category, most buyers demand some form of credible evidence that real progress is being made.
- Governments and donor organizations can link grants, aid, tax breaks and other incentives to the completion of certain phases, even before full compliance with the standard is achieved.
- Financing institutions can provide credits at preferential terms, tying the release of funds with the verified progress on the ground; for them, certification can act as a risk mitigation instrument.

Therefore, the development and use of phased approaches to certification can provide a useful tool to improve forest management and to facilitate access to markets for tropical and other timber.

17.2 Existing models and initiatives

A number of alternative models and parallel initiatives have been developed by various stakeholders in order to put phased approaches to certification into practice (Simula et al, 2003). These can be divided into producer initiatives and purchaser initiatives:

- *Producer initiatives* include initiatives by individual producers, approaches developed by certification bodies and systems, and initiatives developed specifically to support phased approaches.
- *Purchaser initiatives* include both the development of private-sector purchasing policies that are based on a stepwise approach and, more recently, public procurement policies, which also allow for the use of phased approaches.

Each of these is discussed below.

17.2.1 INDIVIDUAL PRODUCER DEVELOPMENT

Many forest organizations which are presently certified achieved full compliance with the certification standard through a 'phased' approach where the requirements were gradually implemented over time. This is very common, but is rarely considered a formal phased approach and it has not been linked with verified statements on the progress made.

17.2.2 PHASED APPROACHES BY CERTIFICATION BODIES AND SCHEMES

Some certification bodies[1] have begun to offer services related to phased approaches. These support programmes consist of two main phases:

1 the initial evaluation (pre-assessment) phase; and
2 the development and implementation phase.

The programme requires companies to make a commitment and develop a plan to ensure full compliance with the standard within a defined time frame. Support programmes have had some success, showing that there is a demand for this type of approach.

Several certification schemes have shown interest in developing a phased approach. None of these proposed phased solutions through existing systems are, as yet, operational; but the approach has significant potential if a credible solution to the problem can be developed within an existing organizational framework, regulating the various aspects of any phased approach.

17.2.3 INITIATIVES TO SUPPORT PHASED APPROACHES

A number of organizations have developed initiatives that actively support phased approaches. These include:

- *The Tropical Forest Trust* (TFT): this organization uses fees collected from timber processors and retailers to work on the ground with forest managers in order to help them implement responsible forest management, with the ultimate aim of becoming certified (TFT, 2004).
- *Global Forest and Trade Network* (GFTN) *Producer Groups*: the World Wide Fund for Nature (WWF) Global Forest and Trade Network has groups in many producer countries who support forest managers in implementing responsible management over a defined period (GFTN, 2004).

17.2.4 MODULAR IMPLEMENTATION AND VERIFICATION

Modular implementation and verification (MIV) is a tool for supporting phased implementation, based on the division of the standard into a series of predefined modules that can be implemented through steps or phases (Nussbaum et al, 2003).

The MIV model consists of two parts: a set of modules outlining actions to be taken, and a framework within which the modules should be used. Completion of all the modules delivers compliance with the requirements of the certification standard.

The advantage of using a defined set of modules, the content of each of which is known, is that it provides a consistent framework for assessing and communicating progress. It also provides the potential for selecting 'minimum' or 'baseline' modules that must be completed prior to recognition.

17.2.5 BUYER INITIATIVES

Numerous companies have made some type of commitment to sourcing their wood and paper products from certified or sustainable sources; but many are unable to implement these commitments immediately. As a result, many active larger buyers have developed some type of phased approach in order to provide a framework for implementing this policy in practice. These buyers are found throughout the supply chain, but particularly in the retail sector.

While the exact approach varies, in general the phases include:

- Establishing the products' forests of origin.
- Ensuring that all products are from legal sources and phasing out unknown or illegal sources.
- Phasing out the purchase of products from various unacceptable sources (for example, endangered ecosystems).
- Preferentially purchasing from suppliers who are progressing towards certification, usually involving these suppliers implementing recognized phased approaches.
- Finally, achieving full certification under a recognized system, although the timing for this may be unspecified.

The buyers' initiatives are a reminder of the shared responsibility of consumers, as well as producers, in addressing the issue of good forest management. This type of approach raises some concerns amongst forest owners since there is a danger that if each buyer develops a different approach with varying requirements, it will become costly and complex to demonstrate compliance. However, early signs are that there is a general willingness on the part of buyers to recognize existing phased approaches, rather than insisting on their own alternative.

A number of publications have been produced to provide support to buyers who want to implement a phased approach to purchasing (TFT, 2003; White and Sarshar, 2004).

17.2.6 PUBLIC PROCUREMENT

During 2000, Group of eight (G8) leaders made a public commitment that participating governments would seek to meet all of their wood and paper requirements from 'legal and sustainable sources'. In practical terms this has led to the emergence of policy commitments to give preference in government procurement to legally (and sustainably) produced timber within the context of the World Trade Organization (WTO) and European Union (EU) procurement rules. The UK government has been spearheading the implementation of this commitment; but France and Germany are also planning to implement similar provisions. The EU has recently issued an *Action Plan for Forest Law Enforcement, Governance and Trade* (FLEGT), which includes public procurement as a key tool to strengthen enforcement in partner countries. This specific focus on the requirement for legal sources has been further supported by several timber trade organizations in importing countries (for example, Japan, the UK, Finland and the US).

In spite of the rapidly increasing attention given to the issue, there is currently no generally applicable system available to verify legality. Certification is the most robust existing mechanism provided that legal compliance is defined as an explicit requirement of the standard. However, where certification is not available, a phased approach that includes verification of 'legal compliance' as the first phase or module could address this need as well.

17.3 Developing a phased approach to forest certification

Phased approaches can be used in a variety of ways; but increasingly the most common use is through some type of organized scheme. Such schemes can be run by certification schemes, certification bodies or other organizations, as discussed above. The general characteristics for such a scheme operating for tropical timber production are outlined in Box 17.1.

The general procedure to be followed for any phased approach is illustrated in Figure 17.1 and is summarized below, while the practical details of implementing the procedure, in practice, are discussed in section 17.4, 'Implementing a phased approach in practice':

- A preliminary audit is first carried out to identify the gaps between the current management and what is required by the chosen certification standard; this audit can be carried out by an internal or external auditor.
- A time-bound plan is prepared by the company in order to close the gaps identified. The auditor can advise on the preparation of the plan; but the main responsibility lies with the forest management unit (FMU) or the forest enterprise.
- Based on the preliminary audit report and the action plan, the FMU can be registered as a participant of the phased approach.
- The forest enterprise then conducts a phased implementation of the action plan.
- Subsequently, there is periodic verification of the progress in implementing the action plan and the performance of forest management. This audit must be carried out by an auditor approved by the scheme (and may be a certification body).
- A full certification audit occurs when all of the gaps have been closed. This audit must be carried out by an accredited certification body.

Two of the key issues that need to be resolved in designing any phased approach based on the procedure set out above are:

Box 17.1 General characteristics for phased approaches for tropical timber production

The following general characteristics are considered essential for phased approaches in the case of tropical timber (Simula et al, 2003):

- Full certification should be the goal of all phased approaches.
- Implementation:
  - There should be mechanisms in place to support forest managers in implementing the requirements of the forest standard.
  - These mechanisms should operate at the forest management unit (FMU) level.
  - They should involve clear commitments from the participating forest managers.
  - There should be a defined time horizon within which all actions that are needed to address gaps in performance should be implemented.
- Verification:
  - Verification should involve independent audits that are based on clear rules and procedures.
  - Conflict of interests in the work of auditors should be avoided.
  - The verification of progress should be transparent.
  - Adequate means to communicate on verified progress should be provided, including chain-of-custody verification if any product claim is made.

Many of the other general principles of forest certification also apply to phased approaches (for example, reliability of assessment, non-discrimination and applicability to all types and scales of forest management).

1 defining the phases;
2 managing communication.

Another important issue is how to ensure that phased approaches do not create barriers for small and community-managed forests. Each of these issues is discussed below.

17.3.1 DEFINITION OF PHASES

There are two basic approaches to defining phases:

1 Each forest organization defines its own steps or phases independently, based on the gaps identified during the baseline review.
2 Phases are predefined through an approach, such as the definition of modules in the MIV tool outlined above.

In practice, implementing either of the approaches is very similar. The advantage of the first approach is that it allows a forest enterprise to focus clearly on its own situation. The advantage of the latter approach is that it provides a good basis for consistent communication and understanding of progress.

In either case, the standard will need to be broken down into a number of legal, economic, environmental and social components, either based on the initial review or as a predetermined set of

FIGURE 17.1 **The general procedure of the phased approach**

components. If there is no locally applicable standard available, the most appropriate international criteria and indicators (C&I) set (see 'Intergovernmental C&I processes' in Chapter 14) or the Forest Stewardship Council (FSC) principles and criteria (P&C) can be used as a basis for defining the components.

With regard to defining the steps and their sequence, there is a general understanding that:

- There should be a set of minimum baseline requirements or components that the action plan should address first (see Figure 17.2).
- Where phased approaches in the forest are linked to phased approaches in purchasing, these baseline requirements should match the basic requirements of purchasing policies, such as legal operations.
- Forest enterprises should have the freedom to choose the order in which the other components are addressed within the specified time period.

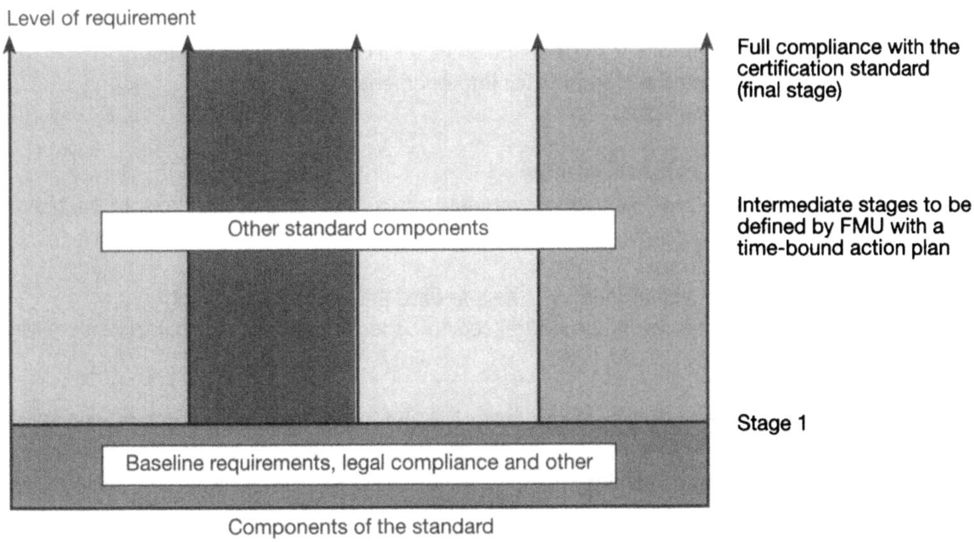

FIGURE 17.2 **Defining stages of the phased approach to forest certification**

17.3.2 COMMUNICATION

Any communication about phased implementation must be truthful and accurate. Many in the forest products sector feel that it should be limited to business-to-business communication and not used for product claims. This is because, during the early part of the process, forest management may, in fact, be quite poor and unless this is clearly communicated there is a risk that a product claim might be misleading.

If communication is made 'on product', it must be made explicit that the product is not derived from a certified forest. In addition, as soon as any market communication related to products and their origin is introduced, there needs to be proper verification of chain of custody.

17.3.3 PHASED APPROACHES FOR SMALL AND COMMUNITY FORESTS

An important issue that needs to be considered in developing and implementing any system for a phased approach is its applicability to small-scale and community forests, which are often managed at a lower intensity and with more diverse objectives than, for example, industrial concessions. The following measures to assist community and small-scale forest owners would be useful:

- simplification of guidelines or the adjustment of standard requirements;
- provision of longer time frames in applying the phased approach than those allowed for large enterprises;
- a group approach to phased implementation and verification using existing organizational structures, wherever possible;
- external funding and assistance, as well as incentives, tied to the progress made;
- establishment of partnerships between communities and the private sector.

17.4 Implementing a phased approach in practice

17.4.1 IMPLEMENTATION

As summarized above, there are three main stages to the process of implementation:

1 a baseline appraisal or audit of the current situation, also called an initial review;
2 development of a time-bound action plan for improvement;
3 implementation of the action plan.

17.4.1.1 Baseline audit or initial review

A good starting point for any process of improvement is an assessment of the current situation. This is important in order to:

- Assess to what extent the requirements of the standard are already being met.
- Identify all of the gaps that must be addressed in order to achieve full compliance with the standard.

The results of the initial review will form the basis for the action plan to guide progress towards implementing the standard.

There are a number of ways in which this review can be carried out, though individual schemes may specify one particular approach:

- *Internal review*: the manager or management team undertakes the review.
- *External review*: the review is carried out by someone external, such as a consultant or a member of staff from a purchaser, donor or investor.
- *Independent review*: the review is carried out by a recognized independent third party, such as a certification body or recognized auditing specialist.

Whoever carries out the initial review, it is essential that they are:

- familiar with the requirements of the standard, including the interpretation of the requirements for the specific location and forest type;
- competent to carry out an audit effectively;
- able to adequately examine technical, social, environmental and economic issues (for medium- and large-sized forests, this will almost always require the use of a team);
- able to report the findings both verbally and in writing in a way that is easy to understand for the forest manager and will ensure that the information is of maximum use for developing action plans.

17.4.1.2 Developing an action plan

At the heart of any implementation of a phased approach is an action plan. This sets out in detail the timetable for addressing each of the gaps identified and, therefore, for achieving compliance with the full requirements of the standard.

It is important that the forest manager or management team is involved in developing the action plan, as they will be the ones who have to implement it. However, it may also be useful to obtain external assistance, such as consultants, academics, government forest departments or non-governmental organizations (NGOs) to provide input on specific issues. It is also useful to include the people involved in undertaking the initial review.

The initial review should have indicated:

- the requirements of the standard where there is already full compliance and no further work is required;
- the areas in which there are gaps between the requirements and current management practices that need to be addressed.

For each of the gaps identified, a plan must be developed to achieve full compliance. All of the separate plans for the different gaps then need to be combined to produce a coherent programme or action plan for full compliance.

For each gap identified, the action plan must set out the details of how full compliance will be achieved, including:

- who is responsible for ensuring that the module is implemented;
- details of all other people who will be involved in implementation;
- the actions that will be required, including any interim targets and the timing for each one;
- what staff, equipment, resources and training will be needed;
- how progress will be monitored, reported and reviewed;
- the timetable for addressing each gap.

Since in many cases the action plan is likely to extend over a period of years, there will need to be some flexibility in order to respond to changes and unforeseen circumstances. This can be provided by developing detailed actions for the first year, together with a less detailed plan for subsequent years, and by linking this to a regular review process, such as once every six months, as discussed below.

The next stage is to put all of the separate plans for addressing different gaps together in a coherent overall programme or action plan.

It is probably best to begin by defining the scale of the total implementation period (for example, three or five years). This may be an internal decision or may be driven by external demands, such as by customers, investors or government. The various actions planned to address different gaps can then be fitted into this overall time frame. When doing this, it is probably helpful to consider the following points:

- Identify all gaps that need to be completed as a basis for addressing other gaps.

> For example, if lack of appropriate operating procedures and lack of adequate training for operators have both been identified as gaps, it makes sense to sort out the operating procedures before training commences.

- Identify issues that need to be completed in order to meet minimum requirements for investors or customers, and timetable them to be undertaken first. This might include legal compliance for customers concerned about the legality of sources or an environmental impact assessment for investors concerned about their own environmental impacts.
- For issues that are going to take some time to address, make sure that activities are begun early enough in the process in order to allow the full plan to be completed.
- Try to ensure that inputs made by a particular person or department, or activities requiring a large input of resources, are staggered.
- Generally, try to divide the costs of implementation reasonably evenly, and ensure that the activities timetabled for any particular year do not exceed the resources available in the budget for that year.

If the timetable covers several years, then it is probably a good idea to incorporate more detail at the beginning and to leave some of the planning for later stages until closer to the time. However, this will only work with two important provisos:

1. Things almost always take longer than planned, so this must be built into any timetable; in particular, it is important not to leave all of the more challenging work until the end of the time available.
2. It is important to have in place a system of regular monitoring and review so that any problems or delays are identified and addressed.

17.4.1.3 Implementing the action plan

Implementing the action plan is the most important part of the entire process since it will deliver improvements on the ground. The way in which this is achieved, in practice, is likely to vary between organizations depending upon organizational culture and preferences.

The most important issue is that there is regular monitoring and review of progress similar to the monitoring and review described in Chapter 8. Based on this, the action plan should be revised and additional or revised activities should be undertaken wherever necessary.

17.4.2 VERIFICATION

Where phased implementation is being undertaken in order to provide assurance on progress to purchasers, investors, donors or other external parties, it is important to verify that the commitments are being met in practice. This may also be useful for internal purposes in order to monitor and maintain progress.

Where verification is being used for external communication, there are three main ways in which it may be used:

1 a baseline assessment of the current level of performance in order to establish that the action plan is based on accurate information;
2 confirmation that any preconditions such as legal compliance have been met;
3 ongoing confirmation that commitments made in the action plan are being achieved in practice.

These assessments will necessitate all of the same considerations discussed in Part Two of this handbook.

Note

1 For example, SGS Malaysia and SmartWood.

18 Capacity-building

18.1 Capacity-building needs

Several weaknesses can be identified in the capacity of many countries to implement certification of forest management. Some are not specific to the forest sector, but tend to be common for standardization, certification and accreditation activities. In order to ensure that the full potential of certification is realized, specific technical capacity needs strengthening, typically in the following fields:

- *awareness-raising* about the pros and cons of certification, its potential role as a policy instrument for promoting responsible forest management and related market requirements;
- *capacity for developing certification standards and procedures*, with supporting participation systems to bring in marginalized actors, especially in developing countries and countries in transition;
- *capacity of producers* to implement responsible management and to comply with related standards, conduct internal audits and establish an effective external audit process (Thornber et al, 1999);
- *capacity of certification and accreditation bodies* to conduct external audits (again, this is especially needed in developing countries, but should not be neglected in developed countries since certification's impacts and credibility revolve around the quality of certifiers' work);
- *capacity to access relevant market information* and to run enterprises in order to serve environmentally sustainable market demands, especially for small- and medium-sized enterprises;
- *engagement/strengthening of existing systems such as government extension mechanisms and rural development banks* in order to address the needs of community forests and small-scale forest owners.

Some of these may best be delivered through training, while others require organizational development.

At the same time, as discussed in Chapter 14, certification can contribute to wider institutional capacity development – for example, in networking, in improving understanding between stakeholders, and in new forms of planning and policy development. Such broader institutional capacities are extremely important in the transition to sustainable management of forests. As Elliott (1999) points out, they can best be developed if certification is set up deliberately as a learning process, as it is in some national processes.

Capacity-building is a crucial part of implementing sustainable forest management and forest certification. Implementing responsible management and forest certification is a cross-disciplinary process that requires expertise in environmental, social and economic aspects at all decision-making levels (see Figure 18.1). Without adequate resources, forest management cannot reach the level required in credible forest certification standards. Certification also sets additional demands – for example, in monitoring, data collection, training and communication – and adequate capacity is needed to undertake these tasks.

Capacity-building can be divided into the development of:

- human resources;
- institutional and organizational framework;
- information systems to provide data for planning, implementation and assessment.

Each of these is discussed below.

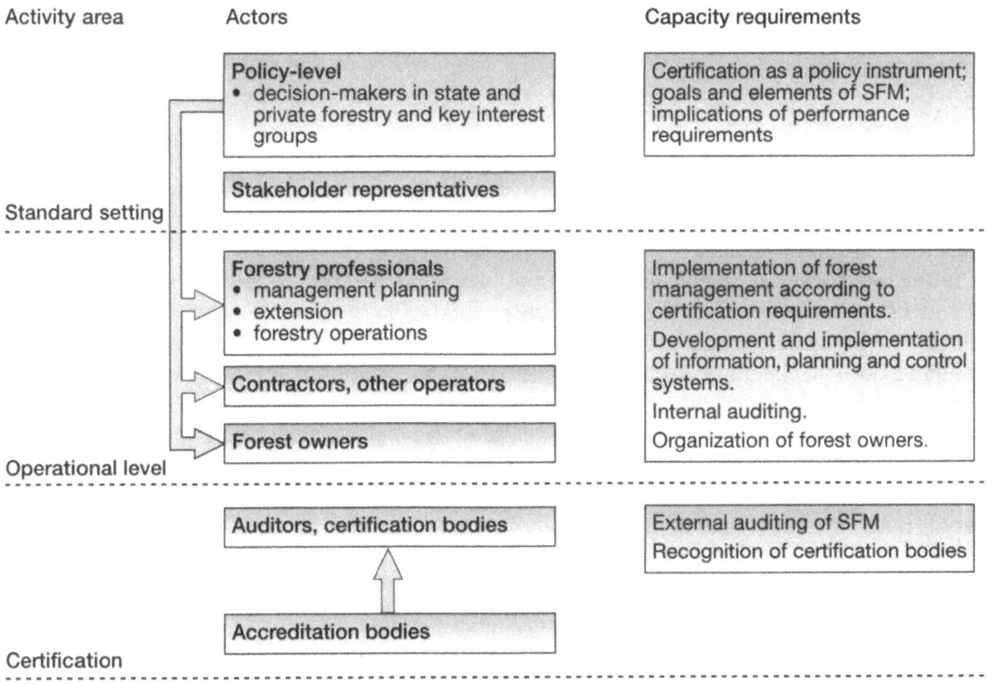

Source: Indufor-Eco (1999)

FIGURE 18.1 **Elements of capacity-building in forest certification**

18.2 Human resources

Forestry is a newly introduced field of activity in the certification sector, and it has a number of aspects that differ from the clearly defined and easily controlled industrial processes in the traditional realm of certification. Auditors need to have an adequate understanding of forests as ecosystems and of the principles of forest management. Each country has its own organizational structures that have been developed to coordinate and share responsibilities in forest management and to promote good forestry by state and private forestry organizations, especially in extension, planning and enforcement. Understanding forest management practices, as well as understanding the management system and the organizational set-up, is essential for reliable audit work.

The necessary types of expertise that need to be established at different levels include the following (Indufor-Eco, 1999):

- Policy-level decision-makers among government agencies and stakeholder organizations (for example, the private sector and civil society) need to be prepared to make informed decisions about how certification as a soft policy instrument can best be used to promote sustainable forest management and how to avoid adverse impacts – for example, on small-scale private landowners or community forests.
- Stakeholder representatives participating in the development of nationally applicable certification criteria and standards should be capable of understanding:
  - how international requirements for certification standards should be interpreted in national conditions;
  - the basic requirements of sustainable forest management and how they can be expressed in terms of relevant criteria and indicators;

- the practical economic, social and ecological implications of the proposed certification standard, including whether or not the requirements are achievable in practical forest management;
- how conflicts between stakeholders can be resolved and consensus achieved in the development process.
- Forestry professionals working as forest managers, extension agents or advisers should be capable of implementing sustainable management in the forests that they manage and of supporting its implementation in state, community or privately owned forests, where they provide management advice or technical services.
- Contractors and other operators carrying out harvesting need to understand the implications of the certification standard for their operations.
- Private forest owners' organizations, which will have to mobilize their members for certification, need to have an adequate understanding of what certification is and what it will mean for the membership.
- Community organizations managing forests need to have the skills and knowledge to upgrade their management systems and their performance in order to make themselves certifiable.
- Auditors of forest management should be trained to verify the certification criteria, drawing on appropriate means of substantiation. There is a need for internal auditors, employed by forest organizations who wish to prepare themselves for certification, and for external auditors to be employed by third-party independent verifiers or certification bodies.

Many developing countries and those in transition have developed a well-qualified pool of forestry professionals with adequate technical skills. However, certification is bringing other stakeholders into the policy process, and they are often poorly prepared to make informed decisions on such complex issues as how sustainable forest management should be interpreted in practice and what the best measures are to achieve it.

Numerous countries have recently experienced policy reforms that have been associated with budget cuts in forest administration. This has led to reductions in staff levels and, in some countries, to outsourcing a range of functions. One specific consequence of this trend is that training budgets have been cut, and in some places the enrolment rates at forest training institutions have declined. At the same time, the tasks of forest managers and authorities have become more complex, requiring a much wider range of expertise and greater efficiency than previously was the case. Therefore, in order to make certification work, there is a need to provide further training.

The main target groups for capacity-building were presented above. The principle areas in which further training is required often include the following:

- *conservation*: the ecological implications of forest management in the country;
- *social order*: the social implications of forest management in the country;
- *management*: integrated forest management planning methods;
- *information systems*: planning and maintenance of appropriate integrated information systems;
- *production*: low-impact harvesting methods;
- *extension methods*: for example, the organization and training of forest owners, and communication;
- *planning*: methods for participatory planning, including stakeholders and cooperation with environmental, social and forestry organizations and authorities;
- *course of action*: certification procedures, with special reference to auditing and verification.

The traditional themes of timber production and silviculture also need to be included in further training, but must be adapted to meet the needs and expectations of responsible forestry standards and certification.

18.3 Institutional and organizational framework

The existing institutional frameworks, such as state forestry organizations and research and training institutes, will often have the necessary resources to provide training for different target groups. Where they do not, they should be helped to develop them. Reliance on the existing organizations, when feasible, strengthens them and contributes to the translation of academic concepts into local forestry practice.

The private forestry sector is developing in many countries as a result of the current tendency to transfer management rights from the government to private entities. Many forms of privatization are being applied, ranging from periodic concession rights to restitution of private ownership on public land. Private investment in forest plantation development is one of the driving forces in this process. In such situations, institutional support from government for training of staff, contractors and forest owners may well be needed.

Private forest owners and other operators will be able to implement good forestry on a large scale and profit from certification only if the level of awareness is high and they are represented and supported by well-established, effective organizations or associations. In many countries, such organizations are weak or lacking.

The current certification industry in the forestry sector is concentrated in a few developed countries. The market leaders are providing services worldwide. In contrast, others are confined to domestic or regional markets and selected countries. There is an urgent need to build up auditing capacity and institutions worldwide in order to broaden the resource base for assessment work.

18.4 Information systems

The traditional information systems in forestry have been quite comprehensive in providing data on timber resources. The concept of sustainable forest management is, however, broader and data collection for certification has to cover the whole range of economic, ecological and social parameters. Such information should be comprehensive, reliable and verifiable. Certification audits principally draw upon the information system of the audited organization; but general information on the forests of the region or country is also important in verifying many certification criteria, particularly those involving landscape-level aspects, such as biodiversity or water management.

The revision of information systems that is required typically includes the following phases:

- Defining the data needs for certification and for monitoring forest management and its implications in general.
- Identifying the data currently collected for forest inventories, surveys, enforcement and other monitoring systems.
- Assessing the quality and coverage of the data (reliability at different geographical scales and in different classes of forest ownership).
- Assessing how the current data collection meets the requirements for certification and forest management monitoring (how representative and reliable the data is and identification of gaps).
- Identifying development needs in data collection at a national level, including links between different data collection systems.
- Identifying technical solutions to be applied in information systems and data collection.
- Assigning resources and responsibilities.

The development of information systems and data collection should be designed so that the systems applied at different levels can be interlinked to the greatest possible extent. This requires a thorough consideration of the principles of data collection, such as sampling or forms of data, in the planning phase of information systems development.

New technologies, such as remote sensing and geographical information systems (GIS), together with geographical positioning systems (GPS), provide cost-efficient ways of improving the monitoring of forest management. The initial investment in new information systems is considerable; but operating costs can be reduced and the quality of data improved compared to using traditional practices. However, there are a variety of technical possibilities for such information systems, many of which are not compatible with each other. Therefore, considerable thought should be given to the possible need to combine information from different organizations when specific information systems are developed.

Certification audits could become one key user of new spatial information technologies in order to maximize benefits from the investment made in them. Remote sensing also has the potential to facilitate certification audits; but practical modalities will have to be developed separately in each case. In particular, surveillance audits could be rationalized and their quality improved through information generated by spatial information systems.

19 Small Forest Enterprises

Increasing attention is being given to the barriers faced by small forest owners and those managing forests at a low intensity when seeking certification. Although the precise definition of both of these categories has been elusive since it tends to vary from one country or region to another, there is, nevertheless, general agreement that there are particular problems that apply to both. These problems stem from two different sources:

1 difficulty in meeting the requirements of the standard;
2 costs and complexity of the certification process.

An additional problem for many managers of small forests and forests managed at a low intensity is in accessing and maintaining markets for certified products.

19.1 Meeting the standard

There are a number of issues that combine to make standards more challenging for small forests and forests managed at low intensity. These are related to the way in which standards are written and the need for interpretation in different situations.

19.1.1 LENGTH AND LANGUAGE

Forestry standards are frequently long and written in complex and formal language. This is difficult to avoid because sustainable forest management is a complicated activity and standards need to be comprehensive and written in precise language. Standards for sustainable forest management have to cover a large range of subjects, including legal, social, economic, and environmental issues, and should ensure that they can be applied to a variety of situations. As a result, the standards often run into dozens of pages.

This can create an immediate barrier for any forest manager who does not have formal forestry training, or simply does not have the time or confidence to spend reading, understanding and interpreting the meaning of a standard. Many of the managers of small forests and larger forests managed at a low intensity fall into this category.

19.1.2 INAPPROPRIATE REQUIREMENTS

Most standards are written to cover a range of forest types and situations. As a result, they need to include all of the requirements that must be met by large forest operations, with their significant impacts. However, many of these requirements are simply not relevant to small or low-management intensity forests, where the impacts are much less.

For example, one of the requirements of the Forest Stewardship Council (FSC) is that forest organizations should provide opportunities for employment and training for the local community: FSC principles and criteria (P&C), criterion 4.1. This is both relevant and reasonable for a company that employs staff or contractors, but inappropriate for a small forest enterprise, which is owner managed and operated.

Similarly, one of the frequent complaints about certification by small forest owners is the amount of documentation required. This is because most standards require detailed documentation of plans, procedures, activities and monitoring. While adequate documentation is important even in small forests, the amount of documentation that is adequate is generally very much less; but this is not always clear.

19.1.3 INTERPRETATION OF REQUIREMENTS

Forest management standards that are written in a generic form in order to be applied to a range of situations (and, sometimes, a range of countries) require interpretation before they can be applied on the ground. National forest management standards often go some way towards providing this interpretation, as they can be fine-tuned to local circumstances, making reference to the appropriate legislation and regulations or national guidelines. However, there is often a great deal of scope for further interpretation before it is clear how a standard should apply to a small forest or low-intensity management.

For example, the Brazilian FSC standard for the certification of forest management for timber in the lowlands of Brazil (September 2000, indicator 4.1.2) requires the forest manager to facilitate local communities' access to the forest for the use of timber and non-timber products, in accordance with the requirements of the forest law, its regulations and subsidiary and connected standards.

However, there is no interpretation provided on what the forest law, regulations and standards might require for small forests (or others), leaving it up to individual small forest owners to establish this for themselves.

In the case of group certification (see Chapter 10), this kind of interpretation may be provided by the group manager. For individual small forests, however, the uncertainty about what is required may be too daunting.

19.2 Costs and complexity of certification

The certification process that must be followed by a certification body (see Chapter 9) is usually defined by the certification scheme or the accreditation body. The cost of certification is affected by the complexity of the certification process, as well as by costs of reporting, reviewing and ensuring the availability of public information. Box 19.1 describes the main components that contribute to the costs of a certification assessment.

Many of these certification costs are fixed, which means that they are the same for a small or a large forest enterprise. For instance, report writing, peer review, translation and travel costs may be similar, regardless of the size or complexity of the forest being assessed. This means that forest certification can be disproportionately expensive for smaller enterprises and forests managed at a lower intensity. Without specific provisions, certification processes are likely to be either too expensive for small forests and forests managed at low intensity, or to be too superficial to ensure a good assessment of larger enterprises.

19.2.1 ADDRESSING THE PROBLEM

The barriers to forest certification that face small enterprises and forests managed at low intensity are increasingly being recognized.

Box 19.1 Cost components of a certification assessment

Pre-assessment	This is not required by all schemes. A pre-assessment visit can add considerably to certification costs, especially where the forest is remote, entailing significant travel costs for a relatively short visit.
Stakeholder consultation	Certification bodies may be required (depending upon the scheme) to carry out stakeholder consultation before an assessment. This can add costs related to evaluating and responding to questions and comments.
Main assessment	Costs of carrying out a main assessment depend upon the sampling intensity required (which may be defined by the certification scheme), transport costs and the complexity of forest management (requiring specialized team members).
Report writing	Reporting requirements differ between schemes; the more detailed the report, the more transparency it is likely to have, but the higher the cost of producing it.
Translations	Reports may be required in one or more specified languages. Where this is not the local language, translation may be required.
Peer review	Peer review increases the transparency and credibility of assessment, but adds a fixed cost to each assessment.
Corrective actions	Actions taken to address major corrective action requests (CARs) may need to be checked on the ground. This adds considerable extra travel and time costs.
Surveillance/monitoring	The frequency and sampling intensity with which certification bodies are required to carry out monitoring visits add to ongoing costs.
Re-assessment	Most certificates are valid for a fixed period, often five years. Reassessment may involve a full re-evaluation or may be similar to a surveillance visit.

The most widely used mechanism to try to overcome these barriers is through the development and use of some form of group certification scheme (see Chapter 10). Several certification schemes have introduced group or regional certification provisions, some of which are summarized in Box 19.2 (see also Chapter 13).

In addition to its group certification programme, the FSC has developed specific 'streamlined' certification procedures that aim to address some of the cost components of certification, specifically for a target group of small and/or low-intensity managed forests (SLIMFs). In order to ensure that the streamlined procedures are available only to target forests, the FSC has developed a definition of SLIMFs, as outlined in Box 19.3.

The revised procedures aim, in particular, to reduce the costs for pre-assessments, reporting and peer reviews, as well as the recurrent costs of monitoring.

Box 19.2

Certification scheme provisions for small forest enterprises (SFEs)

Scheme	Provision for SFEs in standards	Provision for SFEs in procedures
Forest Stewardship Council (FSC)	• Requirements for national initiatives to consider small and low-intensity managed forests (SLIMFs) in national standards come into effect in January 2005. Alternative indicators and means of verification must be developed for small and/or low-intensity managed forests for specific FSC criteria.	• Group certification SLIMF-streamlined procedures.
Programme for the Endorsement of Forest Certification (PEFC)	• Standards can be drafted for individual, group or regional certification. • There are no specific provisions in guidance for standard development; but national PEFC schemes are initiated by national forest owners' organizations or by national forestry-sector organizations, so there is often better representation of SFEs in standards development processes.	• Group certification. • Regional certification.
Finnish Forest Certification Scheme (FFCS)	• Responsibility for implementing the standard is split between regional management and the FMU level, reducing the number of things that small forest owners are expected to do.	• Group certification possible at two levels: 1 regional certification through Union of Forest Management Associations (UFMA); 2 group certification through local forest management association. • Individual forest owners can opt out.
Sustainable Forestry Initiative (SFI)	• There are no specific provisions, although *core indicators* may be substituted if they are not considered relevant, due to scale of operation. • A mutual recognition agreement exists with the American Tree Farm System (ATFS), designed for small non-industrial owners, which enables forests certified under ATFS to be considered as SFI certified.	• No charge for ATFS certification assessments. • ATFS initiated a pilot group certification programme in 2003.
Canadian Standards Association (CSA)	• There are no specific provisions. • The requirements for public participation in setting defined forest area (DFA)-specific values, objectives, indicators and targets could be difficult for an SFE.	• CSA operates on a DFA, which can be a combination of small units or non-contiguous operating areas; group certification is therefore possible.

Box 19.3

FSC criteria defining small and/or low-intensity managed forests (SLIMFs)

The Forest Stewardship Council's (FSC's) small and low-intensity managed forests (SLIMFs) eligibility criteria are intended to allow certification bodies to decide which forests are eligible to be assessed and monitored under streamlined certification procedures. The criteria may be modified at a national or sub-national level by FSC-accredited national initiatives.

Small forest

Forests of 100 hectares (ha) or less may be considered 'small'. This threshold can be increased, up to a limit of 1000ha, by an FSC-accredited national initiative.

Low-intensity management

Forests may be considered to be under 'low-intensity management' when the rate of harvesting is less than 20 per cent of the mean annual increment (MAI) of the total production forest area, *and*:

- either the annual harvest (from the total production forest) is less than 5000 cubic metres; or
- the *average* annual harvest (from the total production forest) is less than 5000 cubic metres over the five-year validity of a certificate.

National initiatives may set a lower rate of harvesting, or may use alternative measurements than the MAI, where this is not available.

Natural forests that are managed for non-timber forest products can also be regarded as under low-intensity management.

20 Selecting a Forest Certification Scheme

As discussed in Part Three, there are a growing number of different forest certification schemes. As a result, anyone in the forest products sector who wants to use certification – whether they are forest managers thinking about becoming certified, purchasers wanting to buy certified products, investors and donors financing forest management, or governments and non-governmental organizations (NGOs) judging the quality of forest management – is faced with the question of which scheme or schemes to use.

This has become a major debate wherever forest certification is discussed or utilized, and one that remains unresolved. This chapter analyses some of the attempts to assess the acceptability of forest certification schemes, reviews some of the tools or approaches that now exist and examines possible options for those faced with the need to chose or assess schemes.

20.1 Existing assessments of forest certification schemes

The Intergovernmental Panel on Forests (IPF) as far back as 1996 (IPF, 1997) concluded that countries should support the following concepts with respect to certification schemes:

- open access and non-discrimination regarding all types of forests, forest owners and operators;
- credibility;
- sincerity;
- cost-effectiveness;
- participation that seeks to involve all interested parties, including local communities;
- transparency in sustainable forest management.

While it is unlikely that any group would disagree with this list, these are very general concepts and do not provide sufficient detail to allow the assessment of certification schemes. Therefore, a range of different organizations and groups have sought to produce their own, more detailed set of criteria for assessing schemes.

In 1997, the Dutch government produced a set of minimum requirements for timber from 'sustainably managed' forests to be eligible for a label on the Dutch market (Dutch Ministry of Agriculture, Management and Fisheries, 1997). These criteria were used as the basis for the Dutch *Keurhout* timber-labelling scheme (which ceased operations in 2003). During 2003, the Dutch government began to revise the criteria with a new version developed for release in 2004.

During 2000 and 2001, five different initiatives proposed criteria sets for global use:

1 Deutsche Gesellschaft für Technische Zusammenarbeit GmbH (GTZ) *Forest Certification Project: Working Paper 2* (Vallejo and Hauselmann, 2000) presents principles, criteria and indicators for assessing the effectiveness of forest certification systems in contributing to sustainable development. The document summarizes and groups relevant hard and soft law, internationally accepted guidelines for standardization, accreditation and certification, and civil society aspirations expressed by representative and non-governmental organizations (NGOs).

2 The Confederation of European Paper Industries' (CEPI's) Comparative Matrix (CEPI, 2000) sets out criteria and indicators for comparing international and national forest certification schemes and presents information collected from the schemes operating in 2000. The document does not state the rationale behind the proposed indicators. Other important limitations to the matrix, which CEPI acknowledged, included:
   - little information on the actual content of forestry performance standards;
   - little indication of a scheme's relative effectiveness and efficiency in actually promoting sustainable forest management on the ground.

   Taking into account a range of comments on the first version of the matrix, CEPI plan to publish a revised version by the end of 2004.
3 A report published by the Australian Department of Agriculture, Fisheries and Forestry (Australian DAFF, 2000) proposes critical elements and potential performance measures for assessing forest management certification schemes and provides a preliminary assessment of existing comparability and equivalence initiatives and certification schemes against the proposed critical elements. The report was prepared for the Australian government to assist their strategic planning and, where appropriate, to inform the ongoing international debate by interested parties about the further development of forest certification.
4 The International Forest Industries Round Table (IFIR) proposed criteria and indicators for credible sustainable forest management (SFM) standards and certification systems in the context of its proposal for an international mutual recognition framework (IFIR, 2001). As with the CEPI matrix, the rationale for the proposed criteria and indicators is not stated and IFIR has not taken the criteria further due to lack of progress in structural recognition between schemes.
5 The environmental non-governmental organization (ENGO) Fern produced a comparative analysis of four certification schemes based on a set of ENGO objectives and criteria. This was followed by a second comparison looking at eight schemes in 2004 (Fern, 2001a, 2004).

There have been a number of other initiatives since 2000. For example:

- During 2000, the World Bank–World Wide Fund for Nature (WWF) Alliance for Forest Conservation and Sustainable Use published a guidance note for its target of improving forest management and certification, setting out 11 criteria for determining credible forest certification systems (World Bank–WWF Alliance, 2000). Based on these, the alliance has prepared a tool for assessing schemes to see whether they meet their criteria, the 'Questionnaire for Assessing the Comprehensiveness of Certification Schemes/Systems' (World Bank–WWF Alliance, 2003). This will be field tested during 2004 and will be used by the alliance in identifying schemes that contribute to its certification target of 200 million hectares, and by the World Bank to guide forestry and forest products investments.
- In 2004, the UK government undertook a process to assess which forest certification schemes deliver its public procurement objectives of purchasing wood and paper products from legal and sustainable sources.

However, none of these initiatives has so far proved able to form the basis for general international dialogue. Each is supported by certain groups and is criticized by others. An analysis of the debate reveals that there are two major reasons why the existence of a number of sets of criteria and comparative analyses has not resolved the debate:

- *Differing objectives*: different interest groups have different goals or objectives which they want a certification scheme to deliver. For each of these different goals, there is a distinct set of attributes that a scheme needs to have, and although the sets are often similar, they are not the same. Therefore, a 'criteria set' that ensures delivery of one group's goals may not be able to deliver the goals of another:

- *Insufficient detail*: many of the sets of criteria and the analyses carried out aim to be straightforward, avoiding too much detail. But, as discussed in Part One, it is the detail in the way that the schemes work which influences the outcome; therefore, a general analysis is often inadequate.

20.2 Developing criteria for assessing forest certification schemes

As noted above, any tool that is to provide an adequate basis for international discussion about certification schemes needs to recognize the importance of identifying a scheme's objectives and must be based on sufficient technical detail. Therefore, any methodology for assessing forest certification schemes must involve the following (Nussbaum et al, 2001):

- *Identify objectives*: firstly, the goals or objectives that an acceptable scheme should deliver need to be clearly identified.

> For example, if the main concern of a purchaser is to ensure that all timber purchased is from legal sources, then this should be an identified objective.

- *Link the objectives to the scheme's characteristics*: the elements or features of a scheme that will be critical in delivering the desired objectives need to be clearly identified; thought must be given to how these elements are designed in order to deliver objectives.

> For example, if legality is an objective, then consideration will need to be given to whether the standard requires legality, whether the certification process checks it is delivered in practice and whether chain-of-custody controls are adequate to ensure that certified only product contains legally sourced material.

- *Develop criteria*: finally, practical criteria need to be developed that allow an assessment to be made of whether the design of a scheme is adequate or not.

This process, which is shown schematically in Figure 20.1, should result in a set of clearly justified criteria against which any candidate certification scheme can be assessed. Each stage in the process is discussed below.

20.2.1 ESTABLISHING OBJECTIVES

Everyone who uses, or is thinking of using, a certification scheme has one or more objectives. Some of these are obvious, some less so. Whenever an assessment of a certification scheme is carried out, it is always against a number of objectives, whether these are explicit or not. Problems tend to arise when objectives are not clearly identified since it becomes difficult to justify the reasoning underlying an assessment.

Therefore, the first step in developing criteria for carrying out an assessment of forest certification schemes is to establish what the user's objectives are.

Box 20.1 provides examples of the objectives of current users of forest certification schemes. Even this short and unsystematic list of objectives collected from reports, presentations and discussions with a range of stakeholders raises two key issues, each of which is discussed below:

1 Some objectives are more directly related to certification scheme design than others.
2 Many users will have more than one objective, and some objectives may conflict with each other.

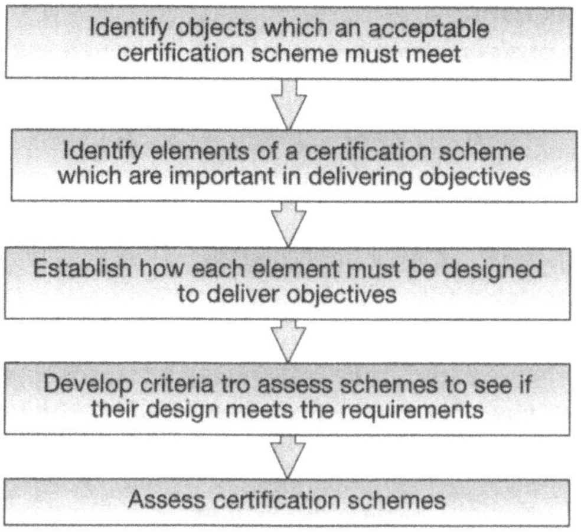

FIGURE 20.1 **The process required to develop and use a methodology to assess whether a certification scheme will deliver a particular set of objectives**

Box 20.1

Examples of the goals and objectives of a range of actual and potential users of forest certification schemes

The goals and objectives set out below have been collected from a range of sources, including reports, presentations, policy documents and discussions:

- We need better market access.
- Our customers demand certified products.
- In order to meet our environmental policy we want to ensure that the products we sell are not contributing to unsustainable forestry.
- Our organization works with wildlife, so for us certification means conservation of rare species.
- I want a scheme that has clear, straightforward requirements and reasonable costs so that I can undergo the process of certification and then get back to doing business.
- The scheme must be accessible to small forest owners.
- As an ethical investment company/government aid agency, we want confirmation that our investment/aid is resulting in sustainable development.
- We want recognition and protection of our right to use the forest as our ancestors have used it for generations.
- As a responsible trading company, we do not wish to buy timber from illegal sources.
- We want a simple message to communicate to our ethically concerned customers.
- We want to be sure that our company will not be the object of an environmental campaign.
- As an international company, if certification is to be useful to us, it must be global and business-oriented.
- We want the scheme that we use to be the most widely accepted.
- Our organization needs to be sure that timber harvesting does not harm critical forest ecosystems or their biodiversity.

Source: Nussbaum et al (2001)

20.2.1.1 Types of objectives

There are two main types of objectives that forest certification scheme users are likely to have:

1 those linked directly to certification scheme design; and
2 those with only an indirect link to certification design.

Objectives that link directly to certification scheme design

These objectives can be further subdivided into objectives that are directly linked to one or two key aspects of the design of the scheme, and objectives that are more complex and require the interaction of several elements of a scheme.

Some objectives link very clearly to the way in which the scheme is designed; therefore, it is relatively easy to establish what elements of the scheme influence them, and the way in which these elements must be designed in order to deliver the objective.

For example, if a key objective is that a scheme must protect and conserve forests of very high biodiversity value, it is essential that there is a requirement for this level of conservation in the standard.

However, in general, the link between the objective and the scheme is more complex and requires that a number of elements are in place in order to ensure that a particular objective is met.

For example, if a scheme is to deliver assurance that products do not come from unsustainably managed forests, this requires a combination of:

- an adequate standard to define good forest management;
- an effective certification process which ensures that forests really meet the standard;
- a reliable process for tracking products from the forest through to the final claim, which ensures that material from uncertified forests cannot enter the product.

In this case, it will be necessary to analyse the way in which each candidate certification scheme addresses all of these needs in order to decide which schemes are adequate.

Objectives that do not link directly to certification scheme design

In addition to objectives that are linked directly to the design of the certification scheme, it is extremely important to be aware that objectives also exist which exhibit no direct link. These are objectives that depend upon other users' perceptions. The two most important objectives that belong to this category are:

1 *Market access*: certification will only provide market access if the scheme chosen is accepted by the marketplace. Certification on its own – however good the certification scheme chosen – will not provide market access unless the market is interested in products certified under that particular scheme.
2 *Managing the risk of negative campaigns*: one of the reasons that many retail companies become involved in forest certification is to avoid the risk of negative campaigns by environmental or social pressure groups. As with market access, in this case it is more important to understand which schemes the campaign groups recognize, than to carry out a systematic assessment of different schemes.

Although there is no direct link, in these cases, between the objectives and the design of the scheme, there is likely to be an indirect link since the other users will probably have their own objectives by which they judge schemes. As a result, it is important to understand what these objectives are; but it is also important to be clear that the delivery of the main objective does not depend directly upon the scheme.

20.2.1.2 Conflicting objectives

Most users have a range of objectives that they want a scheme to fulfil. This is not problematic since schemes can easily meet a number of objectives. However, there is a problem when two or more of the desired objectives conflict with each other. The most common examples of conflicting objectives are:

- *Cost versus just about everything else*: this is one of the key issues that needs to be debated since every additional requirement in a scheme has potential cost implications, and increasing cost, in turn, has implications for competitiveness, sustainability and equity.
- *Equity of access versus the need for rigorous standards*: while it is widely accepted that certification schemes should be accessible to organizations with very limited resources, such as small forest enterprises, community forests and forests in developing countries, there is also a desire to maintain high standards and a rigorous certification process in order to ensure that only well-managed forests can gain access to certification.
- *Confidentiality versus transparency*: an important mechanism for building credibility within a certification system is through requirements for transparency. However, for many companies there is a risk associated with making too much information available since competitors or buyers may be able to use this information for their own benefit.

Much of the debate about forest certification schemes over recent years appears to be the result of different approaches to dealing with conflicting objectives. Therefore, it is very important to identify objectives clearly and to analyse where compatibilities and conflicts occur.

Once conflicting objectives have been identified, it is helpful to prioritize. This is done by identifying those objectives that are essential and without which certification is no longer useful, and by separating them from objectives where some degree of compromise may be possible.

If it is necessary to justify a particular assessment of schemes to other parties, it is useful to be clear about where compromises have been made, and why, in order to allow others to judge whether assessment decisions are acceptable to them.

20.2.1.3 Defining objectives for a specific user

As discussed above, the first stage in carrying out an assessment of forest certification schemes is to establish the objectives of the user carrying out the assessment. This may be relatively simple if objectives are already documented, such as in policy documents.

For example, the Group of 8 (G8) governments have documented policy commitments to:

- the principle of sustainable development through Agenda 21 and related initiatives;
- multi-stakeholder consensus-based processes for developing forest policy and standards, as agreed at the recent United Nations Forum on Forests (UNFF) session;
- measures to oppose illegal logging, as set out in the G8 foreign ministers' 1998 Action Programme for Forests.

Therefore, an analysis of certification schemes carried out by a G8 government might be based on the objectives that the scheme must:

- Contribute to and support sustainable development.[1]
- Require multi-stakeholder, consensus-based standard setting.
- Provide a mechanism for ensuring exclusion of timber from illegal sources.

Similarly, a company or industry association that carries out an assessment is likely to have a number of objectives which are more or less defined in their reports, policy documents or commitments to shareholders.

It may be more difficult to establish objectives if no prior thought has been given to the issue. In this case, it will be necessary for the organization to spend time deciding what its objectives are. This is an essential prerequisite for analysing certification schemes, and is also likely to be a useful and worthwhile internal clarification process.

20.2.2 LINKING OBJECTIVES TO SCHEME DESIGN

Linking objectives to scheme design is probably the most complicated aspect of the process of preparing to assess certification schemes and involves three stages, as shown in Figure 20.1:

1 Identifying which elements are important in ensuring that a particular objective is delivered.
2 Establishing how each identified element must be designed in order to deliver the desired outcome.
3 Developing criteria to assess the adequacy of the elements identified.

20.2.2.1 Identifying important elements

For each of the objectives identified, it is necessary to consider which elements of a certification scheme will be the most important in order to ensure that the objective is delivered in practice. For each objective, it is sensible to consider all four components of a scheme:

1 the standard, including the way in which it is developed and the content (Chapter 3);
2 the certification process, including technical requirements and mechanisms in order to ensure credibility (see Chapter 4);
3 accreditation, including requirements and credibility (see Chapter 5);
4 control of claims, including chain of custody and labelling (see Chapter 6).

For some objectives, there may be only a few key elements, while for others the entire system may be important. In the latter case, it is still useful to identify the most important elements since this will help to prioritize where there is most focus and least flexibility in assessing the compliance of schemes.

20.2.2.2 Establishing how each element must be designed

The review of existing initiatives to assess certification schemes outlined in 'Existing assessments of forest certification schemes' suggests that most of the difficulties arise at this stage of the process. Often, agreement exists between different interest groups about which elements of a scheme are important – so it is easy to reach a consensus about broad principles. The disagreement arises about how to design a particular identified element in order to deliver the required outcomes.

For example, there may be broad agreement that standard-setting should involve a range of stakeholders. However, there is a big difference between a standard-setting process that makes use of inputs from different stakeholders and one which ensures that each stakeholder group has equal rights and influence, though both could be described as 'multi-stakeholder processes'.

Similarly, there is general agreement that standards should include conservation of biodiversity. However, there is a difference between a standard that requires the destruction of biodiversity to be minimized and one that requires active protection and enhancement of biodiversity – though both address 'biodiversity conservation'.

Therefore, it is important to be precise at this stage about exactly how each key element must be designed in order to deliver the required output. The review of certification scheme design provided in Part One of this handbook provides the level of detail required for this type of analysis.

20.2.2.3 Developing criteria for assessment

Once the necessary design of each element has been established, it should then be relatively straightforward for users of certification to develop criteria that can be used to assess whether:

- a certification scheme contains the elements that have been identified as being important;
- the design of each element can deliver the desired outcomes (or objectives).

It is important that the criteria developed are sufficiently precise to be interpreted in the same way by different users. If the process outlined above has been followed, it is likely that the list of criteria will be fairly long. Therefore, there is always a temptation to try to shorten and simplify it. However, this type of simplification may result in a set of criteria that are interpreted differently by various interested parties; as a result, they may no longer provide a basis for common understanding.

At the end of this process, there should be a clear, objective set of criteria that can be clearly explained and justified, and which provide a sound basis for assessing the adequacy of schemes.

20.3 Legitimacy Thresholds Model approach

An alternative approach that has been proposed for assessing certification schemes is the Legitimacy Thresholds Model (LTM) approach (WBCSD, 2003), developed by the World Business Council for Sustainable Development (WBCSD) in response to discussions convened by the Forests Dialogue (Forests Dialogue, 2004).

This model is based on the premise that any assessment of schemes can only be undertaken once it is clear what is required for a scheme to be legitimate to a particular stakeholder group. These legitimacy requirements provide the thresholds for comparison. In practice, this is very similar to the identification of objectives discussed above; but the LTM adds a useful graphic element, as well as a clearly defined context.

The model, which is shown schematically in Figure 20.2, has a vertical axis consisting of criteria or attributes, and measures of legitimate certification schemes. A range of transparent 'thresholds' or benchmarks of legitimacy or credibility are then agreed by relevant stakeholder groups – for example, customers, governments, NGOs and industry. These are represented by the three lines, T1, T2 and T3.

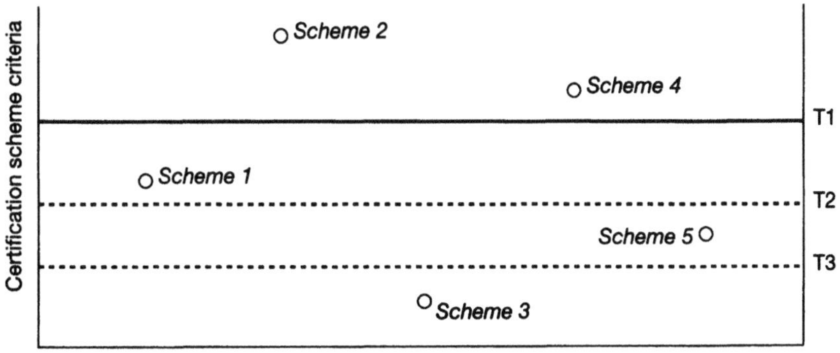

FIGURE 20.2 The Legitimacy Thresholds Model, showing three thresholds and the position of five schemes relative to the thresholds

Schemes (identified in the horizontal axis) can then be assessed against the various thresholds to provide a clear and transparent way of deciding which are legitimate for a particular stakeholder group.

For example, the three thresholds shown could be defined as:

- T1: a 'high' threshold that certification systems must meet in order to be considered by a defined stakeholder group, such as customers for procurement, as delivering sustainable forest management. This might comprise schemes that adequately give effect to one of the recognized international sustainable forest management (SFM) criteria and indicator sets, have been developed via a multi-stakeholder process, are subject to credible third-party verification and provide for chain of custody.
- T2: an agreed 'mid-level' threshold that could, for instance, define the requirements which certification schemes used by small-scale forest owners or developing country suppliers must meet in order to be considered acceptable by a defined group of stakeholders. This threshold could involve a more limited number of attributes compared to T1, reflecting the different capabilities of small-scale forest owners or community forestry operations.
- T3: an agreed 'minimum' threshold, such as legally sourced wood, which must be delivered by a certification scheme before it receives any recognition by a defined group of stakeholders.

In Figure 20.2, schemes 2 and 4 would both deliver credible certification of sustainable forest management, while scheme 1 might be accepted for small forest enterprises or developing countries. Scheme 5 would deliver legally sourced wood only. Scheme 3 did not deliver anything.

The success of the LTM approach will depend upon the ability of stakeholders to:

- *Define thresholds that have an adequate degree of legitimacy*: there are various potential approaches. At one extreme, each stakeholder group could develop and use their own definition, while, at the other, an attempt could be made to obtain wide agreement on thresholds. A compromise between these two may be to reach broad agreement on a number of thresholds, but maintain differences about which of the thresholds is the minimum one for an acceptable scheme.

For instance, in the example above, some stakeholder groups may decide that anything that is legal (T3) is adequate, while other groups support only those schemes that deliver SFM (T1).

While this latter approach would not provide agreement, it might provide a much clearer basis for debate about disagreements.

- *Agree a practical and reliable methodology for assessing certification schemes*: it will also be necessary to develop a methodology to assess schemes. This will need to be sufficiently transparent and rigorous to ensure credibility with all stakeholder groups involved.
- *Identify an agency to undertake assessments*: ideally, an independent agency, recognized and respected by stakeholder groups, will be used to benchmark schemes, employing the agreed methodologies and thresholds.

20.4 Conclusions

It is very likely that there will be an increase in the number of forest certification schemes operating in the short or medium term. Therefore, mechanisms for assessing forest certification schemes in order to allow different groups to choose the ones they want to use will continue to be necessary for some time.

Options available include:

- Using a set of existing criteria developed by various initiatives, such as the Confederation of European Paper Industries (CEPI), the NGO Fern, the International Forest Industries Round Table (IFIR) and the World Bank–WWF Alliance, or a combination of them.
- Developing an agreed set of criteria using the approach outlined in 'Legitimacy Thresholds Model approach' and discussed in detail in Nussbaum et al (2002).
- Using an approach such as LTM, outlined in 'Legitimacy Thresholds Model approach', and discussed in detail in WBCSD (2003).

Note

1 This approach has already been taken in the criteria for assessing certification schemes developed for the German development agency GTZ (GTZ, 2000).

21 New Applications of Forest Certification

The potential of forestry markets to provide a mechanism for raising revenue in order to finance sustainable management has recently received increasing attention (Landell-Mills and Porras, 2002; Pagiola et al, 2002; Scherr et al, 2004). These markets are still incipient but emerging, particularly in carbon sequestration, in which the trade potential will be largest (Katila and Puustjärvi, 2003). Other services, such as biodiversity conservation or water and soil conservation, also offer significant opportunities. Selling an environmental service raises the issue of verification to justify the respective payment. For this reason, forest certification potentially has an important role to play in these emerging markets. This is discussed further below, firstly for carbon and then for other environmental services.

Providing a mechanism for reducing the risks of investment or increasing access to finance also comprise an important and growing use of forest certification, which is examined in more detail in this chapter.

Finally, the potential for the lessons, experience and know-how accumulated through a decade of forest certification to contribute to the development of certification for other natural resource sectors is also explored.

21.1 Carbon sinks

Independent verification or certification of carbon sinks will be relevant to implementing the flexible mechanisms of the Kyoto Protocol. Carbon sinks that occur as a result of afforestation and reforestation[1] qualify as a climate change mitigation measure under the Clean Development Mechanism (CDM) and joint implementation (Auckland et al, 2002).

Certification of sinks and certification of sustainable forest management have a number of potential synergies (see Box 21.1). Linking the two instruments, or applying them in parallel in a coordinated way, could reduce costs and help to position carbon sink maintenance and enhancement within the framework of responsible forest management, even though separate audit protocols would have to be applied.

21.1.1 GENERAL APPROACH

General procedures of existing accreditation bodies for ISO 9000 and the ISO 14000 series standards or products are also applicable for forest and sink verification. Small-scale private forest ownership represents a challenge for both types of audits. Appropriate group certification approaches, which are already in place for forest certification, can help to address this issue.

21.1.2 VERIFIED INFORMATION

Forest certification verifies that a forest is under management, but it does not directly verify carbon stocks and flows in the forest. However, forest certification can verify land-use changes and changes in the growing stock. It can also verify the implementation of positive and negative management measures that exert an influence on carbon sinks.

Under the Clean Development Mechanism, where sinks are provided by afforestation and reforestation projects, socio-environmental impacts within and outside of the project boundary must

Box 21.1

Synergies between certification of forest management and carbon sinks

Areas of potential synergies between certification of forest management and carbon sequestration may include:

- whether common methodologies, definitions and concepts can be developed;
- the building of capacity, which is required for both instruments;
- whether forest management certification and other management tools can contribute towards the preparation of accurate inventories through providing data relating to land-use changes and changes in the growing stock;
- whether forest management certification, if further developed, may verify the implementation of measures or the lack of measures, both positive and negative, that affect sinks;
- whether auditing procedures could be complementary for forest management and sinks certification, even if both instruments require separate protocols and accreditation;
- whether general procedures of existing accreditation bodies (for example, for ISO 9000 and the ISO 14000 series standards) could also be applicable – just as for forest management procedures – for sinks validation, verification and certification systems, after having been augmented to specifically deal with sinks projects;
- how group certification may reduce barriers (such as costs), for individual (small) forest owners, to implement forest management certification schemes and facilitate the implementation of (bundled) sinks activities;
- whether, and to what extent, any sinks credit return may provide additional financial support to the private sector for also implementing forest management certification schemes (for example, cap management), or, the reverse, where forest management certification may give added value and marketing advantages to carbon sequestration.

be analysed. If they are significant, they must be formally assessed according to host country procedures. Forest management certification includes assessment and verification against social and environmental criteria, providing both baseline information and ongoing monitoring of socio-environmental impacts.

In addition, forest certification has the potential to assist in addressing issues that are related to sinks and their verification, such as:

- *Permanence*: commitment at either the level of a forest management unit (FMU) or a group can be arranged to maintain or increase carbon storage as part of the goals of certification.
- *Leakage*: this refers to the production of any greenhouse gases outside of the project boundary which are measurable and attributable to the project. This leakage has to be subtracted from project sequestration figures. Leakage may be due to, *inter alia*, deforestation or excessive harvesting in neighbouring areas of the project if induced by the project's intervention. Forest certification can provide information on such impacts.
- *Additionality*: forest certification could offer a possible baseline for managed forests to help identify whether they will qualify for the CDM in the future.
- *Uncertainty and risks*: there are a number of synergies. Accurate inventories benefit both forest certification and carbon verification; responsible forestry standards address fire and other damage

relevant to sequestration; regional or group certification represents a lower risk level for carbon projects than individual holdings if they are small.
- *Biodiversity impacts*: these are a concern for sink enhancement and are incorporated within certification standards.
- *Transaction costs*: economies of scale can be achieved through combined auditing; group certification may provide a mechanism for allowing smallholders to access sink benefits.

Finally, another potential area of synergy can be identified in harvested wood products (accounting for carbon storage/flows). The chain-of-custody certification, which is part of forest certification-based labelling, verifies the origin of traded products. This could help to provide reliable inventory data on carbon in harvested wood products, and it could also enable the joint labelling of the sustainability and carbon neutrality of products (including bio-energy-based electricity and heat). However, the rules for accounting carbon in harvested wood products have not yet been agreed upon, and this remains an area for future exploration.

One risk is that verification or certification of carbon sequestration will add to the transaction costs and, consequently, will reduce the market opportunities. One way of reducing the transaction costs is to link certification of carbon sequestration by forests with existing forest certification schemes. Two existing forest certifiers, SGS and Smartwood, have developed their own certification schemes to independently verify the existence of sinks and the amount of carbon sequestered.

21.2 Other environmental services

Independent verification of forests' environmental services, other than carbon sequestration, is a potentially powerful tool to raise new sources of funding for forest management. Such services can include a range of possibilities, such as water retention, erosion control or protection of biodiversity. In this case, the certification or verification activity could be market based, regulatory based or project based, and therefore is often likely to be undertaken in conjunction with performance-based certification of sustainable forest management.

The focus of any specific verification must be on establishing a baseline and measuring the project's impact upon the selected outputs. It will probably often also be market oriented, verifying the quantity and quality of services produced in a way that can help to convert the environmental benefits of forests into marketable services.

Certification or verification of environmental services is still in the initial stages of conceptual development; but it seems likely that it will progress relatively rapidly. Costa Rica has been spearheading the process, and other countries are expected to follow.

If verification of the quality of forest management and the quantity of environmental services provided is needed for the same forest area, assessments should preferably be combined to make them cost efficient. In many sets of regional criteria and indicators (C&I), the maintenance of forest cover of adequate quality is used as a proxy for providing environmental services in terms of water and biodiversity. This can be considered, initially, as a feasible way of keeping verification costs low. In the longer run, more specific or refined indicators are likely to be needed.

The driving force for developing a market for forest-related environmental services will be the need to clearly channel the payments made to those who verifiably produce the desired services. Therefore, the current practice prevalent in many countries – where the funds collected from the use of natural resources through taxes or fees are mostly channelled to public authorities or their projects – may no longer be adequate as the main delivery mechanism in the future.

Independent third-party certification could be especially helpful for land managers who require public confidence and credibility when dealing with high conservation values or where payments for services call for transparency and accountability. Private investors or other environmental service beneficiaries will want to know that they will get for the price that they pay.

In the US, both government (tax payers') money and private funds through non-governmental organizations (NGOs) are being used to establish conservation easements to protect outdoor recreation sites and valuable habitats, such as wetlands. In addition, considerable tax advantages are provided to landowners to promote conservation easements. There is, thus, considerable interest in verifying that the conservation easement areas are actually managed and protected so that the desired environmental services are provided in perpetuity and according to the contract (in order to meet specified ecological goals).

With regard to other approaches, such as international transfer payments for environmental services (for example, related to climate change and biodiversity issues) that use mechanisms such as the Global Environment Facility (GEF), it is also important to demonstrate that these funds are being employed to finance *incremental global benefits* that the country in question would not otherwise produce in an agreed baseline situation.

Occasionally, improvements in service delivery – for example, for water services – are sought that will require quantification of changes in the service quantity and quality against the baseline. It is important to establish a strong scientific basis for linking forest and other land management practices to such services, as well as to develop cost-effective and reliable ways of verifying their delivery. In the case of biodiversity conservation, the prevailing level of biodiversity should be maintained or enhanced; this involves comparison with a baseline scenario of biodiversity reduction in the absence of project intervention. The question is: how much biodiversity would have been 'delivered' in the absence of a specific market-based measure to ensure the maintenance of the service? The 'additionality' requirement is partly related to the issue of incremental (marginal) environmental values, which can sometimes be small in relation to incurred costs, including the opportunity costs of forgoing alternative land use (Katila and Puustjärvi, 2003).

21.3 Risk mitigation and access to finance

Forest investments, particularly in natural forest management, are notoriously perceived to be high-risk ventures due to their inherent nature and lack of sectoral knowledge among potential investors (Moura Costa et al, 1999). On the other hand, there is a growing interest among the financing community to diversify their portfolios and prioritize 'Green investments'. The emergence of the sustainability assessment of companies' publicly quoted shares will also place new requirements upon these companies for information on resource use and environmental and social impacts.

Forest certification can be used as a tool to reduce the exposure of forest industries and their customers to environmental and social risks and liabilities. For example, poor forest management can cause damage to hill slopes, trigger mudslides, pollute local streams and rivers and diminish fisheries. This can leave companies open to legal action by local people, and may require costly remedial action.

Potential benefits for forest managers include reduced financing and insurance costs due to lower risk profiles, pre-empting increasing regulations and reducing the liability exposure of the board of directors (Crossley and Points, 1998). Certification is obviously a safeguard measure against publicity risks, as well.

There is a growing trend of making access to international development finance conditional upon certification. For example, the recent World Bank forest strategy has adopted certification as a useful safeguard instrument for risk mitigation (see Box 21.2).

The US-based Overseas Private Investment Corporation (OPIC) is another example of a financing institution that identifies certification as a requirement for forest investments in which it can participate. More recently, a number of European NGOs have been calling for a review of the criteria applied by export credit institutions in the forest sector. These, together with many investment funds, are examples of early action to use certification as a risk mitigation instrument in the financing community. The 'bad publicity' risk is likely to encourage expanded use of certification as a tool to manage environmental and social impacts of internationally operating forest industry corporations.

Box 21.2 **World Bank forest strategy on certification**

The World Bank has accepted the principle of independent monitoring of forest operations. However, the bank has not endorsed any particular certification system, but will assess particular approaches in relation to their compliance with the bank's principles and criteria. The World Bank recognizes the ongoing 'mutual recognition' debate in the international community to harmonize acceptable standards and approaches that these principles and criteria will contribute to discussions.

Planning is needed to produce equitable outcomes and raise the overall social value of forests. In some cases, the need for greater transparency and accountability at the local level will require the use of stakeholder assessment as an alternative to third-party assessment of commercial-scale operations.

The World Bank will encourage national governments to develop standards for natural forest management and forest restoration that are both locally relevant and meet internationally recognized principles and criteria for sustainable forest management (SFM). The institution will also provide support to national governments to create representative, multi-stakeholder, independent forest-monitoring bodies.

Source: World Bank (2003)

21.4 Sustainability certification in other sectors

The forest sector was the first major natural resources sector to embrace the concept of certifying sustainable management. While other sectors have addressed some aspects of sustainability (for example, fair trade for coffee and chocolate and organic production for a range of agricultural commodities), forest certification remains the global leader in the development of certification that addresses the combination of economic, social and environmental factors needed to deliver sustainability.

There is now growing interest in developing standards or certification schemes for other natural resource sectors (for example, oil palm, cotton and soy). The experience from the forestry sector could be invaluable to the progress of these newly emerging initiatives in a number of ways:

- *Requirements of standards*: although each sector will need its own standard or set of criteria, there is likely to be considerable overlap in the range of issues that must be addressed, particularly in relation to social and environmental requirements. Therefore, forestry standards can be used as a source of information for the development of criteria or standards for other natural resource sectors.
- *Standard development process*: all of the lessons learned by the forest sector about the importance of multi-stakeholder, transparent processes in developing standards are equally applicable to other natural resource sectors. Using a similar process is likely to result in the development of widely supported and credible standards.

In addition, forestry has pioneered the development of international standards that contain requirements which must then undergo national interpretation. This is likely to be equally necessary for other natural resource sectors, which can draw on the processes developed for forest standards.

- *Accreditation and certification*: the mechanisms, procedures and issues related to accreditation and certification of forests are equally applicable to other sectors. Therefore, all of the lessons learned

Box 19.2

Certification scheme provisions for small forest enterprises (SFEs)

Scheme	Provision for SFEs in standards	Provision for SFEs in procedures
Forest Stewardship Council (FSC)	• Requirements for national initiatives to consider small and low-intensity managed forests (SLIMFs) in national standards come into effect in January 2005. Alternative indicators and means of verification must be developed for small and/or low-intensity managed forests for specific FSC criteria.	• Group certification SLIMF-streamlined procedures.
Programme for the Endorsement of Forest Certification (PEFC)	• Standards can be drafted for individual, group or regional certification. • There are no specific provisions in guidance for standard development; but national PEFC schemes are initiated by national forest owners' organizations or by national forestry-sector organizations, so there is often better representation of SFEs in standards development processes.	• Group certification. • Regional certification.
Finnish Forest Certification Scheme (FFCS)	• Responsibility for implementing the standard is split between regional management and the FMU level, reducing the number of things that small forest owners are expected to do.	• Group certification possible at two levels: 1 regional certification through Union of Forest Management Associations (UFMA); 2 group certification through local forest management association. • Individual forest owners can opt out.
Sustainable Forestry Initiative (SFI)	• There are no specific provisions, although *core indicators* may be substituted if they are not considered relevant, due to scale of operation. • A mutual recognition agreement exists with the American Tree Farm System (ATFS), designed for small non-industrial owners, which enables forests certified under ATFS to be considered as SFI certified.	• No charge for ATFS certification assessments. • ATFS initiated a pilot group certification programme in 2003.
Canadian Standards Association (CSA)	• There are no specific provisions. • The requirements for public participation in setting defined forest area (DFA)-specific values, objectives, indicators and targets could be difficult for an SFE.	• CSA operates on a DFA, which can be a combination of small units or non-contiguous operating areas; group certification is therefore possible.

Appendix 1
Overview of Existing Schemes

Introduction

As discussed in Chapter 13, in the decade since forest certification first appeared, a number of certification schemes have been developed. This appendix provides detailed information on seven of the main schemes. These are:

1 CertforChile: national certification scheme in Chile;
2 Canadian Standards Association (CSA): Canada's national scheme for sustainable forest management;
3 Forest Stewardship Council (FSC) Scheme;
4 Lembaga Ekolabel Indonesia (LEI): Indonesian sustainable production forest management certification scheme;
5 Malaysian Timber Certification Council (MTCC) Scheme;
6 Programme for the Endorsement of Forest Certification (PEFC) Schemes;
7 North American Sustainable Forestry Initiative (SFI).

The information has been provided by the schemes themselves, giving a unique insight into each scheme as it is seen by those involved in developing and running them. A comparative summary of the main features of each of the seven schemes can be found in Chapter 13.

For each of the schemes, managers were asked to provide details of the following.

TYPE OF SCHEME

There is a wide range of approaches to forest certification schemes, so managers were asked to begin by identifying whether the scheme is a national or regional standard, an international initiative, or a programme for mutual recognition of different schemes.

SCOPE

Something that is often unclear in the discussion about certification schemes is the scope of each scheme. Some schemes can be applied anywhere, while others can only be used in a particular country or region. There are also some schemes that are applicable to all forest types, while others are restricted to one type of forest.

DATE SET UP AND HISTORY

It is often useful in understanding a scheme to know something about its history. How did it begin, who was initially involved and how did it develop? What have been the major successes, challenges or changes over the years?

STRUCTURE AND GOVERNANCE

Who owns and administers the scheme? Is it the national bodies in a particular country or a dedicated organization set up specifically for this task? Are there members and what is the governance structure? Who, ultimately, makes the decisions, and how do they get to be in that position?

STANDARD

This section looks at how the standard was developed and what it requires. Was the standard developed through a multi-stakeholder process or was it the work of specialists? When was it completed and is there any programme for revision? Is it a performance standard or a systems standard? If it is a performance standard, where did the performance requirements come from?

CERTIFICATION APPROACH

How is certification undertaken? Is it through certification bodies or does the certification scheme organization become involved itself? What are the requirements for personnel and process? What is certified? Is it always individual organizations or are group or regional approaches included?

REQUIREMENTS FOR CONSULTATION OR PUBLIC INFORMATION

The issue of consultation and provision of information has been the focus of so much attention that it is examined specifically for each scheme; this section assesses what is required and why.

ARRANGEMENTS FOR SMALL FOREST OWNERS

In many countries a major issue that has been identified is the risk of forest certification becoming a significant barrier to small forest owners who do not have the resources to deal with the cost and complexity of certification. Therefore, this section examines the extent to which schemes have sought mechanisms to minimize the barrier created.

ACCREDITATION ARRANGEMENTS

There are a number of ways in which accreditation can be provided. It may specifically deal with the forest management standard, or be more general in approach, relying on existing arrangements for other standards such as environmental management systems. It may be undertaken by the national accreditation body in a country, by an international accreditation body or by the organization which owns the certification scheme.

CURRENT STATUS

This examines the status of the scheme in 2004. How big is the organization; how many hectares have been certified; what is the ongoing development work currently under way?

CHAIN OF CUSTODY

Most of the schemes include arrangements for chain of custody. This examines exactly what is required. Is chain of custody based on tracing each bit of raw material, or is a percentage approach or a purchasing control approach used? Percentage approaches allow certified and uncertified material to be mixed and rely upon labels and claims that specify the percentage of certified material.

LABELLING AND LOGOS

Many of the schemes have logos and allow the labelling of products. This section examines how this is controlled and how the logos are used.

POLITICS AND PERCEPTIONS

While it is very important to consider each of the schemes as objectively as possible, it is impossible to ignore the politics that surround the different schemes. Therefore, this section briefly summarizes the main supporters and opponents of each scheme, with a suggestion of the causes.

INFORMATION ABOUT THE SCHEME

This section summarizes where further information about the scheme can be found: primarily from websites, but often also from publications.

CERTFOR

A1.1 CertforChile

TYPE OF SCHEME

This is a national scheme run by a purpose-created body.

SCOPE

The CERTFOR standard is specifically designed to apply to plantation forestry in the environmental and social context of Chile.

DATE SET UP AND HISTORY

The development of a series of criteria and indicators for plantation forestry in Chile began in 1999 as a result of a research project conducted by the Instituto Forestal (INFOR), with assistance from Scandiconsult and Matthew Wenban-Smith of the Soil Association, and funded by the European Union (EU). By late 1999 this project had developed a first draft of a series of criteria and indicators (based on Montreal Process criteria and indicators). Despite the fact that this was a very valuable effort, what the industry needed at that time was a sustainable forest management certification standard that was adapted to the reality of the forestry sector in Chile, taking on board the experience gathered by diverse public and private initiatives. This task called for the participation of all the principle national-level stakeholders, as well as some well-known international stakeholders.

For this reason, during 2000, Fundación Chile, INFOR and the Forest Owners Association of Chile (Corporación Chilena de la Madera, or CORMA) established a partnership to combine project knowledge and experience to create a national standard based on the previous work conducted by INFOR. This project was sponsored by the Chilean Development Corporation (CORFO), CORMA and forestry companies, with a total cost reaching US$450,000, 50 per cent of which was from public funds.

Fundación Chile was appointed as the process secretary and was in charge of managing the project. INFOR was a co-partner in this project and responsible for the technical aspects.

The main objective of this project was to develop an 'internationally recognized national forest certification standard for plantations, monospecific *Lenga* (*Nothofagus pumilio*) forests and second-growth *Nothofagus* forest'. The primary goal was to develop a forest certification system that would be recognized worldwide, with its own accreditation system for certifying organizations.

The spirit of the project was founded on the recognition that a certification initiative is principally based on a voluntary and private decision, and therefore requires forest companies to explicitly express a need for the development and implementation of such a system. However, as well as being voluntary and private, any new certification scheme requires as a minimum to comply with the legislative framework in force. In the case of managing the forest resource in Chile, it is important to mention the environmental legislative framework, the DL 701 Forest Law and all the other legislative norms of the sector.

The standards development work began in January 2001. A first draft of the standard was ready for field testing in May 2001. The revised draft standard was published for public consultation in

September 2001. Two public consultation meetings were held in October 2001 and January 2002. The standard was accepted by the superior council in February 2002. The first certificate according to the standard was issued in October 2003.

STRUCTURE AND GOVERNANCE

The owner of the initiative is CertforChile, a non-profit, independent organization, recognized by the state and laws of the nation, ruled by a board, and with numerous members who represent a cross-section of stakeholders from environmental organizations, the university, companies, small owners, large wood-processing companies, buyers groups, research organizations, and other private environmental non-governmental organizations (ENGOs) and NGOs. CertforChile is a membership organization which is free to receive membership applications from all interested parties on a non-discriminatory basis. CertforChile is a properly constituted charitable foundation registered under Chilean law and governed by a board elected by the members.

The superior council has determined that the standard will be revised every five years in order to take into account advancements in science, technological improvements, changes in the social and environmental concerns of stakeholders, and changes in the legal and political situation in Chile. Accordingly, the certificates will have a five-year validity.

The day-to-day running of the scheme is carried out by a secretariat, a function currently carried out by Fundación Chile.

STANDARD

Organization

The standard was developed by a hierarchical structure of political and technical groups.

Four primary multi-stakeholder bodies were created for this purpose:

1 The highest decision-making group was the *superior council*, which consisted of a body of highly respected people representing diverse civil society interests.
2 The second level of grouping was the *technical committee*, which comprised a body of professionals representing ENGOs, the forestry sector, indigenous people, professional societies, research organizations and academics, government environmental and forestry agencies, and small forest owners.
3 The third level was the *working group*, which consisted of a five-member group made up of representatives from Fundación Chile (two members), INFOR (one member) and CORMA (two members), and reporting to the technical committee. This group was responsible for preparing the draft documents to be evaluated and revised by the technical committee, and, upon their recommendation, discussed and approved by the superior council.
4 A fourth, very important, group comprised the *international consultants*. All of these consultants are Forest Stewardship Council (FSC) auditors and have assisted the whole process; key decisions were taken under their guidance.

Basis for CERTFOR standard

The following considerations were taken into account in developing the CERTFOR Standard for Sustainable Forest Management in Chile:

- existing definitions of sustainable forest management (SFM) and international agreements, such as the Montreal Process and the principles and criteria (P&C) of the FSC;
- the draft of the standard developed by the Forestry Institute (INFOR) in December 2000;

- the concerns identified by international consultants during interviews organized with the individual stakeholders visited;
- the concerns of the environmentally sensitive markets;
- the situation of the forestry sector in Chile.

The standard is also based on several international standards and seeks to establish procedures that accord with what is considered to be sound sustainable forest management, and to take into consideration the main concerns that have been raised by the different stakeholders with respect to plantation forestry. The standard aims to generate a positive change in the way in which forest plantations are managed and not to focus on a few bureaucratic obligations. This is the reason why the principles, criteria and indicators developed in this standard can be favourably compared with other respected international SFM standards.

Structure and application of the standard

The standard will guarantee that certified entities conform to a specific level of internationally recognized sustainable forest management practices, adapted to the national context. Additionally, the standard will help to identify forest management entities that present an excellent level of performance.

The hierarchical structure of the standard includes the following categories:

- *Principles*: those elements with a global character that define a conceptual framework, and whose function is to serve as a basis or guideline for activities that contribute to the main objectives of sustainable management. They are seen as the ultimate long-term objectives of forest management.
- *Criteria*: key elements, dimensions and processes that define, limit and permit the practical application of the principles. Through the criteria it is possible to evaluate, periodically, the management of the forests for sustainability and, therefore, to guarantee a true and constant improvement of the management practices.
- *Indicators*: elements that can be described or measured quantitatively and qualitatively in order to evaluate, periodically, their trend and to verify in this way the level of compliance with a specific aspect of a criterion.
- *Minimum level of compliance*: quantitative or qualitative value of the indicator that is considered to be the bare minimum required to demonstrate an adequate level of compliance.

Standard-setting process

During 2000, a two-tier organization for developing and ratifying the standard was agreed. The first tier consisted of a higher council of certification made up of well-respected members of Chilean society who were selected to represent a wide range of environmental, social and economic interests. The second tier is the technical committee for certification, a larger group of technical experts in a variety of fields.

The task of drafting a working standard was delegated to a task team consisting of representatives of CORMA, the forest industry association Fundación Chile and INFOR, under the chairmanship of an external consultant, H J van Hensbergen, with the assistance of Borje Drakenberg of SvenskSkogsCertifiering (SSC). The task team began work in January 2001.

The work of this task team was informed by the previous INFOR process; however, in addition to this a first round of consultation with a variety of stakeholders (environmental and social NGOs, government agencies, forestry companies and indigenous peoples) was carried out by the external consultants in order to identify key issues for developing a standard.

A draft standard and auditor's manual were prepared by mid May, and these were field tested by two teams of auditors led by H J van Hensbergen and Matthew Wenban-Smith. Following the field test the draft was refined, and in September 2001 a public meeting was held at which the draft standard was released to the public as part of a more detailed consultation with interested and affected parties. This was followed up by a First Public Consultation Workshop, which was held on 26 October 2001 with the participation of representatives from a wide spectrum of interests. These included several NGOs (Defensores del Bosque Chileno, Greenpeace and Concepción, Econativa, or CODEFF); entrepreneurs; independent professionals; academics; international organizations such as the United Nations Food and Agriculture Organization (FAO); forestry workers' representatives; native peoples; members of the Technical Committee of the Forest Certification Initiative; and technical delegates of the National Standards Institute (the Instituto Nacional de Normalización, or INN), the Agriculture and Livestock Service (Servicio Agrícola y Ganadero, or SAG), the Chilean Safety Association (ACHIS), the Chilean Forest Service (Corporación Nacional Forestal, or CONAF), the Forest Owners Association of Chile (CORMA), the Ministry of Agriculture, the Chilean Development Corporation (CORFO), and the Forestry Institute (INFOR). The inputs from this consultation round were carefully documented and the standard was revised in the light of the comments received. A second public consultation was held in January 2002 and the standard was ratified by the superior council after all consultations had been exhausted and all important issues had been resolved to their satisfaction.

As can be seen from the above, the standard for plantations was developed by a consultative process in which stakeholders were repeatedly asked to comment on the standard proposals. However, final approval of the standard was carried out by the superior council without formal participation by the stakeholders consulted.

The standard is strongly performance based; but in common with most standards, it has certain requirements in terms of the management system. The most important of these system requirements focuses on the management plan and the training and responsibility of the staff responsible for executing the plan.

CERTIFICATION APPROACH

Ultimately, the system will depend upon auditing companies accredited by the Chilean National Standards Institute (INN) according to guidelines issued by CertforChile. It is expected that the first accreditation of such auditors will take place during 2004.

As an interim measure, the superior council has defined the following procedure for issuing certificates based on inspections carried out by auditing companies with experience of forest management auditing. The audit teams must have members who have been trained by CertforChile in auditing according to the CERTFOR standard. The lead auditor should be an experienced forest management auditor with ample experience of forest certification auditing in performance-based certification systems.

Audit, reporting and evaluation procedure

Auditors carry out inspections of the office-based records of the company and of their performance in the field. A wide stakeholder consultation is carried out in relation to each forest management unit inspected. A report is prepared according to a predefined format that is submitted to CertforChile for peer review and evaluation.

This report will be revised by two independent experts and by the forest management unit (FMU) to be certified, which will have the right to make some comments on the scores given and on other elements of the report. The auditing team will have to publish a final report that takes into consideration the comments received, and recommend or refuse the certification to the FMU.

A minimum mean score of 3 for each principle will be required in order to certify the FMU. However, no indicator must have been marked with a nil score for the approval of the corresponding criterion or principle, and where a criterion has been marked with a score of 2, the certification can be endorsed only if the FMU shows its commitment to correct all of the main non-compliances that have been observed.

Decision-making procedure

The superior council issues certificates as explained below. In order for these to have the highest level of credibility, all decisions taken by the superior council should be by consensus. Furthermore, it is necessary to have a very high level of openness about the certificate and the accompanying reports. Since these certificates are destined to operate in the international market, it is necessary to include international auditors and peer reviewers in the process.

The ownership and administration of the CertforChile system rests with the superior council of CertforChile. The superior council has the power and responsibility to decide the manner in which certificates can be issued. It can also give itself the power to issue CertforChile certificates for plantation management.

In order to ensure that these certificates are internationally credible, the auditing process should include an international presence on the audit team. The lead auditor should be international and should have extensive experience as a lead auditor in other forest management certification systems.

In order to ensure that the certificates issued do not put the reputation of the CertforChile system in danger, all of the decisions should be taken by the superior council.

Furthermore, so that all relevant considerations are taken into account in making the decision, there should be a comprehensive peer review of each certification report involving at least one international expert and two Chilean experts.

Peer review documents should be returned to the lead auditor for comment.

All evidence should be reviewed by the members of the technical council prior to the report being presented to the superior council for a decision. The peer review reports are withheld from the technical council at this stage in order to preserve their independence.

The secretary of the superior council is responsible for the administration of certificates and for auditing the volumes of timber traded under interim certificates.

REQUIREMENTS FOR CONSULTATION OR PUBLIC INFORMATION

The CertforChile audit procedures require extensive stakeholder consultation by the auditors. Auditors must contact as wide a range of stakeholders as possible. Initial contact may be made by letter, telephone or email. All stakeholders have the opportunity to request a meeting with the auditors.

CertforChile requires that certified companies make the basis of the management plan publicly available and that the results of monitoring are publicly available. CertforChile publishes the audit reports on its website.

ARRANGEMENTS FOR SMALL FOREST OWNERS

CERTFOR considers group certification as a viable alternative to individual certification for small- and medium-sized owners for whom the overhead cost of individual certification audits would make certification uneconomic.

The CERTFOR Group Standard has been developed to offer small- and medium-sized forest owners the possibility of becoming certified. Fundamentally, apart from meeting the requirements of sustainable forest management, this group certification system affords the possibility to reduce the costs associated with an audit process. The costs of an audit exclude many small owners from the market for certified products. This standard is applicable to forests in small- and medium-sized landholding whose owners wish to become certified.

Group certification permits the reduction of audit costs through the formation of groups with similar activities and the implementation of an internal control system oriented towards reducing the intensity of external audit inspections, which are needed to verify that the standard for sustainable forest management is being upheld.

One should not, however, expect group certification to imply savings for individual forest owners who lack an internal control system or for those who carry out different forest management activities. In such cases, the intensity of visits and the complexity of the certification report could outweigh potential savings.

It should be pointed out that each group member must comply with all of the requirements of the CertforChile standard independently, and that it is not possible to share responsibility for any of the standard's aspects among the members of a group. For example, it is required that at least 10 per cent of the area of each member's forests be maintained as native vegetation.

Nevertheless, there is a certain degree of flexibility within the group with regard to the division of responsibilities for the management system of the group and of its members. In some cases, the group can be responsible for internal audits and control of the certification inspections only, while in other cases the group may assume responsibility for carrying out management activities, including the physical work.

The aspects considered in this group certification scheme deal with the systems needed to ensure a uniform level of performance in the forest management of each of the members, and with the evidence that must be presented to auditors to confirm that the level of the standard is upheld in each of the forests belonging to the group.

ACCREDITATION ARRANGEMENTS

Current arrangements

At present there is no formal system for the accreditation of auditors. The superior council has determined that auditors should have experience of auditing according to performance-based forest standards and that there should be an international presence on the team. The lead auditor should have wide experience of sustainable forest management auditing according to an internationally recognized scheme. The final decision on adequacy of the audit process rests with the superior council.

Future arrangements

In future, auditors will need to be accredited by the National Standards Institute of Chile (INN). The INN already accredits auditors to perform inspections for a variety of International Organization for Standardization (ISO) standards. The system used by the INN involves an initial inspection of the certifying body before accreditation is granted. There is annual monitoring of the certifying body by the INN in the company of a representative of CERTFOR. The training and experience standards for auditors and the composition of the required audit teams have been determined by CERTFOR.

It is understood that a number of international forest auditing companies have applied to INN for accreditation and it is expected that some will receive accreditation during 2004. As soon as the first certification body has received its accreditation from INN, the current arrangements described above will lapse.

CURRENT STATUS

Hectares certified

A total of 926,900ha have been certified, comprising:

- Bosques Arauco SA (pine plantations): 257,300ha;
- Forestal Celco and Forestal Cholguan SA (pine plantations): 446,100ha;
- Forestal Valdivia SA (pine plantations): 223,500ha.

Major successes

CERTFOR has been a member of the Pan-European Certification Council since November 2002.

In February 2004, the Programme for the Endorsement of Forest Certification (PEFC) schemes announced the start of the public consultation period for the assessment of CertforChile, the Chilean national forest certification scheme. This is the first non-European scheme to be assessed.

Plans

There are plans to develop a standard for natural forests during 2004–2005.

CHAIN OF CUSTODY

The CERTFOR initiative has developed a standard for chain-of-custody compliance. This standard has been approved by the superior council and has been tested. It includes the conventional procedures of chain-of-custody compliance.

The principal objective of a chain of custody (CoC) is to guarantee the connection between the flow of certified timber inputs to the output of products. In order to achieve a chain of custody and to label final products as certified (that is to say, sourced from a forest conforming to the SFM standard), each unit responsible for the timber, from the forest to the retail distribution channels, must possess a CoC certificate to endorse it.

The current standard applies to certified materials according to the CERTFOR standard. However, CertforChile also unilaterally recognizes the PEFC and the FSC systems.

To this effect, the CoC standard of CertforChile takes the three major modalities of accounting for certified timber used around the world into consideration:

1 physical separation;
2 percentage in = percentage out;
3 percentage claim.

The standard makes clear the requirements of these three modalities under Principles 3a, 3b and 3c. Organizations certifying the CoC may opt for any of these modalities, adhering to the exigencies defined by this standard. However, it is not possible to freely mix the timber from these different modalities.

General aspects such as documents, records, and input and output procedures, amongst others, that are valid for the three modalities are referred to in Principles 1, 2 and 4.

Compliance with environmental and social requirements during the processes as certified by this standard is not usually required for a CoC standard; nevertheless, CertforChile, under its Principle 5, endeavours to minimize environmental impacts (for example, toxic waste and river pollution) and social impacts (for example, child labour, workers' health and security).

LABELLING AND LOGOS

All certified companies receive the logo use guide, which contains rules and instructions for the correct use of the logo.

CERTFOR

FIGURE A.1 The CERTFOR logo

POLITICS AND PERCEPTIONS

Since 1999, there have been two major initiatives within Chile in forest certification.

The CertforChile scheme has been supported by government and the larger forestry companies in Chile.

The alternative ICEFI process, which has not yet concluded, follows a participatory process and has in its initial stages concentrated on natural forest management. This process is being supported, in particular, by environmental NGOs such as CODEFF.

In greater detail, major supporters are:

- *The government*: who funded the initiative through a contest in which CERTFOR won over other development projects. The government sees the growth of forest certification as a source of opportunities and challenges – for example, non-tariff trade barriers and market access for Chilean forest products.
- *Foresters*: CORMA strongly supports the initiative because it raises the standard of forest management in the country.
- '*Moderate*' NGOs: these NGOs see certification as a market tool for promoting forest sustainability.
- *Academics*: most academics strongly encourage the development of this initiative, especially young academics who advocate social participation and those who support the incorporation of environmental and ecological concerns within silvicultural management.
- *Industry*: the industrial sector wanted the best possible standard, but adapted to local conditions and with a body of governance that ensures the representation of relevant stakeholders.

There is opposition to the scheme from some NGOs who see CERTFOR's development as a source of competition to the FSC, and who do not believe that the standard adequately deals with their concerns.

Initially, the industry was extremely wary of certification schemes; however, as a result of the process, these schemes have largely become accepted. Most companies have made significant improvements in their forest management as a result of the advent of certification in Chile. In particular, there is a much better understanding of biodiversity and social issues, and systems for managing these have been developed. As a result, NGOs, ICEFI and CERTFOR have become more supportive of a convergence of both initiatives, particularly in the area of natural forests.

INFORMATION ABOUT THE SCHEME

For more information, visit CERTFOR's website on www.certfor.org.

Contacts

Aldo Cerda, executive secretary
Email: acerda@fundacionchile.cl
Tel: +56 2240 0328

Maria Ines Miranda, technical officer
Email: mmiranda@certfor.org
Tel: +56 2240 0349

A1.2 Canadian Standards Association (CSA)

TYPE OF SCHEME

The CSA Sustainable Forest Management Standard, CAN/CSA-Z809-02 *Sustainable Forest Management: Requirements and Guidance*, is a Canadian national standard in terms of its scope and, more formally, because it has been approved as a national standard of Canada by the Standards Council of Canada (SCC). National standard status means it is in full conformance with the National Standards System of Canada and all of its supporting procedures and requirements, including development, implementation and revisions.

The CAN/CSA-Z809-02 standard is the only sustainable forest management (SFM) standard in North America to be developed according to an internationally recognized and accredited standards development process. The standard is further supported through the Canadian National Conformity Assessment Programme, which allows independent third-party certification of individual auditors, certifiers and applicant organizations.

The CAN/CSA-Z809-02 standard has three separate sections for the three key sets of requirements that are essential for sustainable forest management in Canada. Together these comprise the CSA SFM requirements: public participation requirements; performance requirements and SFM system requirements. The CAN/CSA-Z809-02 standard is based on the international Helsinki and Montreal Processes and incorporates Canada's own national SFM criteria, which were developed by the Canadian Council of Forest Ministers.

The standard outlines the requirements that forest managers must follow in order to demonstrate to customers and the public that forests are managed in an inclusive, responsible and sustainable manner.

SCOPE

In Canada, 94 per cent of all forests fall under the jurisdiction of the federal and provincial governments. The CAN/CSA-Z809-02 standard is applicable to all forest types, tenures and sizes across Canada. The context of the standard is the defined forest area (DFA), which is the specified area of forest, including land and water (regardless of ownership or tenure), to which the requirements of the standard apply. Defined forest areas are the parcels of land that become certified to the CAN/CSA-Z809-02 standard.

DATE SET UP AND HISTORY

As an internationally recognized and accredited standards development organization, CSA was approached in 1993 by provincial and federal government agencies, the Canadian forest industry, academia and other stakeholders to develop credible sustainable forest management standards for Canada. As per national standards requirements, a 32-member balanced-matrix sustainable forest management technical committee (SFM TC) was established in 1993 to develop the CSA SFM standard. In October 1996 through the work of the SFM TC, the CSA published two national standards of Canada: CAN/CSA-Z809-96, *A Sustainable Forest Management System: Specifications Document*, and CAN/CSA-Z808-96, *A Sustainable Forest Management System: Guidance Document*.

All national standards of Canada are required to undergo a mandatory review every five years. In 2000, the CSA and the the SFM TC began the revision process for the CAN/CSA SFM standards. During 2002, the revised versions were published as one document, entitled CAN/CSA-Z809-02 *Sustainable Forest Management: Requirements and Guidance*. In 2003, this document was re-confirmed as a national standard of Canada.

Canada's forests make a significant contribution to the quality of life, the integrity of the environment, and the supply of paper and building materials and other forest products, both in Canada and abroad. Throughout Canada, these forests are owned predominantly by the people of Canada, a nation that has hundreds of forest-dependent communities with diverse forest types and particular circumstances.

The CSA was established in 1919 and has a long and rich history rooted in the development of consensus-based standards. The CSA's standards and certified products have had a positive influence on people around the world for over 80 years. It is for these reasons that the CSA worked with a diverse range of stakeholders interested in SFM in Canada in order to develop the CAN/CSA-Z809-02 standard and the Forest Products Marking Programme.

Across Canada, the provinces have rigorous legislation and policies for the protection, conservation and sustainable management of forests. This legislative framework is continuously improved, as is forest management in Canada. In addition to using regulatory tools, organizations benefit from using voluntary tools, such as the CAN/CSA-Z809-02 standard, to help them achieve SFM. This standard gives organizations a system for continually improving their forest management performance and engaging interested parties in a focused public participation process.

STRUCTURE AND GOVERNANCE

The CSA SFM TC writes the standard with input from public consultation, and a CSA standards development staff person is responsible for seeing the development and publication of the standard through to completion. In turn, it is the Standards Council of Canada who ultimately gives a CSA standard higher status by approving it as a national standard of Canada.

Members of the CSA SFM TC are selected by the CSA in cooperation with the technical committee chair and in accordance with the CSA's balanced matrix requirements. The matrix approach ensures that all sectors of society are represented and no one group can control the technical requirements of the document.

The standard was developed using the CSA's accredited consensus-based process. In this context, 'consensus' means 'substantial agreement, but not necessarily unanimity'. In addition to numerical requirements, the process requires that all negative ballots are specifically reviewed and addressed.

Technical committees are structured to include representation from all major stakeholder groups. Minimum and maximum numbers of voting members in each category are set to provide for a reasonable balance of representation and to ensure that one stakeholder group cannot control the outcome of a ballot.

In addition to official five-year reviews of the standard, should anybody have a request for interpretation, a sub-committee of the CSA SFM TC reviews these interpretation requests as they are received and puts forward suggested responses, which are then put forward to the CSA SFM TC to review and potentially approve. If approved, the person issuing the request is made aware of the interpretation, as are other users, as is the Standards Council of Canada, who, in turn, notifies all accredited registration bodies of the interpretation so that they can ensure their audit process includes this new information. In addition, all updates and releases are also posted on the CSA Forest Products Group website at www.certifiedwood.csa.ca.

The CSA project manager is responsible for the administration of the standard, such as responding to queries and meeting with stakeholders to address questions and concerns.

STANDARD

In accordance with CSA standards development requirements, a CSA SFM TC was established to write the standard. The CSA SFM TC consisted of a balanced matrix from four chambers: academia/professional/practitioner sectors; general interest/environmental groups; government/ regulatory authority; and business interests.

In 1996 the CSA SFM TC voted unanimously to accept the standards, CAN/CSA Z809-96 (requirements) and CAN/CSA Z808-96 (guidance), and in October of the same year CSA SFM standards were approved and published as national standards of Canada.

As mentioned above, all national standards of Canada are required to undergo a mandatory five-year review. In December 2002 the CSA SFM TC completed an 18-month review of both CAN/CSA-Z809-96 and CAN/CSA-Z808-96, including public consultation and public review periods. The purpose of the mandatory review was to build on new and current research and implementation experience, and to continually improve the content and applicability of the standard.

By July 2003, the CSA SFM TC had not only unanimously approved the revised version of the standard, CAN/CSA-Z809-02, it was also approved, once again, by the Standards Council of Canada as Canada's national standard for sustainable forest management. Throughout the review process it became clear that adaptive forest management should continue to be the basis of SFM in Canada, and in order to link adaptive forest management to forest certification, three key requirements continue to be necessary: public participation requirements, performance requirements and system requirements.

These three requirements are referred to as the CSA SFM requirements in the CSA-Z809-02 standard, and each of these requirements is addressed in a specific section of the standard. Hence, the CSA SFM standard is founded on a rigorous public participation process, is performance based on national SFM criteria and supporting elements, and has system requirements based on ISO 14001.

Over the past five years it has become clear to forest managers across the country that management systems such as ISO 14001 are very powerful tools. Recognizing this, the CAN/CSA-Z809-02 standard is consistent with the ISO 14001 standard. It is essential to have a management system that can help to fulfil of all the CSA SFM requirements year after year. The management system requirements are viewed as the vehicle to deliver both the public participation and performance requirements in a systematic manner. This ultimately helps to drive adaptive forest management and continual improvement.

It is this integration of the three sets of requirements that sets the CSA SFM standard apart from many other existing forest management standards.

The performance requirements of the CAN/CSA Z809-02 standard are based on the Canadian Council of Forest Ministers (CCFM) SFM criteria, elements and indicators. The CCFM SFM criteria and elements are fully consistent with those of the Montreal and Helsinki Processes, which are both recognized by governments around the world. It is believed that the CCFM criteria and supporting elements represent the best set of broadly accepted Canadian forest values.

In CAN/CSA Z809-02, the CCFM SFM criteria have been adopted verbatim; but the elements have been revised to reflect the need for meaningful application at the DFA level and are now called CSA SFM elements.

The CAN/CSA Z809-02 standard addresses the wide range of issues related to SFM through performance-based actions in three specific ways. Firstly, the mandatory CCFM criteria and related CSA SFM elements must be addressed through the public participation and value, objective, indicator and target (VOIT)-setting processes and applied to the DFA. For example, CSA SFM Element 1.4, Protected Areas and Sites of Special Biological Significance, requires an organization to respect protected areas and identify sites of biological significance within the DFA, and to implement management strategies appropriate to their long-term maintenance. These strategies are defined through the establishment of associated values and the application of on-the-ground objectives, indicators and targets related to protected areas and sites of biological significance.

Section 5.4 (vi) of the standard allows for any value or issue, related to SFM on the DFA, to be brought forward by any individual or interested party. These issues must be addressed through the public participation process through one of several methods provided in the standard.

Also contained in the standard is a mandatory performance matrix that requires the organization to clearly define, set and record the DFA-specific performance requirements. This matrix must be completed through the public participation process, and must describe the DFA-level application of the CCFM criterion; the CSA SFM elements; VOITs; the basis for the targets; any legal requirements; the means of achieving the objectives and targets; monitoring and measurement requirements; and levels of acceptable variance. This matrix will become a key vehicle for addressing and communicating the organization's SFM performance in a transparent and publicly accessible manner.

CERTIFICATION APPROACH

Certification (registration) is conducted by registration bodies accredited by the Standards Council of Canada. A current list of accredited registration bodies can be found at the Standards Council of Canada website: www.scc.ca.

Certification to CAN/CSA-Z809-02 is granted to the 'defined forest area' (DFA) to which the requirements of the standard have been applied and successfully audited. The DFA is a specified area of forest, including land and water (regardless of ownership or tenure), making certification to CAN/CSA-Z809-02 area or region based. The applicant or holder of the certification can be an individual, an organization (for example, a company) or a collection of organizations, all working to meet the requirements of the standard.

All lead-auditors and auditors are required to meet ISO 19011 and CSA PLUS 1134 criteria. One member of team must be a registered professional forester (RPF) or hold Canadian Environmental Auditing (CEA) (SFM) designation from the Canadian Environmental Auditing Association (CEAA). SCC as an environmental management system (EMS) auditor certification body accredits the CEAA and it certifies EMS auditors to ISO/IEC Guide 66. In addition, the CEAA certifies SFM auditors with a designation of certified environmental auditor (CEA) – sustainable forest management.

REQUIREMENTS FOR CONSULTATION OR PUBLIC INFORMATION

The CSA requires extensive public participation in developing its standards. The initial standard was published following years of discussion and work, using an open and inclusive process managed by the CSA. In addition, a Canada-wide public review of this standard generated considerable interest, with CSA distributing over 1500 copies of the draft standard in response to requests for review. Public meetings were held in Montreal, Toronto and Vancouver to seek further input. In 2000, when the CSA set out to review and improve upon the original standard, it sought and incorporated public input once again.

The need for public participation is also strongly emphasized in the standard itself. In fact, this standard requires organizations to seek comprehensive, continuing public participation and to work with aboriginal peoples at the local community level. The public identifies forest values of specific importance to environmental, social and economic concerns and needs. The public also takes part in the forest planning process and works with the organization to identify and select SFM objectives, indicators and targets in order to ensure that these values are addressed. The public participation requirement of this standard is one of the most rigorous of its kind in certification standards in the world today. Because Canadian forests are primarily publicly owned, it is vital that a Canadian forest certification standard involves the public extensively in the forest management planning process. Forest management that meets the SFM requirements of this Standard necessitates a positive relationship between the organization and the local community.

During the audit process, auditors are required to verify that all of the public participation requirements of the standard have been fulfilled. This is done in a variety of ways including (but not limited to) random calls to participants of the public participation process; observing the public advisory group during regularly scheduled meetings; and meeting privately with public advisory group members. Lastly, the standard requires that all SFM plans and audit reports are made publicly available. A listing of certified forests in Canada is maintained and available through www.CertificationCanada.org.

ARRANGEMENTS FOR SMALL FOREST OWNERS

The CAN/CSA-Z809-02 standard is applicable for all types of ownership, tenures and forest types across Canada. As part of the continual improvement process, the CSA SFM technical committee periodically updates the content and quality of the standard in order to reflect scientific advances and experiences gained from the application of the standard. The applicability of the standard to a variety of forest tenures, including private woodlots, is an area that the CSA SFM TC is committed to enhancing.

ACCREDITATION ARRANGEMENTS

The Standards Council of Canada (SCC) accredits standards development organizations such as the CSA. In fact, the SCC oversaw the development of CAN/CSA-Z809-96 and the subsequent 2002 version. The SCC also accredits registrars (certifiers) to register applicant companies under the CAN/CSA SFM series of standards.

In addition, the SCC accredits auditor certification bodies, such as the Canadian Environmental Auditing Association (CEAA), who in turn certify auditors who are hired or contracted by registrars to conduct audits to CAN/CSA-Z809-02.

CURRENT STATUS

As of 31 December 2003, there were 28.4 million hectares of Canadian forests certified to CAN/CSA-Z809. The CAN/CSA-Z809 standard is the leading forest certification standard in Canada, with a forecasted 71.7 million hectares to be certified to the standard by the end of 2006.

CHAIN OF CUSTODY

The CSA's SFM programme's chain of custody and SFM mark requirements were created in 2001 in response to requests from a variety of stakeholders to link CAN/CSA-Z809 to forest products and consumers. Once assurance is provided that forests are being managed according to established principles of sustainable forest management, the next voluntary steps are the recognition and promotion of forest products that have originated from a forest area certified to CAN/CSA-Z809.

The CSA chain of custody requirements are of the most rigorous and credible in the marketplace today. The requirements are outlined in CSA PLUS 1163: *Chain of Custody for Forest Products Originating from a Defined Forest Area Registered to* CAN/CSA-Z809. In addition to critical requirements such as verification of the origin of raw materials and inventory monitoring and control, the CSA chain of custody also includes a requirement for environmental management and product claims. The CSA chain-of-custody programme is the only chain-of-custody scheme to include these requirements, which are based on the belief that certified forest products should be produced by facilities that are environmentally responsible.

The chain-of-custody programme is managed and delivered through CSA International's Forest Products Group. Currently, 44 chain-of-custody certificates have been issued covering 78 sites, with additional certifications underway. Once certified to CSA PLUS 1163, an organization may choose to apply CSA International's SFM mark to their product.

In July 2001, CSA International created the *Forest Products Marking Programme*. Under this programme, CSA International offers organizations the opportunity to further demonstrate their commitment to SFM through a chain-of-custody and product mark, summarized in Figure A.2.

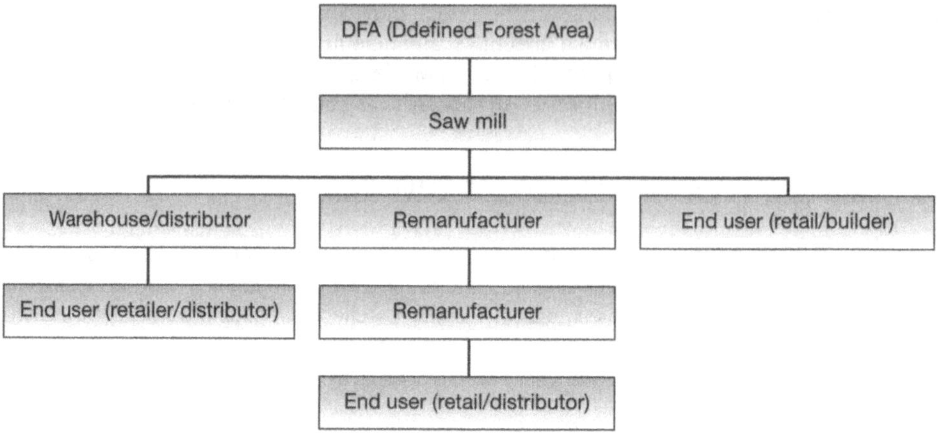

FIGURE A.2 CSA chain of custody for forest products

In the manufacturing of forest products and wood-based products, there may be one or several stages in the transformation process from tree to final product. The ownership of the certified product may change several times throughout the transformation process. These transformation or ownership changes are referred to as 'links' in the chain of custody from forest to consumer. Examples of these links may include defined forest areas; modes of transportation (carriers such as trucks, ships, booms and rail); manufacturing facilities (primary, secondary, re-manufacturing, value added) that use certified forest products; warehouses, stockyards and reloads that repackage certified forest products; and traders/brokers who do not have physical contact with the forest products, but obtain ownership for a period of time.

The scope of a chain of custody will be defined by the implementing organization. Generally, it will cover the certified forest products from the point at which the organization takes ownership or control to the point at which they are delivered to the next 'link' in the chain, or to the end consumer.

Certified forest products can take many forms. Examples include (but are not limited to) the following:

- traditional forest products, such as unprocessed logs, sawn timber or plywood;
- by-products or residues that arise from manufacturing or re-manufacturing processes, such as chips and sawdust;
- products to be re-manufactured, such as pulp and sawn timber;
- products for the end consumer, such as sawn timber;
- composite products, such as paper, doors, window frames and furniture;
- non-traditional forest products, such as blueberries, mushrooms or Christmas trees.

LABELLING AND LOGOS

In 2001, CSA International created a new product mark for companies that are manufacturing forest products originating from a forest certified to CAN/CSA-Z809 and are tracked through a certified CSA PLUS 1163 chain of custody. The CSA SFM mark features both coniferous and deciduous trees and the familiar CSA insignia.

CSA International's Forest Products Marking Programme has brought together decades of product-marking knowledge and experience and combined it with sustainable forest management. The CSA SFM mark demonstrates to customers that forest products bearing this mark have originated from a forest certified to CAN/CSA-Z809 and have been verified through an independent chain-of-custody audit. This creates a powerful opportunity for both suppliers and buyers of forest products to demonstrate and communicate their commitment to sustainable forest management.

FIGURE A.3 Option 1: Minimum average percentage system for solid wood

Within CSA International's Forest Products Marking Programme, there are three product-marking options (see Figures A.3–A.5). These options are based on how the inventory of certified forest products is managed.

The minimum average percentage system approach is based on the premise that the total batch of products can be labelled with the CSA SFM mark when the amount of certified (originating from a forest certified to Z809) wood-based raw material in the input batch meets or exceeds the set minimum average threshold. In order to apply the CSA SFM mark, the minimum average input is 70 per cent (by volume or by weight) of certified (originating from a forest certified to Z809) wood-based raw material. In addition, none of the wood-based raw material (including the remaining input percentage of wood-based raw material) will have come from controversial sources. This mark appears on the product and/or the packaging.

FIGURE A.4 Option 2: Minimum average percentage system for composite products

The minimum average percentage system approach is based on the premise that the total batch of products can be labelled with the CSA SFM mark when the amount of certified (originating from a forest certified to Z809) wood-based raw material in the input batch meets or exceeds the set minimum average threshold. In order to apply the CSA SFM mark, the minimum average input is 70 per cent (by volume or by weight) of certified (originating from a forest certified to Z809) wood-based raw material. In addition, none of the wood-based raw material (including the remaining input percentage of wood-based raw material) will have come from controversial sources. This mark appears on the product and/or the packaging.

In this approach, certified wood, wood raw materials and wood products are received and clearly marked or physically segregated, or otherwise identified as certified (originating from a forest certified to CSA Z809), and remain clearly identifiable as certified throughout the transportation, handling, processing, manufacturing or remanufacturing process or processes. This mark appears on the product and/or the packaging.

FIGURE A.5 **Option 3: Physical separation (segregation)**

POLITICS AND PERCEPTIONS

Major supporters

A growing number of national and international buyers are now specifying 'certified' forest products in their procurement policies. There are a number of reasons for this increased demand. A number of buyers are committed to being social and environmental leaders, and want to demonstrate this commitment through their purchasing polices and practices. Others are simply responding to the wishes of their stakeholders, while yet other groups are responding to pressure and threats by environmental activists to 'clean up their supply chain' or face the consequences of protests and negative exposure.

This shift in purchasing preference and priority is illustrated globally through the growing list of procurement policies that acknowledge a preference for certified products. In essence, buyers are looking for assurance that the products they are purchasing have come from certified forests. Regardless of the motivation, a clear trend has emerged in the marketplace: buyers of Canadian forest products are increasingly looking for products to be certified. This trend has been confirmed in the recent IBM Business Consulting Services report, A *Greenward Shift in the Market for Forest Products from British Columbia*, which concluded: 'there is clear evidence of a greenward shift in the market for forest products, including those from British Columbia. The shift is real, buyers believe it will continue, and we believe it will have a negative impact on forest regions and producers that do not respond to it'.

The CSA SFM standard is increasingly meeting the needs of buyers' purchasing policies as the list of organizations that accept forest products certified through the CSA SFM programme continues to grow. For example, organizations that accept the CSA SFM standard for meeting certification requirements are as follows:

- 84 Lumber;
- Anderson Windows;
- AOL/Time Warner;
- Centex Homes;
- Hallmark Cards;
- Lowe's;
- MASCO Cabinets;
- Masterbrand Cabinets;
- McCoy's;
- Norm Thompson Outfitters;
- Office Depot;
- Pella Windows;
- Staples;
- Wickes Lumber.

INFORMATION ABOUT THE SCHEME

Information and documentation about the CSA SFM programme are available free on the CSA website, www.certifiedwood.csa.ca.

Contact

CSA International Forest Products Group
90 Burnhamthorpe Road, Suite 402
Mississauga, Ontario, Canada
L5B 3C3
Email: csawood@csa-international.org
Tel: +905-275-0284 or +866-CSA-WOOD
Fax: +905-272-2491

By buying products with an FSC label you are supporting the growth of responsible forest management worldwide.
© 1996 Forest Stewardship Council A.C.

FSC

The Global Benchmark for Responsible Forest Management

A1.3 Forest Stewardship Council (FSC)

TYPE OF SCHEME

The Forest Stewardship Council (FSC) is an independent, non-governmental and non-profit organization, registered in Mexico as an association of members (Associación Civil, or AC). The membership consists of a diverse group of representatives from environmental and social groups, the timber trade and the forestry profession, indigenous peoples organizations, community forestry groups and forest product certification organizations from around the world.

The organization operates internationally and provides its services through the FSC International centre, based in Bonn, Germany, as well as through a worldwide network of national initiatives.

The FSC offers an international accreditation programme for independent certification bodies and a labelling scheme for forest products serving as a credible guarantee that products come from well-managed forests – that is, forests that meet the FSC's forest management standards, the so-called principles and criteria (P&C).

SCOPE

The FSC scheme is international in scope. Certification bodies from all countries can apply for accreditation and forest management or manufacturing operations from all over the globe can ask those bodies with an international accreditation to become certified against FSC standards.

The principles and criteria for forest stewardship are intended to apply without discrimination to tropical, temperate and boreal forests or plantations worldwide that are managed for the production of forest products.

DATE SET UP AND HISTORY

During the late 1980s, deforestation found its way onto the agenda of environmental groups, particularly in the UK. In 1985, Friends of the Earth (FoE) launched the first consumer forest campaign regarding deforestation, which included a boycott of retailers that were selling products made out of tropical timber from clear cuts or non-replacement selective logging. That same year, the World Wide Fund for Nature International (WWF-International) hired its first forest conservation officer to develop tropical forest policy campaigns.

Increasing consumer awareness led some retailers to start paying attention to deforestation. The British group B&Q pioneered this trend and revised the company's sourcing policies. For this, the later FSC founding Executive Director, Timothy Synnott, was hired as a consultant to help develop a new approach. After engaging the most vocal local environmental groups – WWF, FoE and the Rainforest Action Group (RAG) – as well as suppliers and others involved in the timber trade, B&Q and Synnott came up with the first programme of wood-sourcing research. However, this remained an isolated case.

A group of timber users, traders and representatives of environmental and human-rights organizations who had identified the need for an honest and credible system for identifying well-managed forests as acceptable sources of forest products met in California in March 1990. The meeting concluded that this system would include a global consensus on what is meant by good forest management, as well as independent audits of the management of forest and a global umbrella organization . It coined the title Forest Stewardship Council (FSC).

In 1991 a 'Certification Working Group' met in San Francisco, US, to discuss certification standards that different organizations could ascribe to, would require for constant 'objective' monitoring of certification programmes in order to protect both the public and certified producers, and would need for some type of organizations to implement the above. The Certification Working Group involved over 30 organizations from 11 countries. This group was instrumental in the development and foundation of the FSC. During the same year, the first buyers group was founded in the UK under the direction of WWF-UK.

1992 saw the United Nations Conference on Environment and Development (UNCED) – the so-called Earth Summit held in Rio de Janeiro. UNCED applied the Forest Principles to natural resources, asserting, 'forest resources and forestlands should be sustainably managed to meet the social, economic, ecological, cultural and spiritual needs of present and future generations'.

The ideas emanating from Rio to have the sustainable development of forests comprehensively defined by the widest possible range of stakeholders were fully taken up when 130 representatives from around the world came together in Toronto, Canada, in September 1993 to hold the Founding Assembly of the Forest Stewardship Council.

In October 1993 an agreement was reached to launch the FSC, and by August 1994 a definitive set of P&C, together with the statutes for the council, were agreed and approved by the votes of the founding members. During the same year the FSC secretariat was established in Oaxaca, Mexico, and Timothy Synnott was hired as first executive director. The first FSC presence outside of Mexico was set up in the UK in 1995 by the endorsement of a contact person as the first stage of a national initiative.

In 1996, the FSC was established as a legal entity under Mexican law. Principle 10 on plantations was completed and the first FSC certified product hit the market in the UK. Four certification bodies (the Scientific Certification Systems (SCS); SGS Forestry Qualifor Programme; the Rainforest Alliance's SmartWood Programme; and the Soil Association) were accredited by the FSC for worldwide forest management and chain-of-custody certification. At the FSC general assembly, the development of new approaches to certification of small landholdings (group certification) was approved.

In 1997 the FSC board approved the 'policy for percentage-based claims,' which allows public recognition for products containing less than 100 per cent FSC-endorsed raw materials.

The first national standard for forest management was endorsed by the FSC in 1998 by approval of the Swedish FSC standard. Later that year, the policy on group certification for small forest landholdings was passed.

FSC members ratified a revision of Principle 9 and the addition of Criteria 6.10 and 10.9 to the P&C in 1999, as well as the definition of the 'precautionary approach' at the second general assembly in Mexico. By the end of the year, the tenth accreditation contract was signed.

In 2000, three additional certification bodies were accredited, seven national initiatives were endorsed, certification standards for Germany were approved, key policies were developed and refined, and more than 100 new members were added.

In September 2001, the first draft of the FSC's group chain-of-custody (CoC) certification policy was circulated for discussion and comment. In November, the FSC received the City of Göteborg, Sweden, International Environment Prize.

During 2002, the third general assembly was held in Oaxaca, Mexico, with more than 200 participants from 44 countries.

In early 2003, the FSC headquarters were relocated from Mexico to Bonn, Germany, according to a recommendation of the change management team. At the end of the year, more than 40 million hectares were certified against FSC standards for forest management.

STRUCTURE AND GOVERNANCE

The FSC is a *membership organization* that defines its policies through a general assembly of both individual members and designated delegates of member organizations. The general assembly is the highest governing body of the FSC, chaired by the chairperson of the FSC board of directors.

The *general assembly* is divided into three chambers, the social, environmental and economic chambers, each of which represents one third of the voting power. The purpose of the chambers is to maintain a balance of voting power between different interests without limiting the number of members. The votes within each chamber are divided equally among its members. At the international level, each chamber is further divided into two sub-chambers, 'northern' and 'southern,' each of which possesses 50 per cent of the voting power within each chamber. The northern sub-chamber comprises countries from high-income countries and the southern sub-chamber is comprises countries from the low-, middle- and upper middle-income countries as defined by the United Nations. The total voting weight of all individual members in each sub-chamber is limited to 10 per cent of the sub-chamber's total voting weight, a feature also meant to preserve a voting balance without restricting access.

Decisions of the general assembly are adopted by an affirmative vote of 66 per cent of the total membership in good standing. Forest industry and private forest owners who join the FSC participate through the economic chamber that they share with retailers, certifiers, consultants and other economic interests. Thus, the interests of forest landowners in combination with other economic interests constitute one third of the voting power in the FSC.

Operationally, the *executive director* and staff carry out day-to-day activities and are accountable to the member-elected board of directors. In addition, a number of board-appointed committees assist in managerial tasks, such as the executive committee and the dispute resolution committee, which handle any complaints – for example, about certification judgements or certified enterprises' behaviour.

The *board of directors* is made up of nine individuals (structured as the general assembly with three chambers and divided in north/south) who are elected by the membership for a three-year term. In order to ensure continuity, three directors retire at the end of each calendar year. They are replaced by postal ballot or general assembly vote.

Directors must either be individual members of the FSC or duly designated delegates of a member organization. They seek to represent the views and concerns of the (sub-) chamber that they represent (for example, south/social) in board deliberations, rather than simply reflecting the views of the organization with which they are affiliated. Directors who are duly designated representatives of member organizations may not be replaced by another representative of that organization. If a director leaves an FSC member organization and becomes affiliated with another organization in the same (sub-) chamber, the director may remain a member of the board if both organizations agree to this in writing. If the director becomes affiliated with an organization in another category, he or she must resign from the board. Commercial interests not demonstratively committed to the FSC, certification bodies and industry associations may not be represented on the board.

The board elects a chairperson and vice-chair by majority vote at the first meeting each calendar year and may also elect a treasurer and secretary. The board will normally seek to take decisions by consensus. If this is not possible, a vote will be taken. For a decision to be taken, six votes in favour are required. The quorum for board meetings is seven.

Below the international level, the FSC is decentralized through a network of national initiatives. The aims of the national initiatives are to:

- Promote, locally and regionally, the FSC and its mission.
- Make the FSC more accessible and more locally adapted.
- Encourage further local participation.
- Develop and test national forest stewardship standards.

- Work effectively with the international membership.
- Support successful implementation and monitoring of certification activities.

STANDARD

Development

In 1991, a working group met in San Francisco to discuss certification standards which different organizations could ascribe to, the need for constant 'objective' monitoring of certification programmes to protect both the public and certified producers, and the need for some type of organization to implement the above. This Certification Working Group involved over 30 organizations from 11 countries.

With the foundation of the Forest Stewardship Council in 1993, the founding members and the board of directors formally decided upon and finally ratified the first set of FSC principles and criteria (P&C) for forest stewardship in September 1994, at that time consisting of Principles 1–9. Principle 10 was ratified by the FSC members and board of directors in February 1996. The revision of Principle 9 and the addition of Criteria 6.10 and 10.9 were ratified by the FSC members and board of directors in January 1999.

Today's P&C are structured in 10 principles and 56 subordinate criteria.

Framework and operational forest stewardship standards

The P&C, by themselves, are not designed to be used as the basis for certification in the field, but constitute broad statements of overarching principles that require interpretation and adaptation in order to become operational. Their purpose is to provide a consistent framework for the development of locally defined forest management standards and to ensure forest certifications with the same results on a global scale.

While the P&C are, by definition, global, forest stewardship standards are national, regional or local. There are two types of such standards: the certification bodies' interim standards, which are applied where approved national or regional standards do not yet exist; and national, regional and local standards defined by national working groups. Both types of standards are approved by the FSC board for compliance with the global principles and criteria. Once a set of national or regional standards has been endorsed by the FSC, all local and international certifiers must, at a minimum, use those standards in their certification processes.

For these national or regional standards, the FSC requires evidence that all three chambers (social, economic and environmental) of the corresponding working group have been involved in developing the standard, that the standard has undergone a wide consultative process, and that it has been harmonized with standards of similar and/or neighbouring regions. Finally, national or regional standards should be reviewed periodically (every three to five years) and revised as necessary.

Performance-based standards

The FSC is a performance-based scheme. Issues such as biodiversity conservation (including set-aside areas), recognition of and respect for local peoples' rights, workers' rights, equal benefit-sharing, use of pesticides and genetically modified organisms (GMOs) must be addressed in all accredited FSC standards.

CERTIFICATION APPROACH

All certifications are carried out by independent certification bodies (third-party certification) who have undergone a thorough evaluation process and have been accredited by the FSC.

Only entities (individuals, enterprises, organizations, associations and authorities) that are responsible for the operational management of the forest resource, and legally entitled to manage the corresponding forest enterprise or forest land, are eligible for certification. An exception constitutes the certification of groups of small forest landholdings under the umbrella of one responsible legal entity, formally acting as group representative. In this case, clear contractual relationships between the group members and the certifiable legal entity, as well as a number of procedures for managing such a group, are required.

The FSC requires that evaluation teams have personnel who are sufficiently qualified to evaluate the social, environmental and economic impacts of the forest being evaluated in order to ensure compliance with the FSC principles and criteria. The minimum requirement is a single person with an appropriate range of experience and expertise.

Generally, some member(s) of each evaluation team of the certification body must have:

- previous experience in the country where an evaluation takes place;
- knowledge of the language of the country where the evaluation takes place.

In addition, for evaluation of forest management enterprises, some member(s) of the evaluation team of the certification body must have knowledge of:

- the forest management system (including silviculture) being implemented in the evaluated forest;
- the local forestry context.

For evaluation of chain of custody, some member(s) of the evaluation team must have:

- training in the certification of chain of custody – for example, as offered by national associations of certification bodies or specialist quality-assurance companies;
- specified in-house training.

Certification audits against FSC standards are made up of the following elements:

- document review, usually during an office visit, of the audited entity;
- field visits, where a variety of operations and other locations are visited to examine how plans are implemented, in reality;
- interviews with people – the audit team will talk to a wide range of people including staff, contractors and stakeholders in order to collect background information.

Audits are carried out by an evaluation team or an auditor of an FSC-accredited certification body and result in an assessment report of the audited enterprise.

The FSC requires a peer review of the assessment report to be conducted by at least two disinterested and credible reviewers with the technical capacity to assess the analytical quality of the report before final certification decisions are made. In addition, the certifying body must consider and document the actions taken in response to comments from the peer reviewers.

Once a certificate has been issued, it is valid for a certifier-specified time period, but may not exceed five years. Annual audits are required, entailing both field and office components, in order to ensure continued adherence to the standard and to monitor compliance with conditions placed on the certificate holder.

REQUIREMENTS FOR CONSULTATION OR PUBLIC INFORMATION

The FSC standards are developed through a participatory process that requires the active involvement of environmental, business and social sectors within its membership, as well as the participation of its members and other stakeholders in openly publicized national or regional standard-setting processes. Beyond standards development, stakeholder consultation is an integral part of the certification process. While protecting confidential information, FSC certification requires public summaries of certification evaluation reports, the certificate holder's management plan and the results of periodic monitoring of the certified forest:

- The FSC requires the certifying body to publish announcement of plans for a pending certification to stakeholders 30 days in advance of the certification audit.
- The FSC requires extensive consultation with stakeholders, experts and interested parties during the audit.
- The award of certification requires a summarized report by the certifying body that must be made available to the public. The contents of the report are itemized by the FSC. Among other prescribed information items, the summary must include an explanation of how stakeholder comments were considered and a list of any 'conditions' on which the certificate has been granted. The summary must be posted on the certifier's website, and the summary and the entire assessment report are also on file at FSC International.
- The certified landowner must make available, upon request, a summary of the key elements of the management plan and the results of monitoring the impacts of the management plan.

Finally, the FSC requires a summary report of positive accreditation of a certifier by the FSC to be made publicly available and includes specific provisions concerning the type of information that must be included in the summary. Likewise, a publicly available annual accreditation report is issued by the FSC providing a status update of all accredited certification bodies.

ARRANGEMENTS FOR SMALL FOREST OWNERS

To date, the FSC has paved two main pathways for small forest owners in order to access forest certification more easily:

1. Encouraging national working groups to define indicators that specifically address the constraints of small forest landholdings in forest stewardship standards ('appropriate to the scale and intensity of forest management' – for example, for environmental impact assessments, safeguards for endangered species, set-aside areas, management plans and monitoring intensity, and assessments of high conservation value forests).
2. Enabling the certification of group management schemes from 1998 in order to:
   - reduce evaluation costs per member;
   - reduce planning, management and other implementation costs;
   - increase opportunities to access new markets; and
   - increase training and educational opportunities for group members.

Through the current Small and Low-intensity Managed Forest (SLIMF) Project, the FSC aims to further facilitate the access of small forest owners – for example, by addressing the following issues:

- the cost of certification;
- the need for information about certification;
- the need to interpret standards and clearly define what is required to become certified; and
- the need for a more flexible evaluation system.

Some certification bodies have already introduced special schemes for small forest owners – for example, the Small Enterprise Scheme (SES) of SGS, which is specifically designed to reduce costs for small enterprises.

ACCREDITATION ARRANGEMENTS

The accreditation process and oversight of accredited certification bodies is internal to the FSC. The FSC accredits certification bodies for the provision of services in two categories. The first category comprises forest management certification, which is independent third-party evaluation of a forest management operation, according to specific environmental, social and economic standards in conformity with the FSC principles and criteria for forest stewardship. The second category is chain-of-custody certification, which is verification of the material flow of forest products from the forest, through processing and trade, to the end user.

The FSC has a comprehensive internal accreditation process that it developed for worldwide application, which is divided into an initial evaluation process and maintained through annual monitoring. It is rooted in International Organization for Standardization (ISO) processes, but is designed to operate under the jurisdiction of the FSC board of directors. The FSC accreditation process is mandatory for firms or not-for-profit entities wishing to provide FSC-endorsed certification services for forest management and/or chain of custody, and the FSC defines comprehensive, explicit requirements to obtain and maintain accreditation. Pending development and acceptance of national standards in a country according to well-defined FSC processes, certifiers develop and employ interim FSC standards, approved by FSC International.

The accreditation contract covers a five-year period and requires an annual review. The contract is subject to the following conditions:

1 payment of all FSC fees and charges;
2 compliance with all FSC requirements for annual monitoring;
3 compliance with all FSC requirements of corrective action requests (CARs);
4 compliance with all FSC disciplinary measures; and
5 compliance with other contractual obligations as specified.

Annual monitoring of accredited certification bodies includes annual desk audits by the FSC personnel at the offices of the certification body, as well as at least one office and field audit of a certificate holder in each category within the scope of accreditation. More audits may be conducted when certification bodies have a high number of certificates issued, and if there are complaints against the certification body.

In the future, the Accreditation Business Unit of the FSC International Centre will operate as an independent organization, thus ensuring full separation of standard-setting processes and the accreditation scheme, as required by ISO regulations.

CURRENT STATUS

Table A.1 presents accurate figures for the current status of the FSC as of 31 December 2003.

CHAIN OF CUSTODY

The FSC requires a chain-of-custody certification to be completed, which is separate and distinct from the certification of forest management operations, if organizations wish to use the FSC logo on a product at any stage of the 'value chain', beginning from the point of harvest to the point of final sale. Each entity controlling certified wood along the chain must be separately certified for chain of custody in order to use the logo on a product. However, the seller of the final product does not have to have chain-of-custody certification if the seller does not change the product's form or packaging and simply resells the final product, and if the buyer does not wish to make claims about an 'FSC product' himself.

TABLE A.1 Current status of the FSC as of 31 December 2003

Area of certified forest management	40,422,684 hectares
Forest management certificates	569
Chain-of-custody certificates	2853
Countries with forest management and/or chain-of-custody certificates	72
Accredited certification bodies	12
Accredited national initiatives	32
Countries with accredited national/regional FSC standards	9
Regional offices (Africa, Asia, Europe, Latin America)	4
FSC members (countries)	561 (61)

Chain-of-custody certification evaluations are conducted against a separate standard, comprising 6 principles and 25 criteria embedded within these principles.

The FSC's policy on *percentage-based claims* allows companies to market products containing less than 100 per cent FSC-endorsed raw materials. The purpose of this policy is to reduce the barriers facing industries that rely on large numbers of suppliers, not all of which are yet certified, and to reduce the perceived disadvantages faced by small forest properties that supply the same markets as larger, more integrated, forest firms.

However, FSC-certified products may not contain *uncertified raw materials* from one of the following categories:

- wood that has been illegally harvested;
- wood from genetically modified trees;
- wood from areas where there is a clear demonstration of violation of traditional, customary or civil rights, or of serious extant disputes with indigenous peoples or other social stakeholders, involving confrontation or violence;
- wood from uncertified high conservation value forests (HCVFs).

LABELLING AND LOGOS

The FSC has three registered trademarks, including:

1 the name Forest Stewardship Council;
2 the acronym FSC; and
3 the FSC logo, comprising the tick (checkmark) and tree symbol and the acronym FSC.

The FSC logo is used both as a generic programme logo as well as a product label. This logo is an integral part of the FSC programme and has not changed since the inception of the FSC.

Trademark services (for example, licensing and monitoring) are provided through trained nominated agents and FSC-accredited certification bodies which act as authorized agents for the administration of the FSC trademarks on the basis of specific contractual agreements.

The accredited certification bodies are required according to the terms of the accreditation contract with FSC to administer the use of the trademark by those individuals and organizations to whom they have issued FSC-endorsed certificates. Nominated agents are responsible for approving and controlling the use of the FSC logo by non-certificate holders in their geographical territory. The FSC secretariat is responsible for approving and controlling the use of the FSC logo by all other users and those of the national initiatives, the nominated agents and accredited certification bodies, and has overall responsibility for monitoring the use of the logo.

The primary use of the FSC logo is to promote products sourced from certified forests. The logo may be used on these products themselves and on their labelling and packaging (*on-product use*). The certification body must ensure that the product falls into one of the following categories:

- solid wood products, of which the wood content is certified by an FSC-accredited certification body as coming from FSC-endorsed certified forests;
- collections of solid wood products, which may display the logo on the collection unit; however, individual units generally cannot carry the logo;
- non-timber forest products that are certified by an FSC-accredited certification body as coming from FSC-endorsed certified forests;
- chip and fibre products for which at least 17.5 per cent by weight of the total chip or fibre used in manufacturing the product line is FSC certified and at least 30 per cent by weight of the new virgin wood chip or fibre used in manufacturing the product line is FSC certified;
- assembled wood products comprised at least 70 per cent by volume of the wood and/or virgin fibre from the FSC-certified wood.

The FSC logo may also be used for *off-product* uses, which comprise three broad categories: product promotion; promoting the FSC-endorsed certified forestry; and promoting the FSC itself and/or their association with and support of the FSC. Examples of these are catalogues, leaflets, advertising and reports.

In nearly all cases, logo use must comply with the *graphic requirements* set forth by the FSC. The logo must be at least 10 millimetres high, and bear the FSC copyright claim: 'The FSC Trademark © 1996 Forest Stewardship Council AC'. The exclusion zone surrounding the logo must be free of all text and graphics. The colour and contrast of the logo must ensure that the image is clear and distinct. There may be no changes to any of the design elements of the logo, including the copyright symbol, lettering, thickness and ratio of height to width. Where appropriate, the certificate registration code must be properly displayed. Whenever possible, an accurate statement must be displayed.

POLITICS AND PERCEPTIONS

Under the numerous supporters of the FSC programme, two major players must be emphasized as they exert considerable political, social or market influence.

The first group consists of environmental non-governmental organizations (ENGOs), both international operating networks as well as national organizations, examples of which are the World Wide Fund for Nature (WWF), Greenpeace and Friends of the Earth (FoE).

The second group is made up of the retail sector, with the leading DIY chains, furnishing houses and mail-order businesses. Examples include B&Q (UK), Home Depot (US), IKEA (Sweden), Migros (Switzerland), Neckermann (Germany), OBI (Germany), OTTO (Germany), and many others.

For many ENGOS, the FSC programme constitutes the only credible certification system because it is based on:

- objective, comprehensive, independent and measurable performance-based standards – both environmental and social;
- equitable and balanced participation of a broad range of stakeholders;
- a labelling system that includes a credible chain of custody;
- reliable and independent third-party assessments and includes annual field audits.

Furthermore, it:

- is fully transparent to the parties involved and the public;
- takes place at the forest management unit level (and not at country or regional level);

- is cost effective and voluntary;
- positively demonstrates commitment from the forest owner/manager towards improving forest management;
- is applicable globally and to all sorts of tenure systems in order to avoid discrimination and distortion in the marketplace.

For the retail sector, the support of the ENGOs is crucial. Apart from this, for marketing purposes, they require a global scheme that can be applied to tropical, temperate or boreal forests and that operates under one logo for the emerging forest products.

INFORMATION ABOUT THE SCHEME

Information and documentation about the FSC are available from the following websites:

- www.fscoax.org;
- www.fsc-info.org;
- www.gtz.de/forest_certification.

Contacts

FSC International Center
Charles-de-Gaulle 5
53113 Bonn
Germany
Tel: +49-228-367-660
Email: fsc@gsc.org

LEMBAGA EKOLABEL INDONESIA©
The Indonesian Ecolabelling Institute

A1.4 Indonesian Eco-labelling Foundation (LEI)

TYPE OF SCHEME

This scheme is a national, voluntary initiative with certification carried out by independent certification bodies accredited by Lembaga Ekolabel Indonesia (LEI). LEI is an independent organization that promotes sustainable natural resource and environmental management through certification.

SCOPE

The system has been established for Indonesian production forest management types. There are three forest management types for production forest in Indonesia, characterized by differences in silviculture and community involvement: natural production forest management; plantation forest management; and community-based forest management. Production forest is one amongst other Indonesian forest categories in a national land use context, which also include protection forest and conservation forest.

DATE SET UP AND HISTORY

The establishment dates of each certification system that LEI has developed are as follows:

- 1998: sustainable natural production forest management (SNPFM) certification system;
- 2000: chain-of-custody certification system;
- 2002: sustainable forest plantation management (SFPM) certification system;
- 2002: sustainable community-based forest management (CBFM) certification system.

LEI was initially established as an independent multi-stakeholder working group in 1994. Chaired by Professor Emil Salim, a distinguished public figure highly respected by NGOs, the private sector, government officials and academic communities, both domestic and abroad, the working group received full support from Minister of Forestry Djamaludin Suryohadikusumo.

The group's task was to develop a national system and institutional capacities for the certification of natural production forest management. The system includes a set of standards (criteria, indicators, norms and verifiers) for sustainable natural production forest management, assessment procedures, minimum requirements, certification decision-making processes and an appeal mechanism. In developing the system, the group has used, as references, the International Tropical Timber Organization's (ITTO's) criteria and indicators (C&I) for sustainable forest management (SFM), the Forest Stewardship Council's (FSC's) principles and criteria, the International Organization for Standardization's (ISO's) environmental management system, and the Tropenbos hierarchical framework for the formulation of SFM standards.

In order to ensure that the system developed is acceptable to all stakeholders, the group has adopted a transparent multi-stakeholder process. A series of multi-stakeholder consultations were held to ensure that inputs from NGOs, the private sector, government institutions and the academic community were accommodated.

On 6 February 1998, the group was officially established as an independent entity named the Lembaga Ekolabel Indonesia (LEI) Foundation. Professor Emil Salim chaired the foundation from

February 1998 until September 2001, and was then succeeded by the current chair, Djamaludin Suryohadikusumo.

In promoting eco-labelling certification, LEI adheres to the following fundamentals:

- A multi-stakeholder process is adopted in all phases of system development, certification implementation and monitoring.
- Certification is undertaken on a voluntary basis by an independent third party.
- Certification is a means for promoting sustainable forest management, not an end in itself.

LEI's vision is to become an independent, constituent-based institution whose main objective is to promote sustainable natural resource and environmental management by the application of a credible eco-labelling certification system.

LEI's aims are to:

- Promote the formulation and implementation of public policies designed to ensure sustainable management of natural resources and the environment.
- Develop and implement a certification system for sustainable natural resources and environmental management.
- Develop and implement an accreditation system to ensure thorough supervision and monitoring of certification implementation.
- Establish national (including human resources and institutional) capacities to implement a certification system for sustainable natural resources and environmental management.

LEI has completed the development of SFM certification systems for natural production forest management, forest plantation management and community-based forest management, as well as a chain-of-custody certification system.

In order to establish international acknowledgement, LEI has been working together with the FSC in a joint certification programme (JCP) for the implementation of a natural production forest certification system since 1999.

STRUCTURE AND GOVERNANCE

The organizational and management structure of LEI is provided in Figure A.6.

The board of trustees is the highest body in LEI. A board of directors, appointed by the board of trustees, is responsible for day-to-day monitoring and overseeing; an executive body, led by an executive director who is responsible to the board of directors, handles the running of the secretariat and the functions of LEI as an accreditation body.

The board of trustees consists of eight members who were appointed by the founding fathers of LEI. Each member is a prominent and credible individual with a well-known longstanding commitment to the environment. The members of the board come from different backgrounds, reflecting the multi-stakeholder nature of LEI certification.

The board of directors consists of three members elected among the members of the board of trustees through a meeting of the board of trustees. While, currently, the board of directors appoints the executive director of LEI, it is important to note that LEI is in a transition period towards a constituent-based organization (due to be completed by the end of 2004).

Major policy decisions, including changes to the system, are presented to multi-stakeholders, to be decided upon through a public consultation process. With transparency as the main principle, information is posted through the website and other media; a public consultation is conducted to invite inputs. Decisions are posted and relayed to all relevant stakeholders through various media. Day-to-day management is handled by the executive body, which is led by the executive director.

FIGURE A.6 Organizational and management structure of LEI

LEI acts as an accreditation body, which accredits legal entities such as certification bodies (CB) to conduct certification using the LEI system. As an accreditation body, LEI monitors and controls the implementation of the LEI certification scheme. One of LEI's functions is to ensure that the national capacity to conduct certification is in place.

The training institution is responsible for securing the availability of qualified assessors and decision-making panel members through the implementation of standardized training. The personnel registration body (PRB) conducts registration and monitoring of personnel involved in the certification as the assessors or as the registered decision-making panel member.

Provincial communication forums (PCF) are LEI's partners in the provincial region; these multi-stakeholder groups provide a source of reliable information concerning unit managements during the certification process. The Certification Review Council (CRC) is an independent institution that handles disputes resolution. CBs are accredited institutions that conduct certification using the LEI system. Currently, there are three CBs operating in Indonesia using the LEI system. Figure A.7 outlines the national structure of LEI certification.

STANDARD

The standard was developed through a multi-stakeholder process that involved government, academics, environmental and social NGOs, forest concession holders' associations, and other forestry practitioners. Until now, LEI has developed four certification systems: a certification system for natural forest management (developed during 1994–1998); a certification system for plantation forest management (1998–2002); a certification system for community-based forest management (2000–2002); and a chain-of-custody certification system (1997–2000). At the end of 1999, the certification system for natural forest was refined after LEI, the FSC and certification bodies under FSC accreditation (SGS Qualifor and Smartwood) conducted a joint assessment based on the LEI standard. All of the systems are required to be reviewed every five years.

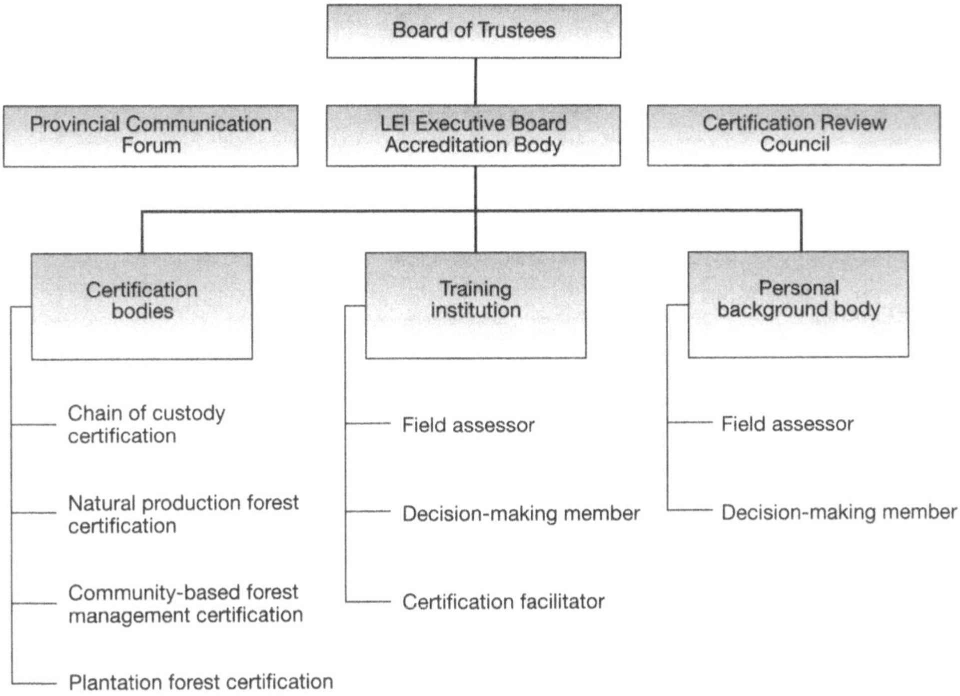

FIGURE A.7 National structure of LEI certification

The standard was arranged in a hierarchical way, with SFM as the primary goal, followed by three major forest function principles (economic/production, environmental and social), the criterion of each principle, the indicators, and the norm of each indicator. There are three types of indicators that may be measured in the field: inputs, processes and outcomes of the forest management activities run by the forest concession holders. In other words, at the indicator level the standard is an actual performance of forest management. Therefore, the standard type is both performance and system. The hierarchical framework of the LEI's certification standard is described in Figure A.8.

The performance requirements of each indicator are defined within norms of indicators, which are classified into three to five classes: excellent, good, fair, poor and bad. Each class represents a combination of outcome or condition, processes or inputs, which must comply with the international standard/convention, government regulations on that particular issue or other widely agreed conditions of forest management measures (in cases where there are no performance standards, either quantitative or qualitative, for a specific issue).

In the decision-making process, or when comparing the actual performance of forest management with performance requirements, the decision-makers also have to consider the local biophysical characteristic of a forest management unit (that is, forest management unit typology). The decision-making process applies the analytical hierarchy process (AHP) as a decision-making tool.

CERTIFICATION APPROACH

An independent certification body that is accredited by LEI, based on LEI's accreditation standard, conducts the certification process.

Certification applies to a forest management unit managed by individual organizations, either government, private or a group of communities (in the case of community-based forest management certification).

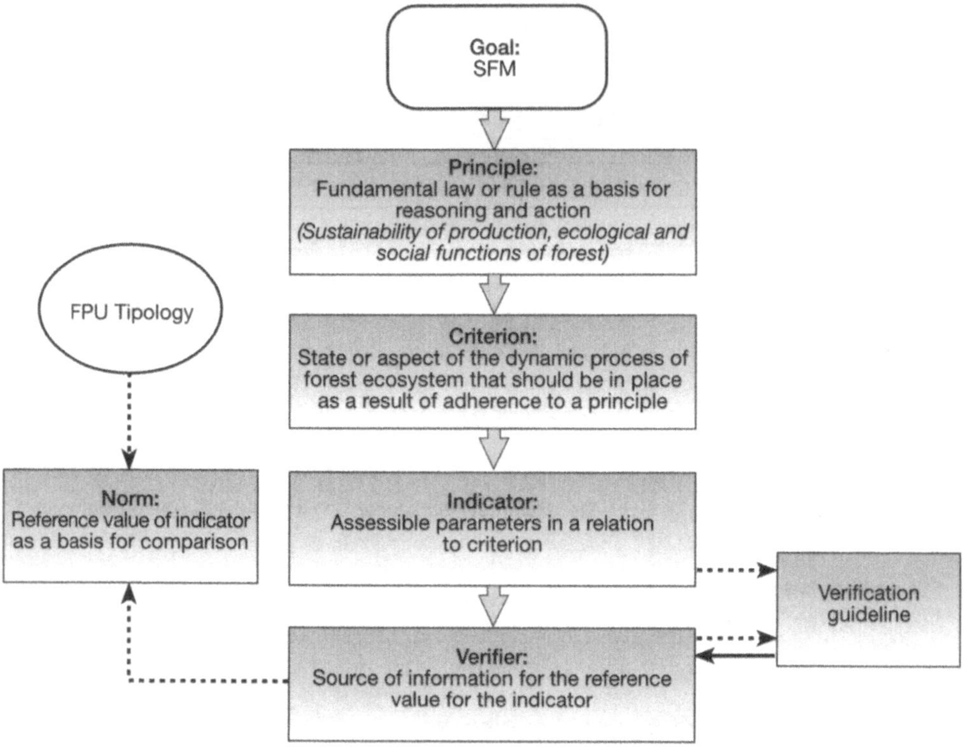

Source: adapted from Lammerts van Bueren and Blom (1997)

FIGURE A.8 Hierarchical framework of LEI's certification system standard

In LEI's certification system, separate parties conduct field assessments and are responsible for decision-making. The requirements for field assessors are:

- They must have at least three years' experience in forestry, on either ecological aspects, social aspects or forest production.
- They must have passed the training for assessors.

There are three grades of assessors, based on their assessment experience and training. All assessors are trained by LEI and registered in the personnel registration body. The decision-makers are qualified experts with a good understanding and experience of forestry. In the final decision-making process, three of the six expert members are local experts who are nominated by a provincial communication forum in which the forest management unit being assessed is located.

Certificates are valid for five years, during which time periodic surveillance audits are conducted. The frequency of surveillance depends upon the performance class:

- *bronze class*: surveillance will occur a minimum of four times;
- *silver class*: surveillance will occur a minimum of minimum three times;
- *gold class*: surveillance will occur a minimum of twice .

The independent certification body accredited by LEI conducts the surveillance.

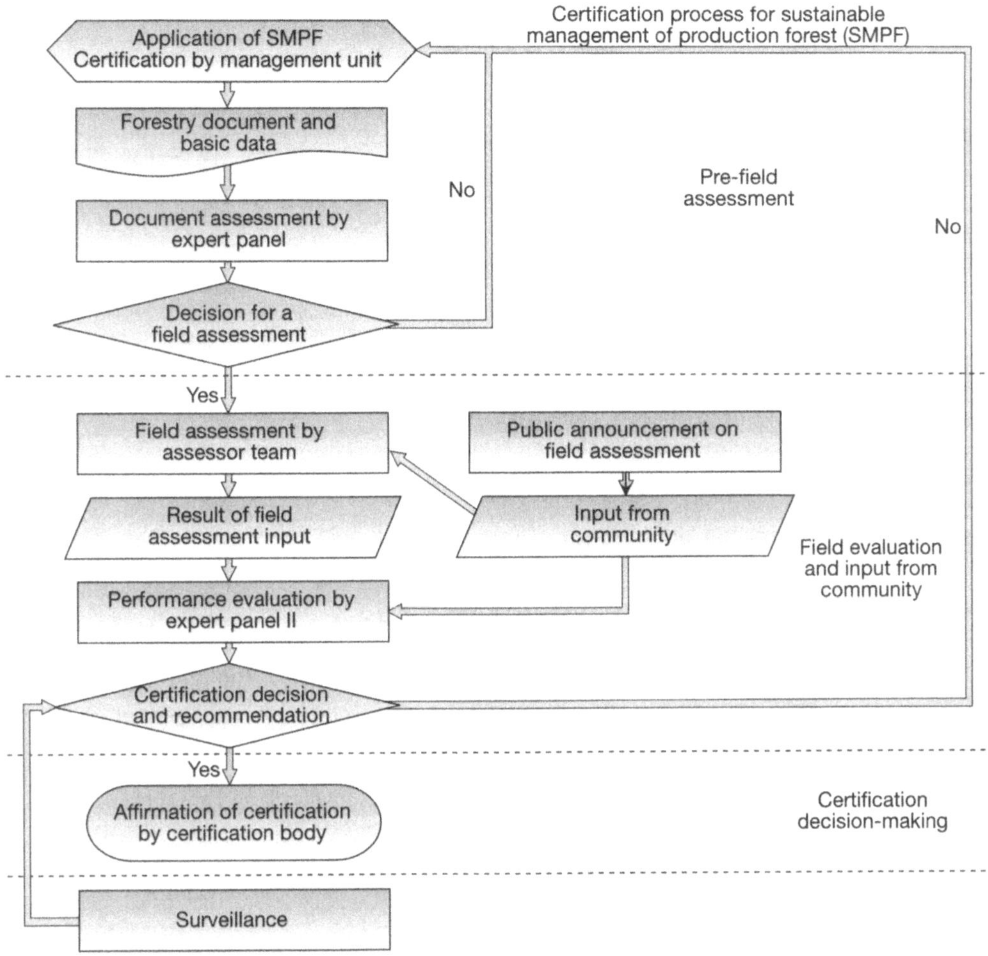

FIGURE A.9 Certification process of sustainable production forest management

REQUIREMENTS FOR CONSULTATION OR PUBLIC INFORMATION

Public consultation is a procedure that must be performed by certification bodies prior to field assessment. The aim is to make the public aware of the certification process of a particular forest management unit, and to invite input from the public for the assessment and decision-making process. The public consultation begins with a public announcement in mass media (newspapers) at both national and local levels, as well as via the internet. This invites the public to deliver information or comments about the forest management unit that is being assessed to the certification body. Furthermore, at least 30 days after the announcement, the forest certification body has to conduct a public meeting that invites relevant stakeholders at a local level, including the provincial communication forum. All relevant input is used for the field assessment and the decision-making process.

A summary of the certification process and general information on the certified forest management unit – for example, location, annual production, forest type and contribution to the local development – are put into the public domain via the websites of either the certification or accreditation body. More detailed information related to the certification is available upon request.

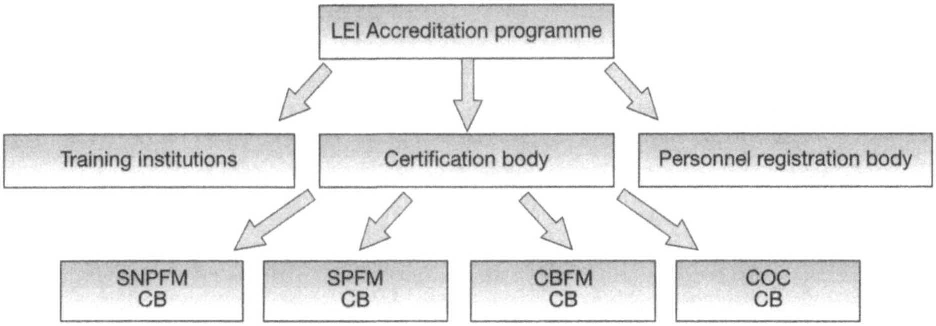

Notes:
CB: certification body
SNPFM: sustainable natural production forest management
SPFM: sustainable plantation forest management
CBFM: community-based forest management
CoC: chain of custody

FIGURE A.10 LEI accreditation scheme

ARRANGEMENTS FOR SMALL FOREST OWNERS

For small-scale forest management, managed by individuals or community groups, LEI has established a specific certification system: the community-based forest management certification system. Due to differences in scale, the type and intensity of community involvement within management, tenure aspects and capital involved, a different set of indicators for measurement and a less complex procedure are required. Requirements for the parties implementing this system are similar to those for natural and plantation forest management.

ACCREDITATION ARRANGEMENTS

These ensure that there is no conflict of interest between LEI and the certification bodies. The LEI accreditation scheme is addressed to ensure the eligibility of certification bodies conducting LEI's system.

It is a prerequisite of the LEI accreditation scheme that the training institution, certification bodies and personnel registration body are independent. Criteria for further evaluation are organizational management, human resources development, financial resources and experience.

CURRENT STATUS

Since 1999, SFM certification for natural production forest management in Indonesia has been jointly conducted by LEI and the FSC, under a joint certification programme (JCP LEI–FSC). According to the JCP protocol, LEI's criteria and indicators are to be used in the certification assessment in Indonesia, and only forest management units (FMUs) that meet both LEI and FSC requirements will be granted SFM certificates.

Three LEI- and two FSC-accredited certification bodies are now in operation for the certification of sustainable natural production forest management in Indonesia.

As of December 2003, LEI has trained:

- 273 personnel as SNPFM assessors;
- 81 personnel as CoC assessors;
- 26 personnel as SPFM assessors;

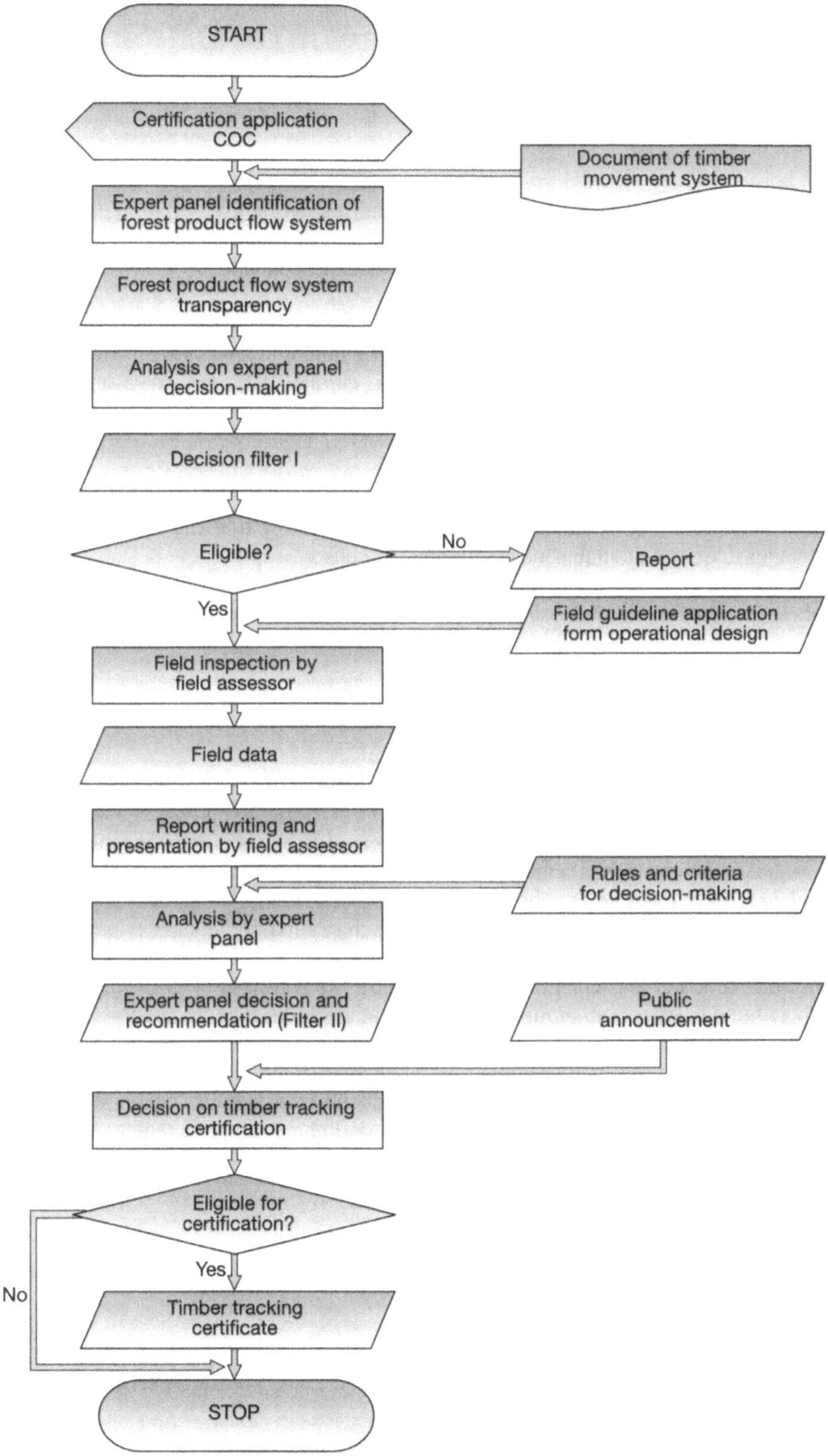

FIGURE A.11 **LEI flow chart of timber-tracking certification procedures**

FIGURE A.12 LEI logo standard

- 81 members for the SNPFM decision-making panel (expert panel);
- 13 members for the CoC decision-making panel (expert panel);
- 13 members for the SPFM decision-making panel (expert panel).

Additionally, LEI has established provincial communication forums in 11 provinces.

As of December 2003, the certification assessment under the JCP has been conducted in 14 FMUs, with a total area of 2,527,700ha, with one FMU being certified (PT Diamond Raya Timber in Riau) with an area of 90,975ha.

CHAIN OF CUSTODY

The chain of custody developed by LEI is aimed at ensuring that the timber used for the wood industry comes from sustainable managed forest (certified).
Chain of custody is based on the principle of assessing timber flow right back to a certified source. The whole process for the chain-of-custody certification is explained in Figure A.11.

LABELLING AND LOGOS

Parties eligible to use LEI's logo are:

- *Certificate holders*: this comprises management units that have been given a certificate by LEI's accredited certification bodies, LEI's accredited training institutions and the personnel registration body. The type of use of logo by certificate holders is categorized into:
  - *on-product use*: the logo is attached to the product itself and on package labels;
  - *off-product use*: the logo is attached to the certificate holder's official report, company profile, press release, promotion, advertisement, brochures, pamphlets, letter paper, name card and other marketing media mechanism.
- *Non-certificate holders*: this comprises institutions/parties that have been given approval from LEI to display LEI's logo in certain media. Non-certificate holders include government institutions, NGOs, educational and training bodies, retailers and wholesalers. The purpose of this logo use is to promote LEI or products attached with LEI's chain-of-custody label. The marketing media includes sales promotions, posters, window-stickers, pamphlets, website/internet sites and leaflets, used under the scope of sustainable natural resource management. In this case, the nature of logo use for non-certificate holders is for off-product use.

LEI's standard logo is illustrated in Figure A.12.
Figure A.13 depicts the LEI logo that may be used in other formats.

POLITICS AND PERCEPTIONS

The LEI certification system was developed through a multi-stakeholder process that involved government, forest entrepreneurs, NGOs and academics who are committed to achieving sustainable forest management.

FIGURE A.13 LEI logo without annotation

This system is strongly supported by the government (Ministry of Forestry), which has made the LEI standard the national standard for sustainable forest management. The association of forest entrepreneurs, the wood industry, academics and NGOs concerned with environmental and social matters also support the implementation of the LEI system as one method of promoting SFM.

To further strengthen the credibility of the certification system, LEI is in the process of formalizing the support from stakeholders by becoming a constituent-based organization (CBO).

INFORMATION ABOUT THE SCHEME

More detailed information about the LEI certification system is provided on the following websites:

- www.lei.or.id;
- www.groups.yahoo.com/group/ecolabelling.

Contact

Secretariat of LEI
Jalan Taman Malabar
No 18 Bogor 16151, Indonesia
Email: lei@indo.net.id
Tel: +251-340-744
Fax: +251-321-739

© 2000 Malaysian Timber Certification Council

A1.5 Malaysian Timber Certification Council (MTCC)

TYPE OF SCHEME

The MTCC timber certification scheme is a national scheme run by an independent purpose-created body, the Malaysian Timber Certification Council (MTCC). However, as preparation for endorsing the scheme under the Programme for the Endorsement of Forest Certification (PEFC), the scheme will be modified to involve the national accreditation body (Department of Standards Malaysia, or DSM) in the accreditation of certification bodies.

SCOPE

The MTCC certification scheme covers the three regions in Malaysia: Sabah, Sarawak and Peninsular Malaysia. The forest management standard – the *Malaysian Criteria, Indicators, Activities and Standards of Performance for Forest Management Certification* (MC&I) – covers the three main native forest types (dry inland forest, peat swamp forest and mangrove forest).

DATE SET UP AND HISTORY

The MTCC was established in October 1998 to develop and operate a voluntary national timber certification scheme in Malaysia. The formation of the MTCC was the result of discussions among government ministries, forestry departments, research institutions, the environment department, universities, timber promotion bodies, the standards institute, timber industry associations and environmental non-governmental organizations (ENGOs) regarding the need to establish an independent and new organization to develop and operate the national timber certification scheme in Malaysia. Formerly known as the National Timber Certification Council, Malaysia (NTCC, Malaysia), it is incorporated under the Companies Act 1965 as a company limited by guarantee. The MTCC began its operation in January 1999. The MTCC scheme started operating in October 2001.

STRUCTURE AND GOVERNANCE

A board of trustees, comprising a chairman and eight other members, is the governing body which decides the overall policy and direction in carrying out MTCC's activities. In addition to the chairman, the members comprise two representatives each from academic and research institutions, the timber industry, NGOs and government agencies and are appointed for a two-year term. The board meets four times yearly, and decisions are made by consensus.

A certification committee, established by the board, has been given the responsibility to:

- decide on applications for forest management and chain-of-custody certification, based on assessment reports submitted by independent assessors;

- decide on the registration of the independent assessors; and
- consider applications for registration as peer reviewers.

The certification committee comprises four members representing each of the stakeholder groups on the board. A committee comprising the other members of the board considers appeals against the decision of the certification committee.

A chief executive officer, who is assisted by a senior manager, two managers and an executive, together with four administrative staff, manages the day-to-day operation of the MTCC. The MTCC management is responsible for implementing the decisions made by the board.

Decisions of the board that are relevant to the stakeholders are communicated to them directly, as well as through the MTCC website and other means, such as press releases. The list of forest management units (FMUs) and companies that have been awarded the MTCC certificate is made available on the website.

STANDARD

The MTCC certification scheme is being implemented using a phased approach. The standard currently being used for assessing permanent reserved natural forests in FMUs is the *Malaysian Criteria, Indicators, Activities and Standards of Performance for Forest Management Certification* (MC&I, 2001). The MC&I (2001) is based on the 1998 International Tropical Timber Organization (ITTO) *Criteria and Indicators for Sustainable Management of Natural Tropical Forests*. It incorporates the corresponding standards of performance (SOPs) for Peninsular Malaysia, Sabah and Sarawak, which were identified during the regional- and national-level multi-stakeholder consultations held in 1999. The stakeholder groups involved in these consultations were from government agencies, research institutions, universities, workers' unions, environmental NGOs, local communities, timber industry associations, women's organizations and timber promotion bodies.

For the next phase of its scheme, the MTCC will use a new standard, the *Malaysian Criteria and Indicators for Forest Management Certification* (MC&I, 2002), which has been developed using the principles and criteria (P&C) of the Forest Stewardship Council (FSC) as the framework. This standard was also developed and adopted in October 2002 through regional- and national-level multi-stakeholder consultations. MTCC has set a target date of January 2005 to use the MC&I (2002). It is envisaged that the standard will be reviewed every five years following its initial use.

Both of these standards are performance-based standards for assessing the FMUs. The performance requirements are expressed as 'standards of performance' in the MC&I (2001) and as 'verifiers' in the MC&I (2002).

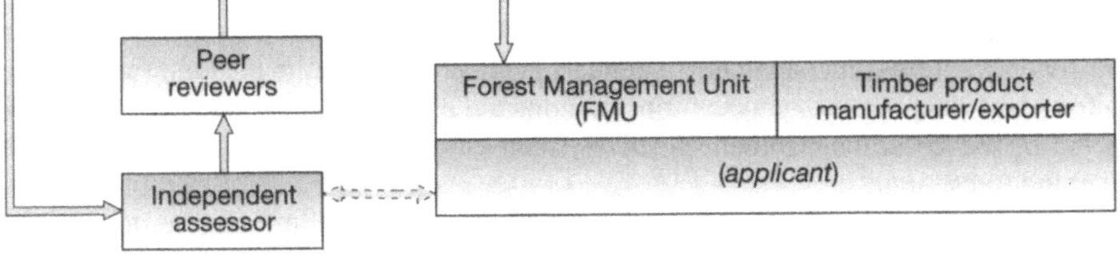

FIGURE A.14 MTCC timber certification scheme

CERTIFICATION APPROACH

Under the timber certification scheme outlined in Figure A.14, the MTCC – as the timber certification organization – receives and processes applications for certification (from FMUs, in the case of forest management certification, and from timber product manufacturers or exporters, in the case of chain-of-custody certification), arranges for assessments to be carried out by registered independent assessors, and prepares evaluation reports based on the assessment reports of the assessors, clients' comments and peer review reports (in the case of forest management certification), which are submitted to the certification committee, who decides on whether an application merits a certificate award.

Requirements for independent assessors

Companies or organizations wishing to be registered with MTCC as independent assessors have to comply with MTCC's terms and conditions for registration. A team comprising a minimum of three qualified and experienced auditors is required for forest management certification. Professional foresters with expertise and a minimum of five years' field experience in forest management, or non-foresters with similar experience in fields related to forest management – such as ecology, environmental sciences, biology, sociology and forest economics – can qualify as auditors for forest management certification.

For chain-of-custody certification, at least two auditors at professional level who have the necessary training, work and audit experience are needed for the assessments. The auditors should also be familiar with the local wood-based industries, as well as with applicable local regulations and documentation relevant to chain-of-custody certification.

Peer review

The peer review process is only necessary for evaluating the assessment report for forest management certification. Normally, two peer reviewers are involved in one assessment and they are appointed by the MTCC, based on an assessment of the FMU. The basic aim of a peer review process is to obtain a second opinion on the compliance of the FMU concerned with the requirements of the certification standard.

Interested individuals are required to meet certain criteria for registration as peer reviewers. They are to have, amongst other requirements, a minimum of five years' experience in forestry or expertise related to various aspects of sustainable forest management. such as ecology, environmental sciences, biology, sociology and forest economics; the necessary training and work experience to assess the adequacy of the assessment reports; and a good understanding of the certification standard and its related assessment procedures.

REQUIREMENTS FOR CONSULTATION OR PUBLIC INFORMATION

FMUs are assessed by the independent assessors to the requirements of the forest management standard (MC&I). The scope of the assessment is limited to the forest management system and practices within the permanent reserved forests (PRFs) of the FMU. The assessment is conducted in order to evaluate current documentation and field practices in forest management and to assess the level of compliance with the requirements of the standard.

During the visits to various sites within the FMU, the auditors consult the relevant individuals who are involved in the forest management practices, such as the forest managers, forest workers and contractors, regarding these practices. Consultations are also held with persons or groups who may be affected by the forest management practices, such as local communities living within or near the vicinity of the forest area.

Therefore, the itinerary must also include time for these consultations. The main aim of such consultations is to obtain feedback on how forest management practices are actually implemented in the field, the practical problems encountered and the results of these practices.

The list of certified FMUs and timber product manufacturers and exporters are put on MTCC's website. The website also provides summaries of the assessment reports of the FMUs that have been awarded the certificate. In addition to ensuring the transparency of the MTCC scheme, putting such information in the public domain serves to encourage feedback from the public regarding the forest management practices of the FMUs concerned, thus assisting MTCC to monitor the FMUs in order to ensure continued compliance to the standard.

ARRANGEMENTS FOR SMALL FOREST OWNERS

The MTCC scheme does not include specific arrangements for small forest owners since the certification standard deals with natural forests and, in Malaysia, the natural forests are owned by the respective state governments.

ACCREDITATION ARRANGEMENTS

Currently, the MTCC serves as the certification body (CB) and the independent assessors are registered with MTCC in compliance with certain terms and conditions. In future, the independent assessors will take on the role of CBs and will be accredited to the DSM. The MTCC will then serve as the national governing body (NGB) for the scheme.

CURRENT STATUS

As of the end of 2003, seven FMUs in Peninsular Malaysia covering 4.1 million hectares of permanent reserved forests (PRFs) have been awarded the *Certificate for Forest Management*. A total of 38 timber companies have been awarded the *Certificate for Chain of Custody*.

CHAIN OF CUSTODY

Applicants for chain-of-custody certification are assessed by the independent assessors for compliance against the requirements as stipulated in the MTCC document entitled *Requirements and Assessment Procedures for Chain-of-Custody Certification* (RAP/COC). Two auditors are involved in the site assessment; one is the lead auditor.

During the site assessment, the assessment team carries out the following activities:

- a review of all pertinent records and reconciliation of the findings, including spot checks of the product on the ground;
- inspection of the mechanisms for identifying certified wood-based material, its utilization, and separation between certified and non-certified products; and
- reconciliation of the documented procedures with actual practices.

There are two systems available for an applicant who wishes to comply with the chain-of-custody requirements:

1 physical separation system; and
2 minimum average percentage system.

The minimum average percentage system allows the applicant to mix certified and non-certified wood in the course of production if minimum average percentages of input of certified wood are met. In this approach, the total batch of products can be labelled as certified when the amount of certified material in the input batch exceeds the set minimum average threshold.

FIGURE A.15 **MTCC logo**

LABELLING AND LOGOS

The MTCC logo on a product provides an assurance that the material from which it is made originated from forests that have been certified by MTCC. The logo is copyrighted and has also been registered as a trademark. The use of the logo has to conform to the rules and procedures specified in the MTCC *Logo Guide for Certificate Holders*.

Only certificate holders are allowed to use the MTCC logo on their products. The use of the MTCC logo requires the inclusion of the following key elements (see Figure A.16):

- the MTCC logo;
- the MTCC copyright claim ('© 2000 Malaysian Timber Certification Council');
- the *Certificate for Forest Management or Certificate for Chain of Custody* number of the certificate holder;
- an approved on-product or off-product statement;
- a mean minimum percentage of MTCC-certified material of the total wood, chip or fibre used in making the product or in the batch manufacturing process (for products using the minimum average percentage system).

POLITICS AND PERCEPTIONS

The phased approach taken by the MTCC in implementing its timber certification scheme has received broad support from most of the stakeholder groups. This is reflected in the large number of organizations that have attended the regional- and national-level consultations coordinated by the MTCC.

However, a small group of environmental and social NGOs have excluded themselves from the standard-setting process when MTCC indicated that it was not in a position to meet certain 'demands' that were made by them because:

- Some of these 'demands' required amendments to the state laws.
- Although the other 'demands' could be included in the certification standard, the extent and manner of their inclusion should still be the subject of discussions and negotiations with the other stakeholder representatives as part of the standard-setting process.

The main reason why these NGOs oppose the MTCC scheme and timber certification, in general, is that they are primarily concerned with protecting the native customary rights of the local communities. One key requirement for certification is security of tenure over the forests, which is normally achieved by gazetting the forests as PRFs. The process of gazetting is, however, seen by these NGOs as extinguishing the native rights of the local communities over the forests.

The MTCC scheme has received good support from the FMU managers and the timber industry. However, many timber product manufacturers and exporters are still reluctant to apply for chain-of-

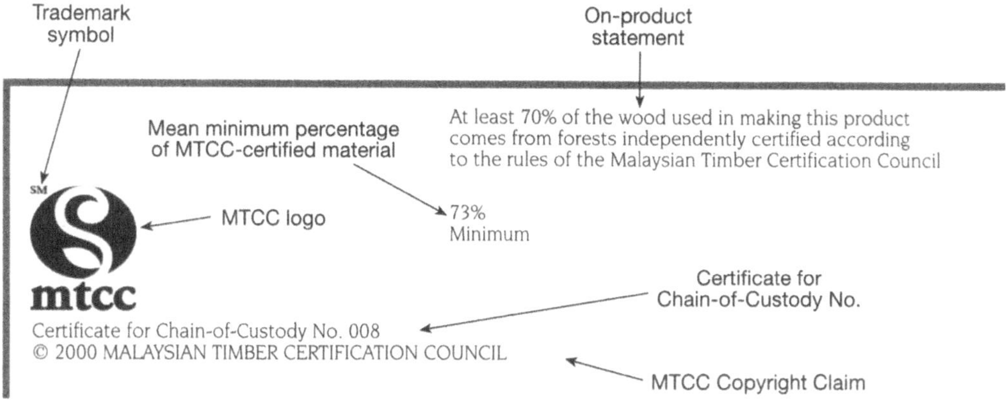

FIGURE A.16 Key elements to be included on a label bearing the MTCC logo

custody certification: they feel that their overseas buyers are not willing to pay a higher price for their certified timber products, despite the extra effort and expenses that they have to put in to achieve certification.

INFORMATION ABOUT THE SCHEME

Further information about the MTCC scheme can be obtained from the MTCC website at www.mtcc.com.my.

Contact

19F, Level 19, Menara PGRM
8 Jalan Pudu Ulu
Cheras
56100 Kuala Lumpur
Malaysia
Tel: +60-392-005-008
Email: mtcc@tm.net.my

A1.6 Programme for the Endorsement of Forest Certification (PEFC) Schemes

TYPE OF SCHEME

The PEFC Council is an umbrella organization for the mutual recognition of independent national or sub-national forest certification schemes. The PEFC Council provides a common PEFC logo for labelling of those products that originate in forests certified according to the PEFC-endorsed schemes.

SCOPE

The PEFC council provides mutual recognition for forest certification schemes from all over the world and does not set any limitation concerning the type of forests covered by participating schemes.

DATE SET UP AND HISTORY

The PEFC Council was established as the Pan-European Forest Certification Council on 30 June 1999 in Paris by 11 national organizations representing or developing national forest certification schemes in Austria, Belgium, the Czech Republic, France, Finland, Ireland, Norway, Portugal, Spain, Sweden and Switzerland. Various European associations representing forest owners and forest industries joined the PEFC Council as extraordinary members. This official launch of the organization followed one year of very intensive discussion amongst all stakeholders, including a series of seminars and meetings at international as well as national level.

One year later, in May 2000, the first forest certification schemes from Finland, Sweden and Norway were officially endorsed by the PEFC Council, followed in the same year by the German and Austrian forest certification schemes. By the end of 2000, the area of PEFC certified forests exceeded 23.5 million hectares.

During 2001, the first non-European countries – the US and Canada – expressed their interest in the PEFC and joined the organization as members. This was an important milestone in the globalization of the PEFC Council and the development of its worldwide framework for mutual recognition of national schemes.

During 2001, the Czech and Swiss forest certification schemes were endorsed and the total certified area exceeded 41 million hectares.

Following the extension of the PEFC geographical coverage, in 2001 the PEFC Council began a comprehensive revision process of the whole PEFC structure and requirements and the general assembly, held in November 2002, accepted a new set of PEFC normative documents.

Growing interest both within as well as outside of Europe also continued throughout 2002 and 2003, and an additional nine countries joined the PEFC Council. Currently, the PEFC Council membership covers 27 countries from 5 continents (Australia, Austria, Belgium, Brazil, Canada, Chile, the Czech Republic, Denmark, Estonia, France, Finland, Germany, Ireland, Italy, Latvia, Lithuania, Luxembourg, Malaysia, Norway, Poland, Portugal, the Slovak Republic, Spain, Sweden, Switzerland, the UK and the US).

The general assembly, held in October 2003, changed the organization's name to the Programme for the Endorsement of Forest Certification schemes, while the PEFC acronym and the PEFC logo remained the same.

STRUCTURE AND GOVERNANCE

The PEFC Council is an international not-for-profit membership organization registered in Luxembourg.

Membership

The PEFC Council recognizes two kinds of membership: ordinary and extraordinary. Ordinary members (PEFC national governing bodies) are independent legal entities which represent national or sub-national schemes operating in single countries. The PEFC national governing bodies are constituted as multi-stakeholder organizations with the participation of a broad range of interests in sustainable forest management.

Extraordinary members represent other interested international organizations which support the objectives of the PEFC Council.

Organization structure

The highest decision-making body is the *general assembly*. Voting rights are distributed amongst the PEFC national governing bodies according to the annual cutting categories given in the official United Nations Economic Commission for Europe (ECE)/Food and Agriculture Organization (FAO) statistics in those countries as a neutral means of determining size and importance of the forestry sector. Delegates to the general assembly therefore represent national forest certification schemes and a broad range of stakeholders involved in their development and administration. The general assembly decides by simple majority. Decisions made by the general assembly are recorded in minutes, which are available to PEFC members. Other interested people can obtain the minutes upon request.

The *board of directors* is the executive body of the PEFC Council and manages the organization between the general assemblies. Board members are appointed by the general assembly so that they represent the major interested parties who support the PEFC, the geographical distribution of the members, the diversity of their annual cutting categories and an appropriate gender balance. The board of directors appoints various working groups and a panel of experts to provide the board with technical advice and assistance.

The *secretary general*, appointed and responsible to the board of directors, runs the work of the secretariat.

PEFC Council requirements

All of the PEFC normative documents and requirements are public and are available on the PEFC Council official website.

STANDARD

Scheme development

The PEFC Council requires that standards and forest certification schemes must be developed in an open and transparent way through a multi-stakeholder process:

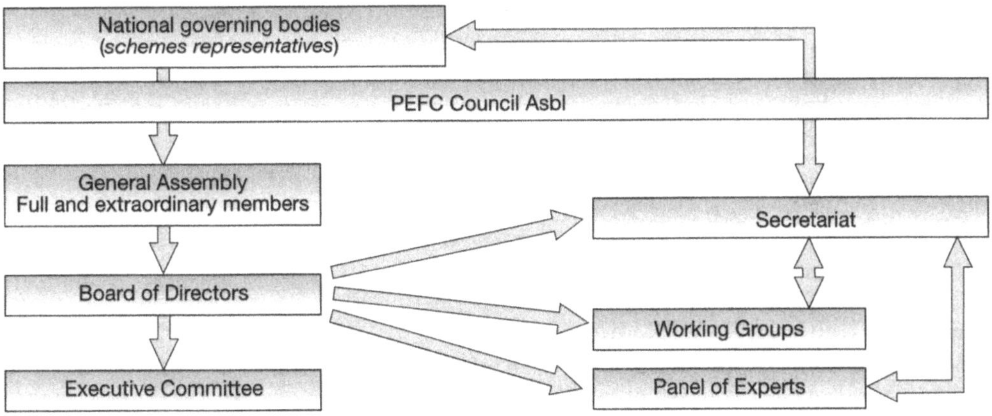

FIGURE A.17 Organization of the PEFC Council

- A forum (committee, council, working group) will be created to which all interested parties representing the different aspects of sustainable forest management are invited – for example, forest owners, forest industry, environmental and social NGOs, trade unions, retailers and other relevant organizations at national or sub-national level.
- The interested parties' participation and views will be documented and considered in an open and transparent way.
- Achieving a consensus will be the objective of the process.
- The start of the standard-setting process will be communicated to the public. Information on the development process will be distributed and discussed and final draft standards will be available to all interested parties.
- The final drafts standards are sent out for a formal national consultation process. Their implementation will be tested and appropriate actions will be taken to incorporate improvements or recommendations.

Forest certification schemes must be reviewed at least every five years to incorporate new experiences and scientific knowledge within standards. The revision process will be participatory, fair and transparent.

Content of the SFM standard

The PEFC Council requires that national standards for forest management must include performance-based criteria, as well as basic management system elements (for example, planning), and must:

- be compatible and consistent with the current Pan-European Operational-level Guidelines (PEOLG) and any deviations, for example, non-adherence on a specific issue, must be explicitly justified;
- include management and performance requirements that are applicable at the level of a forest management unit and optionally also at group and regional levels;
- be in compliance with national laws, programmes and policies; references to these will be indicated in the scheme documentation when relevant (for example, if the requirement of the PEOLG is not addressed in the certification criteria but is included in normative regulations);
- be in compliance with the core International Labour Organization (ILO) conventions, if the requirements of the conventions are not incorporated within national legislation, which is the case when a country has ratified the conventions.

CERTIFICATION APPROACH

The PEFC Council recognizes three basic approaches in forest certification:

1 individual certification;
2 group certification;
3 regional certification.

The latter two approaches enable forest owners to establish a group (if the group is geographically limited, then it is called regional certification) and apply collectively for certification. Forest owners commit themselves to following the certification criteria and their compliance is audited regularly based on random sampling.

Independent third-party certification

Certification of both forest management and chain of custody must be carried out by an independent, technically competent and impartial certification body. All certification bodies certifying against PEFC schemes meet international requirements for certification bodies that are

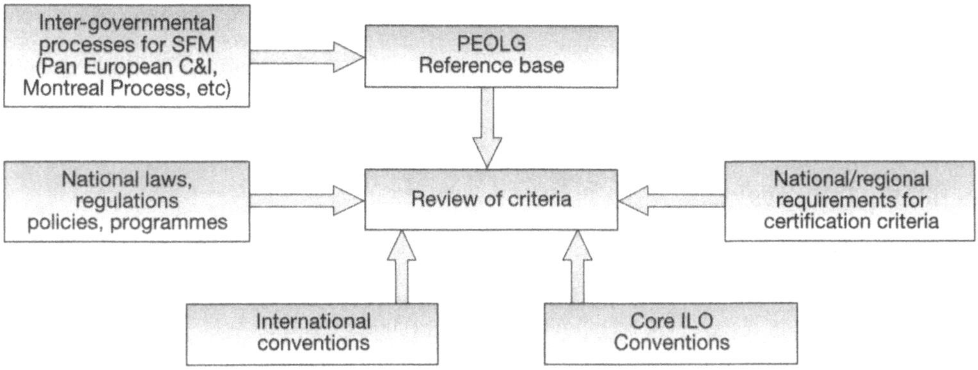

FIGURE A.18 Basis for forest certification criteria

defined by the International Organization for Standardization (ISO) documents. These standards (ISO Guides 62, 65, 66) define the:

- certification body's structure and its external relationships;
- rules for selection of competent auditors and technical experts;
- certification procedures, including the decision-making process;
- rules for complaints, appeals and disputes.

The auditors will fulfil general criteria for quality and environmental management systems auditors, as defined in ISO 19011 (guidelines for quality and/or environmental management systems auditing), and additional sector-specific requirements defined by the respective forest certification schemes.

The certification body's compliance with these stringent requirements is verified through an accreditation process. The accreditation is carried out by national accreditation bodies which are fully independent from the PEFC Council and PEFC schemes and which follow internationally recognized rules for accreditation, as defined by the ISO (ISO Guide 61). The same arrangement and procedures are applied in case of all credible, third-party certifications, such as ISO 9001, ISO 14001 and others.

Internationally applied standards for certification and accreditation and multilateral agreements between national accreditation bodies facilitated by the International Accreditation Forum (IAF) or similar regional organizations ensure that *certificates issued by certification bodies in different countries are equivalent*.

REQUIREMENTS FOR CONSULTATION OR PUBLIC INFORMATION

Standard-setting process

- All interest groups must be invited to participate in the standard-setting process. The start of the standard-setting process must be communicated to the public, and different views must be documented and considered.
- The final draft must be made publicly available for national consultation for a period of at least 60 days.

Certification process

- All certification requirements are made public by the relevant PEFC national governing body and from the PEFC Council official website. The certification body's procedures can be obtained from the certification body upon request.

- All information disclosure will be in compliance with the internationally recognized requirements of ISO Guides 62, 65 and 66.
- Information on certified entities can be obtained from the relevant certification bodies and the PEFC national governing body. The PEFC Council administers an international internet register of all PEFC certificate holders and PEFC logo users.

Accreditation process

- All procedures and requirements of accreditation bodies can be obtained from the accreditation bodies.
- All information disclosure will be in compliance with the internationally recognized requirements of ISO Guide 61.
- Information on accredited certification bodies is publicly available from the relevant accreditation body.

PEFC endorsement process

- An applicant scheme is published on the PEFC Council website.
- All interested people and groups are invited to comment on the submitted scheme during the consultation period, which lasts for 60 days.

ARRANGEMENTS FOR SMALL FOREST OWNERS

The group and regional certification approaches enable the possibility for even the smallest forest owners to obtain 'certified' status in a cost-effective and responsible way.

ACCREDITATION ARRANGEMENTS

The PEFC Council follows the same accreditation requirements (ISO Guide 61) and accreditation arrangements (national accreditation bodies) as most certification schemes and programmes (for example, ISO 9001, ISO 14001 and the European Union's Eco-management and Audit Scheme (EMAS)), which are focused on different types of business and industry. This approach ensures that hundreds of certification schemes and programmes can make use of a unified and harmonized accreditation programme, instead of having in place hundreds of scheme-specific accreditation bodies.

This approach means that national organizations with official status as national accreditation bodies in their countries are responsible for assessing certification bodies and issuing an accreditation diploma.

All of these accreditation bodies are committed to following internationally recognized procedures and requirements for accreditation according to ISO Guide 61.

Membership of the national accreditation bodies in a regional or international umbrella organization (such as the IAF and European Cooperation for Accreditation) and participation in their multilateral agreements guarantees equivalency, compatibility and mutual recognition of accreditations issued in different countries.

CURRENT STATUS

Currently, the PEFC Council has endorsed 13 forest certification schemes (Austria, Belgium, the Czech Republic, Denmark, Finland, France, Germany, Latvia, Norway, Spain, Sweden, Switzerland and the UK). Four schemes are currently undergoing the assessment process (Australia, Chile, Italy, and Portugal).

In January 2004, the total area of forests certified according to these 13 national schemes exceeded 52.5 million hectares, while more than 1204 chain-of-custody certificates have been issued and more than 10,000 organizations or individuals have received the PEFC logo usage licence.

CHAIN OF CUSTODY

The PEFC Council framework also includes rules for chain-of-custody certification, which enable the holder to verify the content of certified raw material during the whole chain, from forests to final consumers. A chain of custody certified by an independent third party (certification body) is a pre-condition for a company to use the PEFC logo for product-labelling purposes.

The PEFC Council chain-of-custody rules provide two methods:

1 physical separation;
2 percentage model.

In the physical separation model, raw material from certified forests must be physically separated from uncertified raw material.

The percentage model allows companies to process certified and uncertified raw material mixed together; but the content of certified raw material must be identified and communicated to consumers. The percentage model includes two options for the use of the certified raw material percentage:

1 The (so-called) minimum average percentage system requires that the certification percentage must be used for the whole production batch for which it was calculated.
2 The input–output method allows the use of the certification percentage for the appropriate part of the production batch (for example, when the certification percentage is 20 per cent, then 20 per cent of the production batch may be considered as certified).

The PEFC Council frameworks also include requirements for uncertified wood and requires that no wood from controversial sources (such as illegal logging and wood from protected areas) can enter the production process.

LABELLING AND LOGOS

The PEFC Council provides its own logo, which is a registered trademark. The PEFC logo can be used only based on the PEFC logo usage licence, which can be issued by the PEFC Council or by the relevant PEFC national governing body on behalf of the PEFC Council. The PEFC logo licence number must be used together with the logo whenever the logo is applied, and this unique code clearly identifies anyone using the PEFC logo, on or off product. The PEFC Council provides an internet database of all PEFC logo users.

The PEFC Council recognizes four types of PEFC logo users groups:

1 national governing bodies (A);
2 forest owners/managers (B);
3 forest-related industries (C);
4 non-commercial users (D).

Only logo users within the logo users group B and C can use the PEFC logo for on-product labelling purposes, and they can obtain their logo usage licence based only upon a valid forest management or chain-of-custody certificate.

Based on chain-of-custody rules, the PEFC logo can be used on product only when the content of certified raw material exceeds 70 per cent.

Usage of the PEFC logo is controlled by a certification body as a part of the forest management or chain-of-custody audits.

POLITICS AND PERCEPTIONS

The PEFC Council is based on the principles of subsidiarity. Therefore, the fundamental decisions influencing certification activities are made at local or national level.

The PEFC Council has received strong support from forest owners and forest industries at both national and international levels, as well as from trade unions, research and the academic forestry community. Many national schemes endorsed by or participating in the PEFC framework have also received the support of local and national ENGOs – for example, in France, Poland, the US, Canada, Malaysia and Austria.

INFORMATION ABOUT THE SCHEME

General information about the PEFC Council can be found on the PEFC official website at www.pefc.org, which includes all of the PEFC normative documents, PEFC brochures (available in various languages) and PEFC-endorsed schemes. The website also provides links to national PEFC websites.

The PEFC Council administers an interactive internet database containing information on all PEFC certificate holders and PEFC logo users.

Contact

17 Rue des Girondinds
L-1626 Luxembourg
Tel: +353-2625-9059
Email: pefc@pt.lu

A1.7 Sustainable Forestry Initiative (SFI)

TYPE OF SCHEME

The Sustainable Forestry Initiative (SFI) is a multinational North American sustainable forestry and certification programme, with the certification standard and certification procedures governed by an independent special-purpose body established solely to carry out these functions.

SCOPE

The SFI programme is designed to fit the legal and cultural context of North America. It is adapted to all types of forest ecosystems, but most significantly those that are managed for the sustainable production of timber and other forest products.

DATE SET UP AND HISTORY

The programme was established in October 1994 and began operation in 1995.

The two watershed events that prompted the creation of the SFI programme were the 1987 Brundtland commission report, *Our Common Future* (WCED, 1987), which set a vision for global sustainable development, and the 1992 Earth Summit in Rio de Janeiro. The Earth Summit focused global attention on the importance of sustainable forestry. Resulting discussions have led to general agreement that a definition for sustainable forestry must include two key concepts:

1 the Brundtland commission's finding that sustainability relies on the commitment of present generations not to limit the ability of future generations to meet their own needs;
2 forest practices must be economically viable, socially acceptable and protective of the environment if they are to be sustainable.

In 1990, the American Forest Council convened a process on behalf of the American Paper Institute (API) and the National Forest Products Association (NFPA) through a Future of Forestry Conference, which resulted in the 1992 adoption of Ten Forest Management Principles. In 1994, the American Forest and Paper Association (AF&PA, an organization that resulted from the merger of API and NFPA) launched a process to 'visibly improve industrial practices and report results'. That process led to the creation of the original SFI principles and implementation guidelines. On 14 October 1994, the AF&PA board of directors adopted the Sustainable Forestry Initiative principles and implementation guidelines, and gave member companies until 31 December 1995 to demonstrate adoption of and conformance with the SFI principles and implementation guidelines.

Thus, the SFI programme was launched during 1995 as a code of conduct for the members of AF&PA. The initial goals did not address forest certification, but were designed to provide a sound basis upon which the industry could improve the management of its free-owned lands and establish a monitoring and reporting programme that would assure investors, customers and the general public that sustainable forest management programmes were an industry-wide norm. In 1996, AF&PA expelled the membership of 15 companies that either refused or failed to meet the SFI programme obligations, and another 10 companies dropped out to avoid the cost of conformance. One hundred and ninety-nine members, representing about 21 million hectares of forestlands, reported conformance with the programme. By 1998, the SFI principles and implementation guidelines had evolved into the SFI standard (SFIS).

As the SFI programme was being designed and established, outside observers were invited to review and critique the programme's content, administration and effects. The Expert Review Panel, initially composed of top executives of environmental, conservation, professional and academic organizations, as well as public resource agencies, was established in 1995. This panel, now called the External Review Panel, has operated under an independent charter since 1997. The panel selects its own membership and agenda, gathers public input from outside organizations and provides a quality control function for the programme. It is now composed of 18 members, and meets regularly to review SFI programme progress reports, the content of the SFI standard and other issues surrounding the programme.

The impact of the SFI programme on forests in North America began with the commitment of all members of AF&PA (primarily industrial forest products companies) to apply the SFIS to their private forests. These lands, representing approximately 14 per cent of the timberlands of the US, produced about 36 per cent of the wood products harvested in 1995. In 1998, non-member private and public organizations were invited to join as SFI programme licencees. By December 2003, some 89 public and private organizations in the US and Canada committed to the SFI programme through licensing, bringing an additional 26 million hectares under the SFIS.

Quality control and programme credibility

The primary role of the External Review Panel has been to ensure that the progress data reported by the SFI programme managers is accurately analysed and presented. During the first years of the programme, the concerns focused on data analysis. AF&PA staff developed methods of data management and analysis to ensure accuracy, while maintaining business confidentiality for individual firms, and the panel spent many hours reviewing methods, results and the interpretation of those results. When all were assured that the information in the annual progress report was as accurately and honestly presented as possible, the text went to press.

As the programme grew and the data began to illustrate trends, the panel's concern shifted to the source of the data: reports from the SFI participants. In 1999, the External Review Panel, AF&PA, the Izaak Walton League of America, the Conservation Fund, and the Rockefeller Brothers Fund cooperated to create a Forest Monitoring Project (FMP). A professional forester hosted by the Izaak Walton League and a forester from the US Department of Agriculture (USDA) Forest Service staffed the project.

The FMP conducted over 50 field reviews in a three-year project, evaluating how effectively practices on the land supported the annual progress reports of SFI participants. These reviews also identified and publicized exemplary problem-solving approaches at the field level. Where participants were encountering problems in documenting conformance with a particular performance measure, the FMP brought those issues back to the External Review Panel and AF&PA for additional research and study. On the basis of the FMP results, the panel could assure the public that the progress reported was an accurate reflection of what was taking place in the participating forests. Today, with most of the land in the SFI programme undergoing independent third-party certification, the FMP has been discontinued and the oversight of quality control in the field has shifted to the independent auditors.

STRUCTURE AND GOVERNANCE

The SFI programme was initially designed and operated within the AF&PA. During 2001, governance of the SFI standard and certification procedures were shifted to a 15-member, independent Sustainable Forestry Board (SFB) in order to improve the effectiveness and credibility of the programme. The SFB is composed of one third conservation and environmental organizations, one third professional and academic experts and one third forest industry leaders. In January 2002, the SFB completed its legal separation from AF&PA, ensuring that management of the SFI standard and verification/certification procedures are now conducted by an independent non-profit organization.

STANDARD

Professional foresters and scientists designed the original 1994 SFI principles and implementation guidelines as part of a multi-year collaborative effort involving consultation with many forest landowners, loggers, conservationists and other stakeholders. In addition, the AF&PA conducted extensive public opinion research in order to ascertain the sustainable forestry issues that were most important to the general public, customers, conservationists and critics of the industry and its management of forestlands.

A voluntary verification and certification process was established in 1998 so that SFI programme participants could engage independent, third-party auditors to verify their conformance with the SFI standard. The process was rapidly adopted, and in 1998 some 283,000ha were verified under the SFI standard. The widespread adoption of the certification process led to the enhancement of the standard in 2001 to align it with the International Organization for Standardization (ISO) procedures and protocols.

During 2001, the newly formed SFB sought public review and comment on the 2001 SFI standard. On the basis of those reviews, the SFB adopted major enhancements to the SFIS, including one that addresses 'forests of exceptional conservation values'. The 2002–2004 SFI standard requires SFI participants to develop plans for the location and protection of known sites with viable occurrences of species and communities that are identified as critically imperilled or imperilled, working with NatureServe or another body with similar expertise (NatureServe is a non-profit organization dedicated to providing knowledge to protect the natural world by working in partnership with the Nature Conservancy and a network of scientific experts). Additionally, this enhancement calls for SFI programme participants to establish procurement policies that promote the conservation of biodiversity hotspots and major tropical wilderness areas outside of North America and that contribute to the elimination of illegal logging.

The SFB also adopted a three-year review cycle for the SFI standard, adopting a formal review process in 2004 for the development of a future 2005 version of the SFIS.

CERTIFICATION APPROACH

The SFIS requires programme participants to adopt policies, plans and procedures through which they will conform to the spirit and intent of the sustainable forestry principles, including meeting all programme objectives through performance measures that are demonstrated by mandatory core indicators. Due to the environmental, legal, economic, geographic and cultural complexity of the programme participants, the SFI standard places the burden of designing specific programme responses to the SFIS on the participants themselves. Participants may then voluntarily choose to obtain independent third-party certification of their conformance to the SFIS (67 SFI programme participants have done so).

Certification audits are conducted by qualified verifiers operating under a published set of verification/certification principles and procedures that were adopted as part of the 2002–2004 SFIS.

The audit team is charged with reviewing evidence to determine whether the organization's performance conforms to the SFIS. Evidence must be compiled by examination of operating procedures, study of materials relating to forestry practices, and on-the-ground examination of field performance, and through meetings with employees, contractors and other third parties (for example, government agencies, community groups and conservation organizations), as appropriate. The audit team grants certification when:

- there are no non-conformances;
- minor non-conformances have been resolved or addressed by a written plan for timely corrective action approved by the verifier; and
- all major non-conformances have been eliminated.

A major non-conformance exists when either:

- one or more of the SFI objectives or performance measures have not been addressed or implemented; or
- several non-conformances exist that, taken together, lead a verifier to conclude that one or more objectives or performance measures have not been adequately addressed or implemented.

Once a programme participant has achieved certification under the SFIS, they must undergo a re-certification audit within three years of the date of the initial certification and every five years thereafter.

REQUIREMENTS FOR CONSULTATION OR PUBLIC INFORMATION

There are several levels of consultation and public information involved in the SFI programme. At the programme level, AF&PA, the SFB and the External Review Panel maintain websites where information about the programme is available. Use of these sites, along with widespread publicity, has invited public input into the 2004 review of the SFI standard. The SFB has also convened an auditor's forum and a customer's forum to help incorporate the views of these important partners into the continuous improvement of the programme.

In adopting the 2002–2004 SFIS following a public review process, the SFB established a more extensive review for 2004, with the goal of adopting enhancements to the standard for the period of 2005 and beyond. The review, which is underway, invites widespread comment from both SFI programme participants and the public at large. The proposed enhancements to the SFIS will be published for a formal 30-day public review period in June 2004, and then presented in a series of public workshops in September. Following another brief public comment period during October, the SFB will consider the recommended enhancements in December 2004.

At the participant level, SFI programme participants who achieve certification are required to file a public summary of their certification audit and its findings. Participants are expected to follow Federal Trade Commission (FTC) guidelines on product advertising and communication in order to ensure that all claims are accurate and consistent with applicable marketing guidelines and requirements. A formal appeals process is open to any individual or group wishing to contest an SFI programme participant's certification. As noted, SFI programme participants are open to claims of inconsistent practices if any member of the public chooses to file such a claim.

At the operation level, the External Review Panel maintains a 'national inconsistent practices access point', where any person who wishes to claim that an SFI programme participant is conducting operations that appear to be inconsistent with their SFIS obligations can make a claim to that effect. Under a strict guarantee of anonymity for the claimant, the ERP works with the SFI implementation committees, programme participants and, if needed, outside experts to review the merits of the claim, ensure that appropriate actions are taken and report results back to the claimant.

SFI implementation committees (SICs) work at the local level to promote the SFI standard as a means of broadening the practice of sustainable forestry and to ensure on-the-ground progress. SFI programme participants provide leadership and invite a wide range of stakeholders to contribute: private landowners, independent loggers, consultants, government land managers, legislators, university scientists and conservationists. These resource professionals volunteer a significant amount of time to ensure that national SFI programme objectives are consistently implemented and adapted to region-specific needs. State and provincial forestry associations often sponsor SICs, which have been formally established in 38 US states and 5 Canadian provinces. Key activities focus on education and training, quality control, communications and reporting. The SICs maintain programmes on a state or provincial basis where anyone may file a claim or raise a question about alleged inconsistent practices. A variety of telephone hotlines, mail access points and other communications are publicized to provide public access to the inconsistent practice programmes.

Additional public and expert input into the programme is obtained through a variety of scientific workshops, widespread publication of the programme's annual progress report, telephone hotlines and the various SFI programme websites.

ARRANGEMENTS FOR SMALL FOREST OWNERS

SFI programme participants in the US face a situation where approximately 58 per cent of the timberland and over 59 per cent of the annual timber harvest comes from some 10 million private ownerships. In Canada, the numbers are smaller; but the importance of private woodland ownerships is high, particularly in the eastern part of the country. Promoting sustainable forest management on those lands is an important part of the SFI programme.

In 2000, the SFI programme entered into a mutual recognition agreement with the American Tree Farm System® to help promote sustainable forestry on family forestlands. The American Tree Farm System is the oldest sustainable forestry and certification programme in the world, currently involving more than 60,000 forest owners who account for 10.5 million hectares of US forest. Additional mutual recognition efforts are being considered within the SFI programme. The goal is to develop a process to recognize other programmes whose sustainable forestry goals and certification procedures provide a functionally equivalent level of sustainable forest management as the SFIS does.

Influencing the quality of management on forestlands that are not involved in any certification programme is a complex and difficult task. The SFI programme addresses this primarily at the level of the primary wood processor (SFI programme participants account for at least 50 per cent of US solid wood products and 87 per cent of US pulp and paper). A primary processor, to meet SFIS obligations, must communicate all aspects of the SFI programme requirements to its procurement staff, buyers and foresters, independent logging contractors, and forest landowners. Since most mills purchase well over half of their timber supply from private landowners over which they have no legal right to dictate land use or land management details, the task is one of communication, education, motivation and encouragement.

The SFIS requires each individual SFI programme participant to establish a written procurement policy; but under US law, the specific content of procurement policies must be left to individual companies. A recent SFIS provision requires that all programme participants must have a verifiable auditing or monitoring system in place to evaluate the results of promoting reforestation, and must employ best management practices (BMPs) within their wood supply. Today, trained loggers produce 92 per cent of wood delivered to SFI programme participants, up from 34 per cent in 1995. This is one way in which the SFI programme is working to make a difference in on-the-ground forestry practices on lands that the participants do not own or control.

ACCREDITATION ARRANGEMENTS

Under the 2002–2004 SFI standard, the SFB established third-party auditor qualification requirements. Lead SFI programme auditors must achieve the highest qualification (environmental management system lead auditor or equivalent) available under the Registrar Accreditation Board (RAB), the Canadian Environmental Auditing Association (CEAA) or the equivalent. In addition to the lead auditor requirements, SFI third-party auditors must meet applicable training, education and experience requirements established by the American National Standards Institute (ANSI/RAB) or equivalent.

The SFB also requires each auditing firm to undergo a formal field peer review of at least one SFI audit annually to maintain approval to conduct SFI certifications. An interpretations committee of forestry experts working with the SFB receives questions from both SFI programme participants and auditors, providing answers and interpretations that form a growing compendium of reference information to assist field practitioners in determining how to meet the SFI standard.

CURRENT STATUS

As of 31 December 2003, there were 206 participants enrolled in the SFI programme. Of those, 118 were AF&PA members and 88 were programme licensees representing a broad range of state and local governments, conservation organizations and other landowners. Of the 55 million hectares enrolled in the programme, 41 million hectares had been third-party certified.

CHAIN OF CUSTODY

The complexity of the wood supply in the US renders a strict 'stump-to-store' chain-of-custody tracking system impossible in many situations. The SFI programme addresses this difficulty in several ways. Certified SFI programme participants wishing to use the SFI on-product label must maintain evidence that 100 per cent of primary sources are independently third-party certified sources – confirmed by a qualified independent third party to be sourced:

- from specific forest tracts managed in conformance with the SFI standard or other acceptable standards (the American Tree Farm System);
- through a procurement system certified to be in conformance with the SFI standard; or
- a combination of those two, but no more than two-thirds (by weight) from other credible sources.

Secondary producers (those that acquire less than 50 per cent of their wood supply directly from the woods) who desire to use one of the three available secondary producer labels (see below), must demonstrate that at least two-thirds (by weight) of the wood or wood fibre in the product(s) or manufacturing unit comes from independent third-party certified sources or from neutral sources, such as recycled or recovered fibre. All non-US and non-Canadian sources must be certified under an acceptable standard, or come from neutral or other credible sources.

LABELLING AND LOGOS

The SFI programme on-product label has been available for qualifying SFI programme participants since 2002. There are four labels available (see Figure A.20):[1]

1. 'Certified participant' is available for use by any SFI programme participant whose manufacturing facility acquires 50 per cent or more of its material from the woods or who sells timber from their own land.
2. 'Participating manufacturer' is defined as a producer of finished forest products, such as plywood, furniture, windows, doors and cabinets.
3. 'Participating publisher' is defined as a producer of magazines, publications and catalogues.
4. 'Participating retailer' is defined as a retailer of wood and paper products.

Each has its own requirements for certification, including being certified to the SFI programme's label-use requirements document. To use the label as a certified participant, for example, a company sourcing its raw materials directly from the forest must have achieved independent third-party certification to the 2002–2004 SFI standard and must also meet the additional requirements of the SFI programme Office of Label Use and Licensing.

These requirements include the following:

FIGURE A.20 SFI labels

- All primary sources must be accounted for as sourced from an independent third-party source, but no more than two-thirds (by weight) from other credible sources.
- The certified procurement system may include material from neutral sources such as recovered wood fibre, and from other credible sources outside the US and Canada, that comes from forest plantations or other well-managed forests that are harvested in compliance with general sustainable forestry practices, and in a manner that does not constitute illegal logging practices as defined in Section 5.32 of the SFI standard.

The SFI label for primary producers denotes that the product is produced by a company which fully subscribes to the principles and practices that comprise the SFI programme. It further represents that this company's practices are subject to a third-party audit in order to ensure that they do live up to their claims. The SFI programme addresses not only the forests owned and/or managed by the company, but also a number of specific steps that the company takes proactively to influence the loggers, family forest owners and other owners in their supply system who do not belong to the SFI or any other certifying agency. Through logger training, education programmes for landowners and on-site auditing of logging practices, the SFI programme participants are influencing the way in which forests are managed, harvested and reforested across the vast forest landscape. In addition to the 100 million acres that they own and manage directly, they are influencing at least another 100–200 million acres (about 40–80 million hectares) of forestlands through their outreach and audit programmes.

For SFI programme participants seeking to use an SFI on-product label, periodic surveillance audits are required to verify the participant's on-going commitment to the SFI standard.

POLITICS AND PERCEPTION

The SFI programme has garnered both a wide range of support and an intense group of opponents. Some 30 private conservation and resource organizations have officially expressed support for the goals of the SFI programme, and 19 states have passed resolutions or proclamations of endorsement. For a listing of those organizations and states, see www.aboutsfi.org.

The supporting organizations generally comprise a spectrum of natural resource bodies (such as the American Bird Conservancy, American Forests and Bat Conservation International), land conservation organizations (such as the Conservation Fund and Ducks Unlimited), professional organizations (for example, the Society of American Foresters and the Wildlife Society), labour unions (such as the United Brotherhood of Carpenters and Joiners of America and the Southern Council of Industrial Workers), and governmental bodies (for instance, the Council of State Governments and the National Association of State Foresters). Some of the supporting organizations have been represented on the External Review Panel and the Sustainable Forestry Board, citing the desire to assist in making the programme effective at promoting the concepts and practice of sustainable forestry.

Opposition has come largely from activist organizations that have been critical of the SFI programme's roots inside the industrial forestry sector, charging that its requirements lacked the independence, environmental rigour and social requirements inherent in other forest certification standards. The 2002–2004 SFI standard and current governance by the SFB address many of these criticisms.

INFORMATION ABOUT THE PROGRAMME

Information about the SFI programme can be obtained from the following websites:

- www.aboutsfi.org;
- www.aboutsfb.org;
- www.abouterp.org;
- www.sfiprogram.info;
- www.afandpa.org.

Contact

1111 Nineteenth Street
NW, Suite 800
Washington, DC
20036 USA

Note

1 All SFI labels are registered service marks of the American Forest and Paper Association (AF&PA).

Appendix 2
Sources of Information

ISO and ISEAL

The ISO website is www.iso.org. It provides information on the organization, together with access to ISO standards and guidelines (most of which must be purchased).

The ISEAL website is www.iseal-alliance.org. The Code of Good Practice for standard-setting is available free from the site.

Certification scheme websites

The websites of some of the main certification schemes are set out below. Most of the sites provide background information, as well as documents, information on certification bodies and information on certified forests and chain of custody.

International schemes	
Forest Stewardship Council (FSC)	www.fscoax.org, www.fsc-info.org
Programme for the Endorsement of Forest Certification (PEFC)	www.pefc.org This website is particularly useful as it also provides links to all participating national schemes.
National schemes	
Australia: Australian Forestry Standard (AFS)	www.forestrystandard.org.au
Chile: CertforChile	www.certfor.org
Canada: Canadian Standards Association (CSA)	www.certifiedwood.csa.ca
Indonesia: Lembaga Ekolabel Indonesia (Indonesian Eco-labelling Foundation – LEI)	www.lei.or.id
Malaysia: Malaysian Timber Certification Council (MTCC)	www.mtcc.com.my
US: Sustainable Forestry Initiative (SFI)	www.aboutsfi.org; www.aboutsfb.org; www.sfiprogram.info

Intergovernmental criteria and indicators processes

Various websites provide further information on intergovernmental criteria and indicators (C&I) processes. One of the most useful is the summary provided by the United Nations Food and Agriculture Organization (FAO), which introduces each process and provides information on content at www.fao.org/forestry/site/16588/en.

Sites specific to particular processes include:

Ministerial Conference for the Protection of Forests in Europe (Pan-European Process)	www.mcpfe.org/
Montreal Process	www.mpci.org/
International Tropical Timber Organization (ITTO)	www.itto.or.jp
African Timber Organization (ATO)	www.focusintl.com/whos0008.htm
Lepaterique Process	www.rds.org.hn/forestal/manejo/criterios_indicadores/zapata.shtml

Critiques and criticisms of certification schemes

Literature that provides a critique of certification schemes includes the following:

Counsell, S and Loraas, K T (2002) 'Trading in credibility: The myth and reality of the Forest Stewardship Council', Rainforest Foundation, London

Fern, (2001) 'Behind the logo: An environmental and social assessment of forest certification schemes', www.fern.org

Fern, (2004) 'Footprints in the forest: Current practice and future challenges in forest certification', www.fern.org

Freris, N and Laschefski, K (2001) 'Seeing the wood from the trees', *The Ecologist*, vol 31(6), www.theecologist.org

Greenpeace and Luonto-Liitto (2001) *Anything Goes? Report on* PEFC *Certified Finnish Forestry* and *Anything Goes? Report on* PEFC *Certified Finnish Forestry*: *Part Two*, 10 April, www.pefcwatch.org/finreport2/index.html

Meridian Institute (2001) *Comparative Analysis of the Forest Stewardship Council and Sustainable Forestry Initiative Certification Programs*, Meridian Institute, Washington, DC

AKU Germany and 46 other NGOs (2003) 'NGOs reject Malaysia's attempts to whitewash its timber practices', Joint NGO statement, 26 May, www.fern.org

Tan, A (2003) *On the Ground: Green Stamp of Approval or Rubber Stamp of Destruction?* Forest Ethics, Greenpeace Canada and Sierra Club of Canada, British Columbia Chapter, Ottawa, Canada

Vallejo, N and Hauselmann, P (2001) 'PEFC: An analysis', WWF Discussion Paper, Pi Environmental Consulting, Pully

Some organizations working with or reporting on forest certification

The list below provides a selection of the many websites with information about forest certification. These frequently change; however, many have links to other relevant sites, so it is relatively straightforward to find more sites after having logged on to the first few:

Greenpeace Canada and Sierra Club of Canada	www.sierraclub.ca
European Forest Institute	www.efi.fi
United Nations Food and Agriculture Organization (FAO)	www.fao.org/forestry
Fern	www.fern.org
Forest Certification Watch	www.certificationwatch.org
Forest Trends	www.foresttrends.org
World Wide Fund for Nature Global Forest and Trade Network (GFTN)	www.panda.org/forestandtrade
Deutsche Gesellschaft für Technische Zusammenarbeit (GTZ)	http://www.gtz.de/forest_certification/english
The Forests Dialogue	www.theforestsdialogue.org
Tropical Forest Trust	www.tropicalforesttrust.com
World Bank–WWF Forest Alliance	www.forest-alliance.org

Illegal logging

For anyone wishing to know more about illegal logging and measures being taken to combat it, the site run by the UK-based Royal Institute for International Affairs (RIIA) is probably the best starting point. This site, at www.illegal-logging.info, provides comprehensive links to other sites related to illegal logging.

Information on implementing purchasing policy

Poynton, S (2003) *Good Wood Good Business: A practical industry-oriented guide to excluding illegal and other unwanted wood from your supply chain*, Tropical Forest Trust (TFT), www.tropicalforesttrust.com

White, G and Sarshar, D (2004) *Responsible Purchasing of Forest Products*, WWF Global Forest and Trade Network, www.panda.org/forestandtrade

Glossary

Accreditation body: an authoritative body which gives formal recognition that a certification body is competent to carry out specific tasks.

Assessment: evaluation of the compliance of a forest and its management with the requirements of a standard.

Assessor: a person trained, competent and approved to undertake assessments.

Audit: see assessment; the two words have the same meaning.

Auditor: see assessor; the two words have the same meaning.

Certification body or certifier: an accredited organization which undertakes the certification of conformity with a standard.

Certification system or scheme: a set of rules for forest certification that includes a standard, certification, certification process, accreditation and rules for making claims. Most certification schemes have been developed by specific organizations.

Chain of custody: the tracing or tracking of certified material from its origin in the forest to the final product.

Closing-out non-compliances or gaps: the process of taking action to resolve identified non-compliances or gaps in order to ensure full compliance with the standard.

Corrective action request (CAR): request to undertake actions to resolve an identified non-compliance with the requirements of the standard.

Criteria: key elements that define principles.

Forest management unit (FMU): a defined area of forest (usually covered by a forest management plan) to which certification is applied.

Gaps: areas where current management does not meet the requirements of a standard.

Indicator: a variable or component of the forest ecosystem or management system used to infer the status of a particular criterion.

International Organization for Standardization (ISO): ISO develops international standards, including guidelines for accreditation bodies, certification bodies and standard-setting.

Non-compliance: a failure to meet a requirement of a standard.

Non-conformance: alternative to non-compliance with the same meaning.

Principle: a fundamental truth or law as the basis of reasoning or action. Principles in the context of sustainable forest management are seen as providing the primary framework for managing forests in a sustainable fashion. They provide the justification for the criteria, indicators and verifiers (source: CIFOR, 1999).

Registration body or registrar: the same as a certification body. Used frequently in North America.

Stakeholders: individuals or organizations with a legitimate interest in the goods and services provided by a forest. They include regulators, owners, managers, employees, contractors, local communities, indigenous people, environmental interest groups, investors, insurers, researchers, customers, consumers and the general public.

Standard: a document containing technical specifications or other precise criteria to be used consistently as rules, guidelines or definitions of characteristics in order to ensure that materials, products, processes and services are suitable for their purpose.

Surveillance: periodic re-examination of a certified forest and its management to ensure continued compliance with a standard.

Verifier: information that enhances the specificity or the ease of assessment of an indicator.

References

ATO/OAB (2001) 'Les ensembles des principes, critères et indicateurs (PCI) de l'Organisation Africaine du Bois (OAB) pour la gestion durable des forêts tropicales naturelles africaines, à utiliser au niveau national de l'Unité Forestière d'Aménagement (UFA), harmonisés avec les C&I de l'Organisation Internationale des Bois Tropicaux (OIBT)', OIST, Commission Européenne, Libreville, Gabon, August

Aukland, L, Moura Costa, P, Bass, S, Saleemul Huq, Landell-Mills, N, Tipper, R and Carr, R (2002) *Laying the Foundations: Preparing the Land-Use Sector – A Quick Guide to the Clean Development Mechanism*, International Institute for Environment and Development, London, www.iied.org

Australian DAFF (Department of Agriculture, Fisheries and Forestry) (2000) *Establishing Comparability and Equivalence Amongst Forest Management Certification Schemes: Critical Elements for the Assessment of Schemes*, Australian DAFF, Canberra, Australia

Baharuddin, H G and Simula, M (1994) *Certification Schemes of All Types of Timber*, ITTO, Yokohama

Baharuddin, H G and Simula, M (1996) *Timber Certification in Transition*, ITTO, Yokohama

Baharuddin, H G and Simula, M (1998) *Timber Certification: Progress and Issues*, ITTO, Yokohama

Baharuddin, H G and Simula, M (2001) *Framework for an Auditing System for ITTO's Criteria and Indicators for Sustainable Forest Management*, ITTO, Yokohama

Bass, S (1997) *Comparing the FSC and ISO Approaches to Forest Certification*, IIED, London

Bass, S (1999) 'Forest Certification – what problems is it solving? What challenges are still to be faced?', Presentation to 7th Papercast Global Conference on Paper and the Environment, Berlin

Bass, S and Simula, M (1999) 'Independent certification/verification of forest management', Background Paper, World Bank/WWF Alliance Workshop, Washington, DC, November

Bass, S, Thornber, K, Markopoulos, M, Roberts, S and Grieg-Gran, M (2001) *Certification's Impact on Forests, Stakeholders and Supply Chains*, IIED, London

Blowfield, M (1999) *Ethical Trade and Sustainable Rural Livelihoods*, Natural Resources Institute, Chatham

Brockmann, K L, Hemmelskamp, J and Hohmeyer, O (1996) *Certified Tropical Timber and Consumer Behaviour: The Impact of Certification Schemes for Tropical Timber from Sustainable Forest Management on German Demand*, Zentrum für Europäische Wirtschaftsforschung GmbH, Mannheim

Bruenig, E F (1997) *Cost and Benefit of Sustainability in Forestry*, University of Hamburg, Hamburg

Cashore, B (2004) 'Learning and forest certification', Paper prepared for the Forests Dialogue, Yale University, New Haven

CEPI (Confederation of European Paper Industries) (2000) *Comparative Matrix of Forest Certification Schemes*, CEPI, Brussels

CICI (2003) *International Conference on the Contribution of Criteria and Indicators for Sustainable Forest Management: The Way Forward*, (CICI–2003), Report, 3–7 February, Guatemala City, www.fao.org/DOCREP/005/Y8694E/TopOfPage

CIFOR (Centre for International Forestry Research) (1997) *Guidelines for Developing, Testing and Selecting Criteria and Indicators for Sustainable Forest Management*, CIFOR, Jakarta

Colfer, C P (1995) 'Who counts most in sustainable forest management?', CIFOR Working Paper No 7, Jakarta

Coulombe, M J (1995) 'Sustainable forest management: Criteria and indicators and certification are not the same thing!!', Discussion paper, USDA Forest Service, Washington, DC

Counsell, S (1999) *Trickery or Truth? An Examination of the Effectiveness of the Forest Stewardship Council*, The Rainforest Foundation, UK

Crossley, R and Points, J (1998) *Investing in Tomorrow's Forests*, WWF Research Report, Gland, Switzerland

de la Rochefordière, A and Mitchell, A (2001) 'National monitoring and certification are both needed to save the tropical forests', SGS (Société Générale de Surveillance), mimeo

DETR (2000) *Green Claims Code*, www.defra.gov.uk/environment/greening/greenpro/gcc/

DFID (Department for International Development) (1999) A DFID *Approach to Certification*, Draft, DFID, London

Dudley, N (1997) *Criteria for Forest Quality*, WWF and IUCN, Gland, Switzerland

Dutch Ministry of Agriculture, Management and Fisheries, Department of Nature Management (1997) *Timber Certification and Sustainable Forestry*, Dutch Ministry of Agriculture, Management and Fisheries, The Hague, The Netherlands

Dykstra, D, Kuru, G and Nussbaum, R (2003) 'Tools and methodologies for independent verification and monitoring', *The International Forestry Review*, vol 5(3), pp262–267

Dykstra, D, Kuru, G, Taylor, R, Magrath, W, Nussbaum, R and Story, J (2003) *Technologies for Wood Tracking: Verifying and Monitoring the Chain of Custody and Legal Compliance in the Timber Industry*, World Bank–WWF Alliance, Washington DC

EA (European Accreditation) (2001) www.european-accreditation.org

Eba'a Atyi, R and Simula, M (2002) 'Forest certification: Pending challenges for tropical timber', Background paper, ITTO International Workshop on Comparability and Equivalence of Forest Certification Schemes, Kuala Lumpur, 3–4 April

Elliott, C (1999) *Forest Certification: Analysis from a Policy Network Perspective*, PhD thesis, Ecole Polytechnique Fédérale de Lausanne, Lausanne

Elliott, C (2001) 'Roles and responsibilities of environmental NGOs', *The ATIBT Newsletter*, no 15, winter, pp29–30

Ervin, J (1996) 'The consultative process', in Viana, V et al (eds) *Certification of Forest Products: Issues and Perspectives*, Island Press, Washington, DC

FAO (United Nations Food and Agriculture Organization) (2001a) *Criteria and Indictors for Sustainable Forest Management: A Compendium*, Forest Management Working Paper, FAO, Rome

FAO (2001b) *Global Forest Resources Assessment 2000: Main Report*, FAO, Rome

FAO (2003) *Evaluation des Besoins en Formation dans le Secteur Forestier en Afrique Centrale*, FAO, Reseau des Institutions de Formation Forestière et Environnement d'Afrique Centrale, UICN, Rome

Fern (2001a) *Behind the Logo: An Environmental and Social Assessment of Forest Certification Schemes*, May, Fern, www.fern.org

Fern (2001b) *Behind the Logo: The Development Standards and Procedures of the Pan European Forest Certification (PEFC) Scheme in Finland, an Introduction*, www.fern.org

Fern (2003) Eco-labels, forest certification and the WTO. Fern Briefing Note, July, www.fern.org.

Fern (2004) *Footprints in the Forest. Current Practice and Future Challenges in Forest Certification*, Fern, Moreton-in-Marsh, UK, www.fern.org

FSC (Forest Stewardship Council) (1994) FSC Statutes, Forest Stewardship Council, Bonn, Germany, www.fsc.org

FSC (1997) *Guide for Applications for FSC Membership, Document* 5.2.1, March, 1997, Oaxaca, Mexico

FSC (1998) FSC *Accreditation Manual*, Oaxaca, Mexico

FSC (1999) *Improvements in Forest Management due to Certification*, Unpublished report to IIED, London

FSC (2000) FSC *Policy on Percentage-based Claims*, FSC *Document* 3.6.3, Oaxaca, Mexico

FSC (2001) www.fscoax.org

FSC (2002) *Report on the Questionnaire To Assess the Impacts of the* FSC *Percentage-based Claims Policy*, www.fscoax.org.

Forests Dialogue (2004) www.theforestsdialogue.org.

Frost, B, Mayers, J and Roberts, S (2003) *Growing Credibility: The Impact of Certification on Forests and People in South Africa*, IIED, London

GFTN (Global Forest and Trade Network) (2004) www.panda.org/forestandtrade

Gray, I, Nussbaum, R, Higman, S and Jennings, S (2002) *Involving Harvesting Contractors in Forest Certification*, UK Department for International Development Forestry Research Programme, ProForest, Oxford, UK, www.proforest.net

Grayson, A J (1995) *The World's Forests: International Initiatives Since Rio*, Commonwealth Forestry Association, Oxford, UK

Ghazoul, J (2001) 'Barriers to biodiversity conservation in forest certification', *Conservation Biology*, vol 15, pp315–317

Gretzinger, S (2003) *Modular System to Achieve SFM Standards*, ITTO Regional Workshop on Phased Approaches to Forest Certification (Latin America), Panama City, 9–10 May
Gullison, R (2003) 'Does forest certification conserve biodiversity?', *Oryx*, vol 37, no 2, pp153–165
Hansen, E and Juslin, H (1999) *The Status of Forest Certification in the ECE Region*, UN/ECE/FAO Timber Section, General Timber and Forest Discussion Papers ECE/TIM/DP/14, New York and Geneva
Hansen, E, Forsyth, K and Juslin, H (1999) *Forest Certification Update for the ECE Region*, UN/ECE/FAO Timber Section, General Timber and Forest Discussion Papers ECE/TIM/DP/17, New York and Geneva
Hauselmann, P (1997) *ISO Inside Out: ISO and Environmental Management*, Discussion Paper, WWF International, Gland, Switzerland
Higman, S, Bass, S, Judd, N, Mayers, J and Nussbaum, R (1999) *The Sustainable Forestry Handbook*, Earthscan, London
Higman, S, Nussbaum, R, Arguilar, F, Nardelli, A and Scrase, H (2002) *How Standards Constrain Certification of Small Forest Enterprises*, UK Department for International Development Forestry Research Programme, ProForest, Oxford, UK, www.proforest.net
Higman, S, Mayers, J, Bass, S, Judd, N and Nussbaum, R (2005) *The Sustainable Forestry Handbook, second edition*, Earthscan, London
IAF (International Accreditation Forum) (2000) *The IAF Charter*, www.iaf.nu
IAF (2001) www.iaf.nu/default.asp
IFIR (International Forest Industries Round Table) (2001a) *Proposing an International Mutual Recognition Framework*, Report of the Working Group on Mutual Recognition between Credible Sustainable Forest Management Certification Systems and Standards, February, IFIR
IFIR (2001b) *Report of the Working Group on Mutual Recognition between Credible Sustainable Forest Management Standards and Certification Systems*, IFIR
IFOAM (International Federation of Organic Agriculture Movement) (2001) www.ifoam.org/accredit/index.html
IIED (International Institute for Environment and Development) (1996) *The Sustainable Paper Cycle*, IIED, London, and WBCSD, Geneva
Indufor (1997) *Options for International Institutional Arrangements in Certification of Forest Management and Implications for ACP Countries*, Prepared for the European Commission DG VIII/A/I, Helsinki
Indufor-Eco (1999) *Implications of Land Restitution for Achieving World Bank/WWF Alliance Targets in Eastern Europe and the Central Asian Region*, Main report prepared for the World Bank/WWF Alliance, Helsinki and Oberaula
IPF (Intergovernmental Panel on Forests) (1997) *Report of the Ad Hoc Intergovernmental Panel on Forests on Its Fourth Session*, E/CN.17/1997/12, United Nations, New York
ISEAL Alliance (Social and Environmental Accreditation and Labelling Alliance) (2000) www.isealalliance.net
ISEAL (2004) ISEAL Code of Good Practice for Setting Social and Environmental Standards, ISEAL Alliance, Bonn, Germany, www.isealalliance.org
ISO (International Organization for Standardization) (1990) ISO 10011-1 – *Guidelines for Auditing Quality Systems, Part 1: Auditing*, ISO, Geneva
ISO (1991) ISO 10011-2 – *Guidelines for Auditing Quality Systems, Part 2: Qualification Criteria for Quality Systems Auditors*, ISO, Geneva
ISO (1992) *Certification and Related Activities: Assessment and Verification of Conformity to Standards and Technical Specifications*, ISO, Geneva
ISO (1994) ISO/IEC Guide 59 – *Code of Good Practice for Standardization*, ISO, Geneva
ISO (1995) ISO/IEC Directives Part 1 – *Procedures for the Technical Work*, ISO, Geneva
ISO (1996a) ISO/IEC Guide 2 – *Standardization and Related Activities: General Vocabulary*, ISO, Geneva
ISO (1996b) ISO/IEC Guide 61 – *General Requirements for Assessment and Accreditation of Certification/Registration Bodies*, ISO, Geneva
ISO (1996c) ISO/IEC Guide 62 – *General Requirements for Bodies Operating Assessment and Certification/Registration of Quality Systems*, ISO, Geneva
ISO (1996d) ISO/IEC Guide 65 – *General Requirements for Bodies Operating Product Certification Systems*, ISO, Geneva
ISO (1996e) ISO 14012: 1996 *Guidelines for environmental auditing – Qualification Criteria for Environmental Auditors*, ISO, Geneva

ISO (1998) *Information to Assist Forest Organizations in the Use of* ISO 14001 *and* ISO 14004 – *Environmental Management Systems Standards*, Technical Report 14061, ISO, Geneva

ISO (1999a) ISO 14021 – *Environmental Labels and Declarations: Self-declared Environmental Claims (Type* II *Environmental Labelling)*, ISO, Geneva

ISO (1999b) ISO 14024 – *Environmental Labels and Declarations: Type* I *Environmental Labelling – Principles and Procedures*, ISO, Geneva

ISO (2000a) ISO/TR 14025 – *Environmental Labels and Declarations: Type* III *Environmental Declarations*, ISO, Geneva

ISO (2000b) ISO 14020 – *Environmental Labels and Declarations: General principles, second edition*, ISO, Geneva

ITTO (International Tropical Timber Organization) (1990) *Guidelines for the Sustainable Management of Natural Tropical Forests*, ITTO, Yokohama

ITTO (1998) *Criteria and Indicators for Sustainable Management of Natural Tropical Forests*, ITTO, Yokohama

ITTO (2002) *Forest Certification and Biodiversity: Opposites or Complements?* Discussion paper prepared for the GEF by the ITTO Secretariat, Yokohama, Japan

Kanowski, P, Sinclair, D and Freeman, B (1999) *International Approaches to Forest Management Certification and Labeling of Forest Products: A Review*, Prepared for Department of Agriculture, Fisheries and Forestry, Australia

Kanowski, P J, Sinclair, D, Freeman, B and Bass, S (2000) *Establishing Comparability and Equivalence Among Forest Management Certification Schemes*, Department of Agriculture, Fisheries and Forestry, Australia

Katila, M and Puustjärvi, E (2003) *Impact of New Markets for Environmental services on forest products trade* (unpublished draft), Document prepared in collaboration with Ecosecurities Ltd, FAO Impact Assessment of Forest Products Trade in The Promotion of Sustainable Forest Management (GCP/INT/775/JPN)

Lammerts van Bueren, E M (1997) *Hierarchical Framework for the Formulation of Sustainable Forestry Management Standards*, Tropenbos Foundation, Wageningen, The Netherlands

Lammerts van Bueren, E M and Blom, M (1997) *Hierarchical Framework for the Formulation of Sustainable Forest Management Standards*, Tropenbos Foundation, Wageningen, The Netherlands

Landell-Mills, N and Ford, J (1999) *Privatising Sustainable Forestry: A Global Review of Trends and Challenges*, IIED Instruments for Sustainable Private Sector Forestry Series, IIED, London

Landell-Mills, N and Porras I T (2002) *Silver Bullet or Fools' Gold? A Global Review of Markets for Forest Environmental Services and Their Impacts on the Poor*, Instruments for Sustainable Private Sector Forestry Series, IIED, London.

Lay Cheng, T (ed) (1996) *Initiatives on Assessing Sustainability: Status and Future Directions*, CIFOR/CATIE Special Publication, Jakarta/Turrialba

LEEC (1993) *The Economic Linkages Between the International Trade in Tropical Timber and the Sustainable Management of Tropical Forests, Final Report and Technical Annexes*, London Environmental Economics Centre, IIED, London

Lindhe, A (2004) *Conservation through Management: Cut Wood as Substrate for Saproxylic Organisms*, PhD thesis, Swedish University of Agricultural Sciences, Sweden

Maini, J (2001) 'International policy: Where do we stand today?', in Söderlund, M and Pottinger, A (eds) *The World's Forests: Rio +8 Policy, Practice and Progress Towards Sustainable Forest Management*, Commonwealth Forestry Association, Oxford, pp273–281

Mallet, P (1999) *Analysis of Criteria Addressed by Forestry, Agriculture and Fairtrade Certification Systems*, Falls Brook Centre, Canada

Mäntyranta, H (2002) *Forest Certification – an Ideal that Became an Absolute*, Metsälehti Kustannus, Finland

Markopoulos, M (1998) *The Impacts of Certification on Community Forest Enterprises: A Case Study of the Lomerio Community Forest Management Project, Bolivia*, Forestry and Land Use Series No13, IIED, London

Markopoulos, M (1999) *Community Forest Enterprise and Certification in Mexico: A Review of Experience with Special Reference to the Union of Zapotec and Chinantec Forestry Communities* (UZACHI), *Oaxaca*, Oxford Forestry Institute, Oxford

Markopoulos, M (2000) *The Role of Certification in Supporting Community-based Forest Enterprise* (CFE) *in Latin America*, PhD thesis, Oxford University, Oxford

Mayers, J and Bass, S (1999) *Policy that Works for Forests and People*, IIED, London

Mayers, J, Bass, S and Macqueen, D (2002) *The Pyramid: A Diagnostic and Planning Tool for Good Forest Governance*, Prepared for the World Bank WWF Alliance for Forest Conservation and Sustainable Use, www.iied.org

Maynard, B and Robinson, D (1998) *Ethical Trade and Sustainable Rural Livelihoods: Quintana Roo Forest Certification Case Study*, Draft, Natural Resources Institute, Chatham

Meridian Institute (2001) *Comparative Analysis of the Forest Stewardship Council and Sustainable Forestry Initiative Certification Programs. Executive Summary Consensus Statement on Salient Similarities and Differences between the Two Programs*, Meridian Institute, Virginia Beach, VA, US

Merino, L (1996) *Social Aspects of Sustainability and Social Certification*, Universidad Nacional Autónoma de México and Consejo Civil Mexicano para la Silvicultura Sostenible, Mexico

Molnar, A (2003) *Forest Certification and Communities: Looking Forward to the Next Decade*, Forest Trends, Washington, DC

Moura Costa, P, Salmi, J, Simula, M and Wilson, C (1999) *Financial Mechanisms for Sustainable Forestry*, UNDP/SEED, Helsinki and Oxford

Muthoo, M (2001) 'Certification and sustainable forest management', Paper presented at the International Workshop of Experts on Financing Sustainable Forest Management, Oslo, Norway, 22–25 January

Nsenkyiere, E O and Simula, M (2000) *Comparative Study on the Auditing Systems of Sustainable Forest Management*, ITTO, Yokohama, Japan

Nussbaum, R (2000) 'Forest certification: Verifying sustainable management', Paper presented at the Workshop on Streamlining Local-level Information for Sustainable Forest Management, University of British Columbia, Canada, 28–30 August

Nussbaum, R (2001) 'Contractors and certification: How does forest certification impact the use of contractors?', Paper presented at the South African Institute of Forestry Symposium: Outsourcing in Forestry – Opportunity or Threat?, Sabie, South Africa, 3 May

Nussbaum, R (2002) *Group Certification for Forests: A Practical Guide*, UK Department for International Development Forestry Research Programme, ProForest, Oxford, www.proforest.net

Nussbaum, R, Bass, S, Morrison, E and Speechly, H (1995) *Sustainable Forest Management: An Analysis of Principles, Criteria and Standards*, IIED, London, UK

Nussbaum, R, Bass, S, Morrison, E and Speechly, H (1996) *Sustainable Forest Management: An Analysis of Principles, Criteria and Standards*, Towards a Sustainable Paper Cycle Sub-study Series No 4, IIED/WBCSD, London

Nussbaum, R, Garforth, M, Scrase, H and Wenban-Smith, M (2001) *An Analysis of Current FSC Accreditation, Certification and Standard-setting Procedures Identifying Elements which Create Constraints for Small Forest Owners*, UK Department for International Development Forestry Research Programme, ProForest, Oxford, www.proforest.net

Nussbaum, R, Gray, I and Higman, S (2003) *Modular Implementation and Verification* (MIV): *A Toolkit for the Phased Application of Forest Management Standards and Certification*, ProForest, Oxford, www.proforest.net

Nussbaum, R, Jennings, S and Garforth, M (2002) *Assessing Forest Certification Schemes: A Practical Guide*, UK Department for International Development Forestry Research Programme, Proforest, Oxford, www.proforest.net

Pagiola, S (2003) *Mecanismos de Cobro y Pago por Servicios Ambientales*, Presentación en el Taller Internacional de Experiencias Aplicadas en el Monitoreo y Seguimiento del Mercado de Servicios Ambientales EXPO FORESTAL 2003, Guadalajara, México, 7–8 August

Pagiola, S, Bishop, J and Landell-Mills, N (eds) (2002) *Selling Forest Environmental Services: Market-based Mechanisms for Conservation and Development*, Earthscan, London

PEFC (Programme for the Endorsement of Forest Certification) (2001) PEFC *Logo Usage Guide for Wood Products*, Annex 7 to the PEFC Technical Document, www.pefc.org/Tech-annex7.htm

PEFC (2002) PEFC *Council Technical Document Annex 1: PEFC Terms and Definitions*, PEFC, Luxembourg, www.pefc.org

Poore, D (2003) *Changing Landscapes*, Earthscan, London

Poynton, S (2003) *Good Wood Good Business: A Practical Industry-oriented Guide to Excluding Illegal and Other Unwanted Wood from your Supply Chain*, Tropical Forest Trust, www.tropicalforesttrust.com

Prabhu, R (1997) *A Criteria and Indicator Tool-Box for the Assessment of Sustainable Forest Management*, CIFOR Discussion Paper, Jakarta

Prabhu, R and Colfer, C (1999) *The Criteria and Indicators Toolbox Series*, Centre for International Forestry Research, Bogor, Indonesia

Phrabu, R, Colfer, C P J and Dudley, R G (2002) *Guidelines for Developing Testing and Selecting Criteria and Indicators for Sustainable Forest Management*, CIFOR. Jakarta

Prabhu, R, Colfer, C, Venkates Warlu, P, Cheng Tan, L, Soekmadi, R and Wollenberg, E (1996) *Testing Criteria and Indicators for the Sustainable Management of Forests: Phase I Final Report*, CIFOR Special Publications, Jakarta

Rametsteiner, E (1999) *Sustainable Forest Management Certification – Framework Conditions, Systems Design and Impact Assessment*, MCPFE Liaison Unit, Vienna

Rametsteiner, E (2000a) 'Sustainable forest management certification: Frame conditions, system designs and impact assessment', Paper presented at the Ministerial Conference on Forests in Europe, Liaison Unit, Vienna

Rametsteiner, E (2000b) *The Role of Governments in SFM Certification*, Institut für Sozioökonomik der Forst und Holzwirtschaft, Discussion Papers No P/2000-1, Vienna

Rametsteiner, E (2002a) *Promoting Political Commitment for the Use of Criteria and Indicators as Tools for Sustainable Forest Management*, International Conference on Criteria and Indicators to Sustainable Forest Management: The Way Forward, Background Paper no 2

Rametsteiner, E (2002b) 'The role of governments in forest certification a normative analysis based on new institutional economics theories', *Forest Policy and Economics*, vol 4, no 3, pp163–173

Rametsteiner E (2002c) 'Markets for certified forest products', in UNECE/FAO *Forest Products Annual Market Review* 2001–2002, *Timber Bulletin Vol* LU,ECE/TIM/BULL/2002/3, UN Publications, Rome, pp157–154

Rametsteiner, E and Simula, M (2001) 'Forging novel incentives for environment and sustainable forest management', in Simula, M, Rametsteiner, E, Blasten, A, Green, T, Pajari, B (eds) *Forest Certification: Forging Novel Incentives for the Environment and Sustainable Forest Management*, EFI Proceedings No 43, European Forest Institute, Joensuu, Finland

SAI (2004) www.cepaa.org

Saile, P (2001) *Training Needs and Capacity Building Challenges in the Development of Certification in* ATO *Countries*, ATO/ITTO Regional Workshop on the Implementation of Principles, Criteria and Indicators for Sustainable Management of African Forests, Yaounde, 25–26 May

Scherr, S, White, A and Khare, A (2004) *For Services Rendered: The Current Status and Future Potential of Markets for the Ecosystem Services Provided by Tropical Forests*, ITTO Technical series No 21, Yokohama, Japan

Scrase, H and Lindhe, A (2001) *Developing Forest Stewardship Standards: A Survival Guide*, Taiga Rescue Network, Jokkmokk, Sweden

Scrase, H et al (1999) *Certification of Forest Products for Small Businesses: Improving Access – Issues and Options*, Report to Forest Research Programme of DFID, London

Simula, M (1999) *Certification of Forest Management and Labelling of Forest Products: Discussion Note on Main Issues*, The World Bank Group, Forest Policy Implementation Review and Strategy Development: Analytical Studies, Indufor Oy, Finland

Simula, M (2003) 'Criteria and indicators for sustainable forest management: Overview of progress and issues: Scene-setting paper', International Conference on the Contribution of Criteria and Indicators for Sustainable Forest Management: The Way Forward, CICI, Guatemala City

Simula, M, Eba'a Atyi, R and Nussbaum, R (2003) *The Potential Role of Phased Approaches to Certification on Tropical Timber Producer Countries as a Tool to Promote Sustainable Forest Management*, ITTO, Yokohama, Japan, www.itto.or.jp

Simula, M, Rametsteiner, E, Blåsten, A, Green, T and Pajari, B (eds) (2001) *Forest Certification: Forging Novel Incentives for the Environment and Sustainable Forest Management*, Proceedings of the International Workshop, EFI Proceedings 43, Brussels, Belgium, 6–7 September

Söderlund, M and Pottinger, A (eds) (2001) *The World's Forests: Rio +8 Policy, Practice and Progress Towards Sustainable Forest Management*, Commonwealth Forestry Association, Oxford

Synnott, T (2000) *Forest Stewardship Council: Position on Mutual Recognition*, Second International Seminar on the Mutual Recognition of Credible Certification Systems, Brussels, Belgium, 28–29 November

TFT (Tropical Forest Trust) (2004) www.tropicalforesttrust.com

Thang, H C (1996) *Formulation and Implementation of Criteria and Indicators for Sustainable Forest Management in Malaysia*, Workshop on Forest Management Certification, Forest Research Institute Malaysia, Kepong

Thang, H C (2000) 'Malaysia's experiences on criteria and indicators for sustainable forest management and timber certification', Paper presented at the Workshop on Criteria and Indicators for Sustainable Forest Management and Timber Certification, Yangon, Myanmar, 21–22 February

Thornber, K (1999) *Impacts of Certification on Forests, Stakeholders and Markets – Case Study: Bainings Ecoforestry Project*, Instruments for Sustainable Private Sector Forestry Project Series, IIED, London

Thornber, K, Plouvier D and Bass, S (1999) *Certification: Winners and Losers: A Discussion of Equity Implications*, IIED, London

UKAS (2001) *Criteria for Selection of UKAS Technical Experts for Accreditation of Forest Certification*, UKAS Internal Document, Document Reference 215, UKAS, UK

UKWAS Steering Group (2000) *Certification Standard for the UK Woodland Assurance Scheme*, UK Forestry Commission, Edinburgh, UK

UNECE (2001) *Markets for Certified Forest Products: Forest Products Annual Market Review 2000–2001*, United Nations Economic Commission for Europe, Geneva, Switzerland

Vallejo, N and Hauselmann, P (2000a) *Institutional Requirements for Forest Certification: A Manual for Stakeholders*, Deutsche Gesellschaft für Technische Zusammenarbeit (GTZ) GmbH, Forest Certification Project, Working Paper 2, Germany

Vallejo, N and Hauselmann, P (2000b) *Forest Certification Project: Working Paper 2, Institutional Requirements for Forest Certification: A Manual for Stakeholders*, GTZ, Germany

Vallejo, N and Hauselmann, P (2001) PEFC: *An Analysis*, WWF Discussion Paper, Gland, Switzerland

Viana, V (1996) *The Role of Public Participation in the Development of Certification Programs: Lessons from Brazil*, International Conference on Certification, Case Study 4 – Brazil, Brisbane

Viana, V M, Ervin, J, Donovan, R, Elliott, C and Gholz, H (eds) (1996) *Certification of Forest Products Issues and Perspectives*, Island Press, Washington, DC

Vilhunen, L, Hansen, E, Juslin, H and Forsyth, K (2001) *Forest Certification Update for the ECE Region, Summer 2001*, Geneva Timber and Forest Discussion Papers, ECE/TIM/DP/23, New York and Geneva

Vogt, K A, Larsen, B C, Gordon, J C, Vogt, D J and Fanzeres, A (2000) *Forest Certification Roots, Issues, Challenges and Benefits*, CRC Press, Boca Raton, Florida

WBCSD (World Business Council for Sustainable Development) (2003) *Forest Certification Systems and the 'Legitimacy' Thresholds Model*, Discussion paper prepared by James Griffiths, www.wbcsd.ch (follow links to forest products)

WCED (World Commission on Environment and Development) (1987) *Our Common Future* (Brundtland Report), Oxford University Press, Oxford, UK

Wenban-Smith, M, Nussbaum, R, Garforth, M and Scrase, H (2001) *An Analysis of the Barriers Faced by Small-scale Farmers and Communities Producing Timber outside a Conventional Forest Matrix, and Recommendations for Progress*, UK Department for International Development Forestry Research Programme, ProForest, Oxford, www.proforest.net

White, A and Martin, A (2002) *Who Owns the World's Forests? Forest Tenure and Public Forests in Transition*, Forest Trends, Washington, DC

White, G and Sarshar, D (2004) *Responsible Purchasing of Forest Products*, WWF Global Forest and Trade Network, www.panda.org/forestandtrade

World Bank (2003) *Sustaining Forests: A World Bank Strategy*, World Bank, Washington, DC

World Bank–WWF (World Bank–World Wide Fund for Nature) Alliance for Forest Conservation and Sustainable Use (undated) *Guidance Note for Improved Forest Management and Certification Target: Achieving the Independent Certification of 200 Million Hectares of Well-managed Production Forests by the Year 2005*, World Bank–WWF, Washington, DC

World Bank–WWF Alliance for Forest Conservation and Sustainable Use (2003a) 'Questionnaire for Assessing the Comprehensiveness of Certification Schemes/Systems', World Bank–WWF Alliance, Washington, DC

World Bank–WWF Alliance (2003b) 'Review of Learning and Capacity Building Tools to Promote Sustainable Forest Management (SFM) and Certification', Draft Report, World Bank, Washington DC, www.forest-alliance.org

World Commission on Forests and Sustainable Development (1999) *Our Forests, Our Future*, Cambridge University Press, Cambridge, UK

Index

Page numbers in *italic* refer to tables and boxes

THE EARTHSCAN FORESTRY LIBRARY

provides a collection of publications that address the key issues and innovations in policy, practice and theory that are shaping Forestry. Combining a wealth of knowledge and expertise from internationally renowned figures in the field, this Library is an essential resource for forestry practitioners, policy makers, conservationists, researchers and students worldwide.

JAMES & JAMES

Browse and buy on-line at
WWW.EARTHSCAN.CO.UK

[a]Erratum

This second edition is based on the *The Forest Certification Handbook* First Edition
© Steve Bass and Chris Upton, 1995

[a]Acknowledgements

The authors of this edition of *The Forest Certification Handbook* would like to acknowledge …

For Product Safety Concerns and Information please contact our EU representative GPSR@taylorandfrancis.com Taylor & Francis Verlag GmbH, Kaufingerstraße 24, 80331 München, Germany

Batch number: 10726322

Printed by Printforce, the Netherlands